Systems Engineering for Projects

Achieving Positive Outcomes in a Complex World

Best Practices and Advances
in Program Management Series

Series Editor
Ginger Levin

RECENTLY PUBLISHED TITLES

PgMP® Exam Test Preparation: Test Questions, Practice Tests,
and Simulated Exams
Ginger Levin

Managing Complex Construction Projects: A Systems Approach
John K. Briesemeister

Managing Project Competence: The Lemon and the Loop
Rolf Medina

The Human Change Management Body of Knowledge (HCMBOK®),
Third Edition
Vicente Goncalves and Carla Campos

Creating a Greater Whole: A Project Manager's Guide to Becoming a Leader
Susan G. Schwartz

Project Management beyond Waterfall and Agile
Mounir Ajam

Realizing Strategy through Projects: The Executive's Guide
Carl Marnewick

PMI-PBA® Exam Practice Test and Study Guide
Brian Williamson

Earned Benefit Program Management: Aligning, Realizing,
and Sustaining Strategy
Crispin Piney

The Entrepreneurial Project Manager
Chris Cook

Leading and Motivating Global Teams: Integrating Offshore Centers
and the Head Office
Vimal Kumar Khanna

Project and Program Turnaround
Thomas Pavelko

Systems Engineering for Projects

Achieving Positive Outcomes
in a Complex World

Lory Mitchell Wingate

CRC Press

Taylor & Francis Group
Boca Raton London New York

CRC Press is an imprint of the
Taylor & Francis Group, an **informa** business

AN AUERBACH BOOK

"College of Performance Management," "PMI," "PMP," and "PMBOK® Guide" are registered trademarks of the Project Management Institute.

"Complex Systems Management" and "CSM" are trademarks and service marks of Lory Mitchell Wingate.

"Gore-Tex" is a trademark of W. L. Gore & Associates.

"JetBoil" is a registered trademark of Johnson Outdoors, Inc.

"IPMA" is a registered trademark of the International Project Management Association.

"ISO" is a registered trademark of the International Organization for Standardization.

"Karitek" is a registered trademark of Kari-Tek.

"Microsoft" and "Microsoft Project" are registered trademarks of Microsoft Corporation in the United States and/or other countries.

"OCSMP" and "Certified Systems Modeling Professional" are trademarks of the Certified Systems Modeling Professional.

"OMG" and "Object Management Group" are registered trademarks of the Object Management Group.

"Red Bull" is a registered trademark of Red Bull GMBH.

"Reed Chillcheater" is a registered trademark of Reed Chillcheater Ltd., North Devon, UK.

"SAE" is a registered trademark of Th e Society of Automotive Engineers, International.

"TQM" and "TOTAL QUALITY MANAGEMENT" are trademarks of Exar Corporation.

"Virgin Atlantic" is a trading name of Virgin Atlantic Airways Limited and Virgin Atlantic International Limited, with its head office in Crawley, United Kingdom.

CRC Press
Taylor & Francis Group
6000 Broken Sound Parkway NW, Suite 300
Boca Raton, FL 33487-2742

First issued in paperback 2022

© 2019 by Lory Mitchell Wingate
CRC Press is an imprint of Taylor & Francis Group, an Informa business

No claim to original U.S. Government works

ISBN 13: 978-1-03-247602-5 (pbk)
ISBN 13: 978-0-8153-6295-1 (hbk)

DOI: 10.1201/9780429464577

Visit the Taylor & Francis Web site at
http://www.taylorandfrancis.com

and the CRC Press Web site at
http://www.crcpress.com

Dedication

Dedicated to all the believers in the world who know that appropriately applying processes can make a difference in achieving desired outcomes.

Contents

Dedication   vii

Contents   ix

List of Illustrations and Tables   xvii

Preface   xix

Acknowledgments   xxi

About the Author   xxiii

Chapter 1 The Discipline of Systems Engineering   1

   Chapter Roadmap   2

   1.1   Origination of Systems Engineering   2

   1.2   Evolution of the Systems Engineering Discipline   4

      1.2.1 Terminology   5

      1.2.2 Standards   6

   1.3   Organizational Implementation of Systems Engineering   7

      1.3.1 Organizational Structure and Culture   7

   1.4   Overview of the Systems Engineering Discipline   8

      1.4.1 System Approach   8

      1.4.2 System-of-Systems Approach   10

   1.5   Case Study: Trans-Greenland Expedition   13

      1.5.1 Structure and Purpose   13

      1.5.2 Project Charter and Plan   14

      1.5.3 Stakeholder Engagement and Communications Management   16

      1.5.4 Scope Management   16

      1.5.5 Schedule Management   18

      1.5.6 Acquisition Management   18

      1.5.7 Risks and Opportunity Management   18

      1.5.8 Quality Management   20

      1.5.9 Test, Verification, and Validation   20

	1.5.10 Governance	21
	1.5.11 Outcomes and Lessons Learned	21
	1.5.12 Case Analysis	21
1.6	Key Point Summary	22
1.7	Apply Now	23
	References	23

Chapter 2 Systems Engineering as a Project Enabler — **27**

	Chapter Roadmap	28
2.1	The Structure of Projects	28
	2.1.1 Basic Definition of a Project	29
	2.1.2 The Project Management Discipline	30
	2.1.3 Types of Project Management Methods	36
	2.1.4 The Systems Engineering Discipline	37
2.2	Process Alignment Between Two Disciplines	49
	2.2.1 Stakeholder Management	50
	2.2.2 Communications Management	51
	2.2.3 Scope Management	51
	2.2.4 Schedule Management	52
	2.2.5 Procurement Management	52
	2.2.6 Risk Management	52
	2.2.7 Quality Management	52
	2.2.8 Governance	53
2.3	Using a Combined Method to Drive Project Outcomes	53
2.4	Case Study: FIRST Robotics Competition—University of Detroit Jesuit High School and Academy Robocubs Team #1701	54
	2.4.1 Project Charter and Plan	55
	2.4.2 Stakeholder and Communications Management	55
	2.4.3 Scope Management	55
	2.4.4 Acquisition Strategy and Integration	57
	2.4.5 Risks and Opportunity Management	57
	2.4.6 Quality Management	58
	2.4.7 Test, Verification, and Validation	58
	2.4.8 Governance	58
	2.4.9 Outcomes	59
	2.4.10 Lessons Learned	59
	2.4.11 Case Analysis	59
2.5	Key Point Summary	60
	2.5.1 Key Concepts	60
2.6	Apply Now	60
	References	61

Chapter 3 Application of Complementary Processes   **63**

Chapter Roadmap   64

3.1   Stakeholder-Focused Complementary Processes   64

    3.1.1   Stakeholder Engagement   65

    3.1.2   Communications Management   69

3.2   Solution-Focused Complementary Processes   75

    3.2.1   Scope Management   76

    3.2.2   Schedule Management   82

    3.2.3   Procurement Management   84

    3.2.4   Risk Management   87

    3.2.5   Quality Management   90

    3.2.6   Governance   93

3.3   Case Study: Simon Beck's Snow Art   96

    3.3.1   Project Charter and Plan   96

    3.3.2   Stakeholder and Communications Management   97

    3.3.3   Scope Management   97

    3.3.4   Schedule Management   98

    3.3.5   Procurement Management   99

    3.3.6   Risks and Opportunity Management   99

    3.3.7   Quality Management   100

    3.3.8   Governance   100

    3.3.9   Outcomes   100

    3.3.10 Lessons Learned   101

    3.3.11 Case Analysis   101

3.4   Key Point Summary   101

    3.4.1   Key Concepts   101

3.5   Apply Now   102

References   104

Chapter 4 Application of Unique Processes   **105**

Chapter Roadmap   106

4.1   Stakeholder-Focused Unique Processes   106

    4.1.1   Stakeholder Engagement   107

4.2   Solution-Focused Unique Processes   110

    4.2.1   Scope Management   112

    4.2.2   Schedule Management   114

    4.2.3   Financial Management   117

    4.2.4   Human Resource Management   119

    4.2.5   Risk Management   121

    4.2.6   Quality Management   123

4.3   Case Study: Jill and Julia—Singers, Songwriters, Musicians   125

4.3.1	Project Charter and Plan	126
4.3.2	Stakeholder and Communications Management	126
4.3.3	Scope Management	127
4.3.4	Schedule Management	128
4.3.5	Procurement Management	128
4.3.6	Risks and Opportunity Management	129
4.3.7	Quality Management	129
4.3.8	Test, Verification, and Validation	130
4.3.9	Governance	130
4.3.10	Outcomes	130
4.3.11	Lessons Learned	130
4.3.12	Case Analysis	131
4.4	Key Point Summary	131
4.4.1	Key Concepts	131
4.5	Apply Now	131
References		133

Chapter 5 Success Measurements   **135**

Chapter Roadmap		136
5.1	Defining Success	136
5.1.1	Successful Outcomes	136
5.1.2	Project Management Successes	137
5.1.3	Systems Engineering Successes	138
5.2	The Use of Measurements	139
5.2.1	Supporting Foundations	139
5.2.2	Measurements Watch Items	143
5.3	Types of Measurements	145
5.3.1	Trend and Variance Analysis	146
5.3.2	Statistical Sampling Measurements	146
5.3.3	Key Performance Indicators	147
5.3.4	Technical Readiness	147
5.4	Essential Measurements	149
5.4.1	Baseline Control	149
5.4.2	Resources Control	155
5.4.3	Risk Control	156
5.4.4	Quality Control	157
5.4.5	Governance Measurements	162
5.5	Communicating Measurements	164
5.6	Case Study: Alva-Green Coaching Group	167
5.6.1	Project Charter and Plan	167
5.6.2	Stakeholder and Communications Management	167
5.6.3	Scope Management	168

5.6.4   Acquisition Strategy and Integration   168
5.6.5   Risks and Opportunity Management   169
5.6.6   Quality Management   169
5.6.7   Test, Verification, Validation   169
5.6.8   Governance   170
5.6.9   Outcomes   170
5.6.10 Lessons Learned   170
5.6.11 Case Analysis   170
5.7   Key Point Summary   170
5.7.1   Key Concepts   171
5.8   Apply Now   171
References   174

Chapter 6 Tailoring   **175**
Chapter Roadmap   176
6.1   Definition of Tailoring   176
6.2   Applying Tailoring   177
6.2.1   Tailoring by Requirements   178
6.2.2   Tailoring Risks   179
6.2.3   Tailoring by Life Cycle Phase   180
6.2.4   Tailoring for Different Management Methods   181
6.3   Tailoring Project Management Processes   183
6.3.1   Stakeholder Engagement   183
6.3.2   Collaboration and Communications   184
6.3.3   Total Project Scope   184
6.3.4   Total Project Schedule   185
6.3.5   Financial Management   185
6.3.6   Risk Management   185
6.3.7   Quality Management   186
6.3.8   Integrative Management   186
6.4   Tailoring Systems Engineering Processes   186
6.4.1   System Stakeholder Engagement   187
6.4.2   Communications and Decision Support   187
6.4.3   Technical Scope Management   188
6.4.4   Interface Management   188
6.4.5   Technical Schedule Management   189
6.4.6   Acquisition and Procurement Management   189
6.4.7   Risk Reduction   190
6.4.8   Quality and Measurements   190
6.4.9   Test, Verification, Validation   190
6.4.10 Governance   191
6.5   Communicating About Project Tailoring   192

6.6   Case Study: Greenland to Scotland Challenge— In the Wake of the Finnmen   195
      6.6.1   Project Charter and Plan   196
      6.6.2   Stakeholder Engagement/Communications Management   196
      6.6.3   Scope Management   196
      6.6.4   Schedule Management   199
      6.6.5   Cost Management   200
      6.6.6   Risks and Opportunity Management   200
      6.6.7   Quality, Test, Verification, Validation   202
      6.6.8   Governance   202
      6.6.9   Outcomes   203
      6.6.10   Lessons Learned   203
      6.6.11   Case Analysis   204
6.7   Key Point Summary   204
      6.7.1   Key Concepts   205
6.8   Apply Now   205
References   206

Chapter 7 Methodology Synthesis and Application   **207**
Chapter Roadmap   208
7.1   Step 1: Understand the Environment   208
      7.1.1   Organizational Considerations   208
      7.1.2   Complex Project Organizational Considerations   211
7.2   Step 2: Structure the Project   212
7.3   Step 3: Apply the Processes   213
      7.3.1   Stakeholder-Focused Complementary and Unique Processes   214
      7.3.2   Solution-Focused Complementary and Unique Processes   218
7.4   Step 4: Choose Measurements   223
      7.4.1   Essential Measurements   224
      7.4.2   Performance Metrics   226
7.5   Step 5: Tailor the Processes   227
7.6   Step 6: Implement the Project   228
7.7   Case Study: Carolina Soap Market   228
      7.7.1   Project Charter and Plan   229
      7.7.2   Stakeholder Engagement   229
      7.7.3   Communications Management   229
      7.7.4   Scope Management   230
      7.7.5   Schedule Management   234
      7.7.6   Acquisition Strategy and Integration   234
      7.7.7   Risk and Opportunity Management   235
      7.7.8   Quality, Test, Verification, Validation   235

7.7.9   Governance   236
7.7.10   Outcomes   236
7.7.11   Lessons Learned   236
7.7.12   Case Analysis   237
7.8   Key Point Summary   237
7.8.1   Key Concepts   237
7.9   Apply Now   238
References   241

Chapter 8 The Future of Systems Engineering   243
Chapter Roadmap   244
8.1   Systems Engineering Origins   244
8.2   Adapting to the Changing World   247
8.2.1   Societal Responsibilities   248
8.2.2   Human Sustainment   255
8.2.3   Human Interactions   264
8.2.4   Human Accomplishments   266
8.3   Systems Engineering Process Evolution   269
8.4   Key Point Summary   270
8.4.1   Key Concepts   270
8.5   Apply Now   271
References   272

Acronyms   273

Glossary   275

Bibliography   283

Index   289

List of Illustrations and Tables

List of Illustrations

Figure 1.1	System Representation	9
Figure 1.2	System-of-Systems Representation	11
Figure 2.1	Basic Project-Level Work Breakdown Structure	34
Figure 2.2	Measuring Technical Performance	40
Figure 2.3	Sample Technical Work Breakdown Structure	44
Figure 3.1	Overarching Stakeholder Complementary Processes	65
Figure 3.2	Sample Stakeholder Matrix	67
Figure 3.3	Methods of Communications	71
Figure 3.4	Sample Project Communications Plan	72
Figure 3.5	Sample Document Tree	74
Figure 3.6	Solution-Focused Complementary Processes	76
Figure 3.7	Example of a Functional Diagram	79
Figure 3.8	Example of Requirement Specifications	80
Figure 3.9	Example of a Risk Register	88
Figure 3.10	Example of Technical Performance Measurements	92
Figure 3.11	Overarching Complementary Model	94
Figure 4.1	Stakeholder-Focused Overarching Processes and Tools	107
Figure 4.2	Example of an Issues Register	109
Figure 4.3	Solution-Focused Overarching Processes and Tools	111
Figure 4.4	Example of N^2 Charts	113
Figure 4.5	Example of Standard Schedule Reporting	116
Figure 4.6	Example of Earned Value Reporting	117
Figure 4.7	Example of Budget Tracking Spreadsheet	118
Figure 4.8	Example of Earned Value Reporting	119
Figure 4.9	Example of Requirements Verification Matrix	124

Figure 5.1   Perception Model   137

Figure 5.2   Example of Roles and Responsibilities Matrix   140

Figure 5.3   Types of Measurements   145

Figure 5.4   Additive Metrics   148

Figure 5.5   Continuums of Reviews   159

Figure 5.6   Example of Foursquare   165

Figure 5.7   Example of Stoplight Chart   166

Figure 6.1   Additive Measurements   178

Figure 6.2   Example Life-Cycle Tailoring Decision Matrix   181

Figure 6.3   Management Methods   182

Figure 6.4   Example Tailoring Risks   193

Figure 6.5   Example Tailoring Treatments   194

Figure 6.6   Example Tailoring Alignments   195

Figure 7.1   Two Organizational Structures   209

Figure 7.2   Project Structure Risk   213

Figure 8.1   The Evolution of Standards   246

Figure 8.2   Impact Areas   247

Figure 8.3   Societal Responsibilities   248

Figure 8.4   Human Sustainment   255

Figure 8.5   Human Interactions   264

Figure 8.6   Human Accomplishments   267

List of Tables

Table 3.1 Apply Now Checklist   **103**

Table 4.1 Apply Now Checklist   **132**

Table 5.1 Apply Now Checklist   **172**

Table 7.1 Apply Now Checklist   **239**

Preface

Civilization in the 21st century is technology driven. Project complexity is expected to continue to increase as a globally connected society incorporates more interconnected technology into the mainstream. Companies in today's environment find themselves in a constant mode of adaptation. To survive into the future, they must be adaptable, creative, and fast.

Many organizations around the world have realized the benefit of applying systems engineering processes to important projects in order to control complexity and achieve the desired outcomes. The systems engineering discipline first emerged in the 1950s as a method of addressing complicated and cross-disciplinary project activities. It was seen as a holistic approach that fundamentally would provide a whole system in its complete form. Two industries, defense and space exploration, required systems that could deliver with precision and in predictable ways and so drove the initial implementations of the systems approach.

As projects continued to get more complex with the introduction of projects that crossed disciplines—such as hardware/software—it became clear that simply decomposing the system to its individual elements and sub-elements, designing and producing them, and then combining them back into a system would not provide predictable system performance owing to a phenomenon called *emergence*. Emergence is expressed when the behavior of the sum of the parts is greater or less than expected for reasons that cannot be easily deduced. This drives risk in a project, which can quickly turn into impacts to project performance and expected outcomes. Projects that required careful integration of project elements from multiple disciplines therefore needed a better way to ensure that they achieved the expected results within the cost, schedule, and scope anticipated by the stakeholders.

Standards-generating organizations have provided a significant body of knowledge to process disciplines supporting design, development, manufacturing, production, and construction. The project management and systems engineering disciplines, for example, have been consistently applied to ensure that for the cost, in the time needed, a viable product or service is provided. New standards-generating organizations have emerged over the last decade, and there has been a continued evolution. At this time in history, these organizations are finding that there are many intersecting and overlapping, but also complementary and enabling, processes across the disciplines. It is the right time to look at how the best of the processes can be brought together into a framework that can facilitate the next generation of projects.

Systems engineering has continued to evolve in the midst of the evolution of other complimentary and overlapping disciplines. Some areas of overlap will cause undue oversight and tension between multiple responsible parties, and some areas will lack any attention at all. There is an increasing need to deconflict and tease apart these overlapping processes so that any ambiguity is removed, increasing the number of accepted standard processes, which when followed will decrease risk. Users of systems engineering have joined forces to bring systems engineering and conflicting or overlapping processes from other disciplines into alignment.

Systems engineering can be used in projects of any type: simple or complicated, homogeneous or complex; and within any industry: science, technology, arts, adventure, etc. Depending on the size and complexity of the project, different levels of systems engineering process rigor is applied (tailoring), which makes the discipline applicable to all projects. This book provides a methodology for optimizing results in all projects, particularly those that are complex. The approaches established here can easily be applied to projects of all sizes and from across projects in any discipline or field. Chapter case studies provide a look into how the methodology can easily fit into any type of project—the more creative and non-traditional the better. To ensure that the reader can apply what is learned, each chapter offers case studies, a "Key Point" summary section, and an "Apply Now" section to practice what is discussed in the chapter.

What makes the methodology described in the book powerful is the flexibility and adaptability that it provides, while lifting the burden off organizations that do not want to invest heavily in implementing a significant number of conflicting standards. This book will also be invaluable to individuals looking for just enough lightweight, flexible structure to help them meet their most important goals, no matter how complicated or complex. Specific guidance and templates are incorporated strategically throughout the chapters, providing immediate access to a powerful toolkit. Anyone wanting to apply new thinking to projects in order to increase the success potential will want to consider this book.

This book is a comprehensive, go-to manual on the application of systems engineering processes to projects of all types and levels of complexity. It provides a comprehensive look at the systems engineering concepts found within the current literature. It also describes the part systems engineering plays in project management, where overlaps occur with this discipline as well as where the systems engineering processes fit with other downstream activities, such as engineering and software development.

To provide the best value from this book, it is structured in a manner that allows readers to understand how to implement systems engineering methods to projects, no matter how simple or complex. The systems engineering discipline will be described in a progressive manner, starting with a background of where the discipline originated and how it evolved and is implemented in organizations. Chapter 1 therefore provides the background on the origination and evolution of the systems engineering discipline.

Chapter 2 provides a view of the overarching disciplines of project management and systems engineering and explores the process alignment between the two disciplines. Chapter 3 provides a thorough explanation of the concepts for applying the Complex Systems Methodology$^{\text{TM SM}}$ (CSM$^{\text{TM SM}}$) complementary processes of integrating systems engineering and project management to a project. Chapter 4 describes in depth the Complex Systems Methodology's unique processes of systems engineering and project management within a project.

Chapter 5 focuses on measurements that assist in the active management of the project by using repetitive assessments that evaluate the progress against the baselines and technical parameters that have been set. And Chapter 6 discusses the important topic of tailoring.

Chapter 7 presents all the processes associated with implementing the Complex Systems Methodology into context so that the direct application steps of the processes within a project is clear. Finally, Chapter 8 investigates the future of systems engineering. Case studies, summary key concepts, and Apply Now exercises, available in each chapter, will allow immediate application of the systems engineering concepts and will provide the reader with the background and experience to approach projects of all types with confidence.

I invite you now to explore the discipline of systems engineering for projects.

— Lory Mitchell Wingate

Acknowledgments

This book was inspired by the incredible individuals who I interact with on a daily basis. My family, team, my friends, and all of my colleagues have played a role in helping me to mold my thinking. In particular, thank you to Christopher McLaughlin, who was instrumental in inspiring me to pursue this challenge.

It would not have been possible to write this book without the support and contributions of many individuals. I would like to thank the individuals who spent countless hours reviewing and editing the manuscript. In particular, special thanks to John Wingate, Ginger Levin at Taylor & Francis, and Susan Culligan at Derryfield Publishing Services, who contributed their time in editing, providing keen insight and solid encouragement, and providing recommendations that enhanced this book tremendously. And to my husband John, the most thanks for his continued support and dedication to quality.

My deepest gratitude to the leaders that I have had the opportunity to work for, and who have also served as my mentors over the last two decades. Your generous support and encouragement has allowed me to become the person I am today. To all the individuals that have helped me along the way, but in particular, thank you Tony Beasley, Scott Bailey, Bob Milburn, and Dan Crowley. Your leadership continues to inspire me.

About the Author

 With over 25 years of experience in both for-profit and non-profit companies, Lory Mitchell Wingate is a successful leader with a history of exceptional performance on complex technical, scientific, and engineering projects that incorporate leading edge technology in all phases of the program and project life cycle. Wingate possesses detailed knowledge and expertise in systems engineering and has developed a strong method for combining the best practices from several disciplines into a winning formula for the management of complex projects. She has ideated and implemented a unique blend of standard project management and systems engineering processes to achieve optimal science and engineering outcomes through the appropriate process rigor applied to business and proposal development and project management across all disciplines.

Lory has a Master of Business Administration (MBA) in Information Technology Management and is both a Certified Project Management Professional (PMP®) and an Expert Systems Engineer (INCOSE® ESEP). Wingate's area of expertise is in project management, program management, and systems engineering, and she actively pursues opportunities to present training workshops and materials associated with her areas of expertise. *Systems Engineering for Projects: Achieving Positive Outcomes in a Complex World* provides a set of tools and techniques valuable in managing the technical scope of any project.

Chapter 1

The Discipline of Systems Engineering

The elements of systems engineering that will be explored in this chapter include:

- origination
- evolution
- impact on the business
- organization

Systems engineering is a discipline that provides structured processes necessary to manage projects that require performance from many disciplines in order to achieve their objectives. In Sections 1.1 and 1.2, this chapter first explores the origination and evolution of the discipline of systems engineering, the resulting terminology changes, and standards development. There is a long history associated with the development of the processes, with a convergence of techniques that have led to a formal designation of an integrated discipline. This evolution is continuing. The future of systems engineering will be fully discussed in Chapter 8.

In Section 1.3, the application of the systems engineering discipline within an organization is discussed. Section 1.4 describes an overview of systems engineering, including the system and system-of-systems approaches. A case study demonstrating the application of systems engineering processes is provided in Section 1.5. The purpose of each case study in this book is to demonstrate that systems engineering processes and practices are a natural controlling mechanism associated with projects across all fields and disciplines. And awareness of the natural alignment of activities to the established systems engineering processes, when appropriately applied, can increase the probability of achieving a successful outcome. Systems engineering processes can be used in most every scenario to develop and implement a roadmap that will lead to successful outcomes and that meets the desires of the key stakeholders. The stakeholder group includes the customers as well as any other key individuals who have a vested interest in the project. Although specific systems engineering activities were not consciously planned into the project, the case study in this chapter demonstrates the natural inclination to categorize and organize in a methodical way in order to increase the potential of achieving success. The focus of these cases provides an opportunity to review an activity that would not typically be chosen to represent systems engineering such as is found in other texts; however, these cases well represent standard systems engineering processes in use. This provides the reader with a greater understanding of the applicability of these processes to their own situations, which may fall outside of the typical organizational examples.

Finally, a summary checklist of the key concepts is available in Section 1.6 to assist learning. Apply Now exercises, which allow for the immediate application of the information in this chapter, are included in Section 1.7.

Chapter Roadmap

Chapter 1 focuses on systems engineering as a discipline. It specifically:

- explains how systems engineering emerged as a discipline
- describes the evolution of the systems engineering discipline
- explores the impact of systems engineering within businesses
- uses a case study to demonstrate the application of systems engineering processes
- provides a summary checklist of the key concepts to assist learning
- provides Apply Now exercises to assist in the application of the concepts from the chapter to a real situation

1.1 Origination of Systems Engineering

This section explains how the systems engineering discipline originated and how it overlaps systems thinking.

- Systems engineering originally emerged from a need to achieve better project outcomes.
- It began as a way to view the interactions of various engineering disciplines as they were related to a single project or product.
- It takes into formal consideration the stakeholders' vision.
- Systems engineering forms a holistic and synergistic perspective of the product.
- It has led to formal education programs and implementation standards.

One of the ways that ideas have been brought to life in the world has been through invention. Interested persons throughout time have attempted to create new products through trial and error. They had a great idea and put it to the test. Ideas that were successful in a prototype form were then produced in larger quantities. Early in the technology age, these inventions were seen as remarkable and interesting creations, mostly unavailable to the average person. Communications about the new object and what it could do were limited to the media reach and the budget of the individual inventor. In some cases, even if the new product was of interest outside a small and local community, the costs to produce the item would limit its production and distribution to a small population. As the technological age advanced, engineering took over the majority of duties from the individual inventor. Inventors were expected to focus on early research and development (R&D), whereas engineers took the design and created production versions that could be mass produced at a lower cost.

For the greater part of the technological age, inventions typically were of the mechanical and analog electronic forms. The move to digital technology was revolutionary. Where analog technology simply converted a signal in its original form, digital technology sampled the signal, converted it to numerals, and then stored it in a digital device. During the output phase, the numbers are transformed into voltage waves that approximate the original signal. Conversely, the analog signal is read as recorded, amplified and projected directly through a speaker or other output device. With the conversion to digital technology, the signal was not subjected to degradation and could be compressed. This allowed more information to be stored on smaller media, which opened up a new paradigm and radically changed the nature of invention and production. Opportunities for new products that could be designed, mass produced, and then made available to the general population at reasonable cost meant that the focus would change from the initial state of invention to production efficiencies.

Although the processes that ultimately make up systems engineering were practiced in various forms throughout history, the first documented use of an approach that took the system into consideration was in Bell Telephone Laboratories. Their use of "systems engineering" as a term to reflect their methods was documented in the 1940s.[1] Bell Laboratories was involved in military action optimization studies during World War II. Scientists and engineers there were using operations research methods, specifically optimization modeling using calculus, linear algebra, and other techniques, as well as stochastic processes such as queuing theory and probability theory. It was not until 1962, however, that "Arthur Hall published his first book on systems engineering: *A Methodology for Systems Engineering.* Hall was an executive at Bell Laboratories and was one of the people who were responsible for the implementation of systems engineering at the company."[2] During the same era, the RAND Corporation, ideated by a newly formed United States Air Force, developed a process for systems analysis that would become an important part of the systems engineering discipline. Systems analysis is an approach that reviews a problem in logical steps, and describes the system thoroughly and explicitly. Using computing resources to perform systems analysis and optimization modeling provides a solid scientifically based approach for performing systems engineering.

From the 1950s through the 1970s, the digital revolution spawned unexpected and inconceivable growth in industry. The act of inventing proliferated through every discipline, and those who utilized technology, particularly digital technology, found tremendous opportunities. Evolution over time from simple systems engineered primarily from mechanical hardware and electronics to complicated systems integrating multiple engineering disciplines became the norm. Complicated, but not complex, designs were developed and used to manufacture the majority of products. The speed at which products were brought to market increased exponentially. The ability to think of an idea; conceptualize a design; develop a prototype; test, verify, and validate the performance of the capability; and produce the design to specifications all formed the basis for movement to the design-build-test method of engineering that was popular during this era.

The design-build-test method of engineering provided the structure for the engineer to research the area of interest, develop a solution based on known requirements, create the prototype, and test it. However, this method did not, as part of its standard process, provide for consideration of the contribution of other disciplines (such as software development that was inserted into hardware), or provide a way for the stakeholders to articulate their needs and validate that the product met their needs. It was not until project development started crossing disciplines on a regular basis that the limitations of a traditional engineering process started to emerge. A systems engineering concept was beginning to be developed. There was a "growing sense of a need for, and possibility of, a scientific approach to problems of organization and complexity in a 'science of systems' *per se.*"[3]

The United States Department of Defense and the National Aeronautics and Space Administration (NASA) were early adopters of this integrated approach to development, driven out of necessity to complete large, complicated products to support war fighting and space exploration. Indeed, most people think of the NASA Apollo Program when they think of "classic systems engineering." "The task . . . was daunting and complicated, it involved breaking the underlying goal into multiple sections or manageable parts that participating agencies and companies could work with and comprehend. These various parts then had to be reintegrated into one whole solution, and as a result, careful attention and management involving extensive testing and verification was necessary. The complex nature of these tasks made systems engineering a suitable tool for designing such systems. It was the principles of systems engineering that resulted in the rigorous system solution, which contributed to Apollo's overall success."[4]

The first attempts to teach systems engineering concepts occurred in the 1950s by Bell Laboratories, but with limited success. By the 1980s, strong demands to meet the needs of the industries interested in developing and implementing an integrated systems approach drove universities in the United States to offer the requisite courses that ultimately produced a pipeline of specialist engineers. Graduates of these courses were following a rigorous approach designed to facilitate major programs. Military standards were also being implemented within government-funded programs, helping to form the foundation for the systems engineering concepts that are in place today.

Once the digital age evolved into what is known as the information age during the 1980s, and projects became, more often than not, interdisciplinary and complex, systems engineering truly became a necessity for successful project execution. These projects generally involved the heavy integration of software. Many of the engineering projects that covered the mechanical, civil, and safety, disciplines or software-enabled systems required integration over networks. Businesses had to rethink how they operated. Stakeholders had more choices and had the ability to shift their loyalties and purchase power from one company to another and from one product to another. They demanded more participation in the development as well. Systems engineering became sought after by these organizations as a way to help meet their stakeholder's needs and improve their organizational performance.[5]

Projects that required careful integration of project elements from multiple disciplines needed a better way to ensure that they achieved the expected results within the cost, schedule, and scope anticipated by the stakeholders. Projects were becoming more and more complicated and spanned all areas of business, such as platform-related products (e.g., aircraft), infrastructure products (e.g., bridges and buildings), information-dense products associated with command and control activities, and enterprise systems distributed across organizations and, in some cases, other companies and countries. All of these independent pieces of the system needed to be brought together at the right time, in the right place, to maximize the effectiveness of the system as a whole. The approach needed to be holistic and synthesizing. And it needed to focus on the desires of the stakeholder. With the application of systems engineering comes the stakeholder focus throughout the project.[6] The more complicated the project became, the more it was liable to be diverted (unintentionally) from the stakeholders' desired vision.

The need to understand and correctly identify what stakeholders wanted, to convert that vision into a straightforward functional architecture, and then to be able to decompose that functional architecture into a physical architecture that could be modeled and tested in order to ensure the best design to meet the stakeholders' expectations were met was an imperative. This top-down approach was called "systems" engineering for organizational-level implementation, or "system" engineering for activities associated with a single project. This approach envisioned incorporating work from across all contributing disciplines instead of through individual downstream disciplines, such as mechanical or electrical engineering, with a focus on ensuring that stakeholders received what they needed from a project. It was seen as a holistic approach that fundamentally would provide a whole system in its complete form to the stakeholders.

Systems engineering added tremendous value to the project management methods. It was holistic in nature and spanned the life cycle of the system. It focused on stakeholders—both their needs and their perceptions of the provided holistic system. And perhaps most importantly, it took into consideration all disciplines that contribute to the performance of the system as a whole, regardless of in which disciplines the elements of the system originate.[7] The key processes in this new discipline included understanding the end state desired by the stakeholders, synthesizing, optimizing, implementing, proving, and then sustaining the solution. These activities are generally divided into stages of a life cycle so that each stage can be practically organized and managed to successful conclusion.

1.2 Evolution of the Systems Engineering Discipline

This section describes how the systems engineering concept evolved and obtained a formal status as a defined discipline.

- The descriptions of the terms "system," "system engineer," "systems engineer," "system engineering," and "systems engineering" were formalized to ensure a common reference and understanding across disciplines.
- Systems engineering has developed into a certificate- and degree-awarding formal discipline.

Once the need for a concerted effort in understanding and effecting projects from a system perspective was acknowledged, it was a natural next step for a disciplinary approach to be crystalized and formalized. The desire was to create a formula that could be used repetitively to minimize risk and increase the probability of successful outcomes for all systems projects.

Over time, papers were written and presented; books outlining techniques became common; and organizations formed to codify terminology, set standards, and define approaches. In addition, more and more undergraduate- and graduate-level degree programs, educational coursework and certifications, conferences, training seminars, and internships became available. The formal job title of System or Systems Engineer was instituted as a valid title for recruitment, although there was, and still is, a wide variation of descriptions as to the role that is performed. And even though the discipline of systems engineering has come far and is now considered part of the mainstream job market, the actual position descriptions can still be far removed from the descriptions found in the standards documentation. There is a long way to go before convergence occurs, where a clear position description in the workforce matches the description of the anticipated work of a systems engineer that is documented in the standards.

As it is anticipated that projects will continue to become more complex, systems engineering is expected to become more important over time. The anticipated future evolution of the discipline will be outlined in Chapter 8.

1.2.1 Terminology

In order to effectively communicate the requirements of performance in this emerging discipline, definitions needed to be established that would be accepted by the broader engineering community and would appropriately convey the intended roles of the systems engineer on the project. A common reference and understanding was needed. "The language used by engineers and the language used to describe what they conceive, make, and do is changing. Engineers used to talk about things like force, shape, size, tolerance, modulus, voltage, temperature, precision, construction, and speed. Of course, they still do, but today, they are even more likely to talk about scale, scope, state, complexity, integration, architecture, resilience, evolution, affordability, and social context."[8] The necessity to speak a common language led to a common description of the terms "system," and "system engineering." These have been formalized and documented in systems engineering standards.

Whenever more than one item is connected in some way (through process, physical connections, or wireless connections, for example), the definition of a system can be used to describe the whole entity.[9] Further, this entity may often achieve greater performance as a whole system, than what can be achieved individually by the parts.[10] Components can refer to elements, sub-elements, or actual component-level products that interact to create a system desired by the stakeholders. "The results include system-level qualities, properties, characteristics, functions, behavior, and performance. The value added by the system as a whole, beyond that contributed independently by the parts, is primarily created by the relationship among the parts; that is, how they are interconnected."[11] A description of a system will include everything that is required to make the system function in its intended environment, and nothing more.

System(s) engineering is described as a multidisciplinary approach, based on established quantitative principles, that provides for addressing and meeting stakeholder needs in the formation, design, and production of a solution within established costs and schedule and within acceptable risk parameters.[12]

It is an interactive, cross-disciplinary, top-down, iterative, and methodical approach to the life cycle of a systems project. Systems engineering "is an approach and a discipline that puts an activity into context and reviews all aspects of performance from a systemic perspective. As a discipline, it provides structure and methods to define and organize projects, to integrate activities and ensure that interfaces are correctly identified and addressed, ensures testing of components and systems are completed, manages risks and reviews, and performs configuration management to ensure that design changes are tracked and implemented methodically so the current configuration is always known."[13]

1.2.2 Standards

Standards, which can be used as established methods for accomplishing tasks within a discipline and for which quality (compliance) can be measured, are important in promoting effectiveness, managing risks, and increasing the probability of success in projects of all types. Implementation of sound systems engineering practices is meant to minimize unexpected emergent behaviors and reduce complexity. There are many applicable standards across multiple disciplines that could be applied to systems engineering processes. Some routinely are used for that purpose, including those from such standards-generating organizations as the Institute of Electrical and Electronics Engineering (IEEE), the International Organization for Standardization (ISO), the American National Standards Institute (ANSI), the Department of Defense Military Standards (DoD MIL-STD-xxx), the International Electrotechnical Commission (IEC), the Electronic Industries Alliance (EIA), the Society of Automotive Engineers, International (SAE®), and the International Council on Systems Engineering (INCOSE). INCOSE was founded in 1990. It hosts the largest membership of systems engineering practitioners to date. It is also a certification-granting organization. "INCOSE champions the art, science, discipline, and practice of systems engineering."[14]

Other standards-generating organizations have their own or similar terminology and processes, including the following:

- International Association of Project Managers (IAPM)[15]
- Association for Project Management (APM)[16]
- International Project Management Association (IPMA®)[17]
- Project Management Institute (PMI)[18]

Some organizations combine efforts to set standards that are overarching as well, such as:

- "ISO/IEC JTC1/SC7 (Information Technology, Systems and Software Engineering)"[19]
- An international standard, ISO/IEC/IEEE 15288, which has been established as an inclusive set of process standards that can be used across engineering disciplines.[20]

These standards, along with many others, have been developed that offer sometimes consistent, but often conflicting, overlapping, and sometimes incompatible or inconsistent guidance. This overlap may cause risk in systems projects because different discipline practitioners are implementing what appear to be the same processes. In particular, there are noted overlaps between software engineering (IEEE SW Engineering standards) and information technology (IEEE CS) vocabulary and processes, and the project management and systems engineering standards-generating organizations' vocabulary and management processes. As systems engineering is an overarching, top-down effort, this can be problematic in that processes can drive action and design at the lowest levels of the project before the project is ready. This in turn drives risk and increases the potential for catastrophic failure of the project.

Efforts are being made to try to bring standards in line to minimize the risk wherever an over-lap with the systems engineering discipline has been noted. In 2011, an initiative sponsored by the United States DoD and led by the Stevens Institute of Technology, in collaboration with the Naval Postgraduate School and other professional agencies, created a wiki-based collection of materials called the *Guide to the Systems Engineering Body of Knowledge* (SEBoK). This documentation was viewed as "a living authoritative guide that discusses knowledge relevant to Systems Engineering."[21] "The three SEBoK steward organizations—the International Council on Systems Engineering (INCOSE), the Institute of Electrical and Electronics Engineering Computer Society (IEEE-CS), and the Systems Engineering Research Center (SERC) provide the funding and resources needed to sustain and evolve the SEBoK and make it available as a free and open resource to all."[22] The SEBoK governance and Editorial Board took the lead on outlining the explicit issues with the overlaps and is actively working with the stakeholders to drive a positive change, which could ultimately benefit all projects.

1.3 Organizational Implementation of Systems Engineering

This section describes the following considerations for organizational implementation of systems engineering:

- Implementation of systems engineering into organizations requires commitment to the discipline as a whole to be successful.
- Determining the best strategy for implementing and performing systems engineering is an organizational leadership decision.
- Systems engineering can increase the probability of project success in any domain.

1.3.1 Organizational Structure and Culture

Depending on the existing structure of the organization, different levels of autonomy in implementing systems engineering will be available. Generally, the larger an organization is, the more difficult it is to implement significant changes to the structure because the number of stakeholders that are involved increases exponentially. Obtaining stakeholder buy-in and support becomes a major challenge to overcome. In addition to size of the organization, the way it is funded and legally structured may add additional constraints to restructuring. For example, organizations that obtain their funding from governmental bodies may be subjected to more stringent organizational controls than companies that obtain their funding from the private sector. And organizations in highly regulated fields, such as the medical or financial domains, may be required to obtain additional approvals for changes to their organizational structure.

Organizational structure and culture can impact the implementation of systems engineering and the resulting successes. "Virtually every significant business or enterprise that creates products or services benefits from performing a wide variety of systems engineering activities to increase the value that those products and services deliver to its owners, customers, employees, regulators, and other stakeholders."[23] However, it is often the case that more established organizations that saw great successes with the traditional design-build-test downstream engineering often find it difficult to implement changes that were necessary to implement a cross-discipline systems engineering function. As systems engineering, by its nature, spans disciplines,[24] the execution often becomes problematic, because there are firmly established and often embedded processes that must be replaced.

Implementing systems engineering into an organization requires an understanding of the business processes, governance model, supplier management, etc. If an organization is new to systems engineering, an evaluation of all the interfaces associated with each systems engineering process must be mapped against each process owned by another department/division, to ensure that the hand-offs between the systems engineer and those departments that do the detailed work are known and addressed before the official implementation begins.

Although the systems engineering discipline has been established to be applicable across all industries and includes processes written independent of any domain, its application in different specialized business areas (e.g., information technology firms, financial industry, medical industry, etc.) might require tailoring of the process rigor to address specialized languages and vocabulary.

Taking the time and effort to install a systems engineering discipline into an organization is well worth the effort. In trying to quantify the return on investment (ROI) from implementing systems engineering processes on projects, research has shown clear and compelling evidence that systems engineering efforts have a definite, discernible impact on the probability of the achievement of successful project outcomes. Indeed, research on impacts on success rates on projects that implement systems engineering shows clear evidence that the value achieved is worth the investment.[25] Associations and institutions from various disciplines have conducted research during the last decade that confirms the existence of evidence that the use of systems engineering makes a valuable contribution to project success.[26]

1.4 Overview of the Systems Engineering Discipline

This section provides a high-level overview of systems engineering to orient the reader on the basic structure of the discipline, as a starting point. The way that system engineering fits and adds value to a project will be discussed in depth in Chapter 2. The specific processes that are used to perform systems engineering will be reviewed in Chapters 3 and 4. Chapter 5 provides information on applying success measurements to track progress. Chapter 6 discusses how to tailor the processes to assure that the appropriate rigor is applied to different project types.

- Systems engineering is an overarching, top-down, integrated approach to managing a project.
- Systems engineering views a project from the life cycle perspective—that is, the end-to-end life of the project from conception through to closure, disposal, or restoration to its natural or original form.
- With experience on more complex projects, systems engineers started referencing "systems thinking" ideas to explain emergence.
- The discipline of systems engineering, when applied to a complex project, requires additional discipline and rigor.
- A system-of-systems approach provides the strongest defense against undesired emergent behavior.

1.4.1 System Approach

The system approach is a holistic, top-down approach. This means that instead of starting from a known capability and designing upward to a more integrated capability, one must look at the intended purpose of the design and decompose the parts of the design into manageable sections—all while keeping a firm perspective on how everything fits together. Systems engineering integrates all required design disciplines and specialty groups by creating and implementing a structured process that overarches the entire life cycle. Figure 1.1 shows a representation of a system.

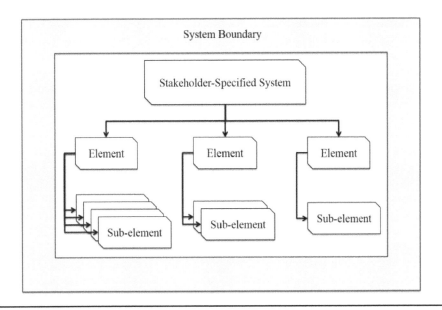

Figure 1.1   System Representation

The design disciplines include software and hardware engineering, reliability and maintainability, human factors (usability), environmental, safety and security, manufacturing/production, support, and many others. It is an interdisciplinary approach, combining resources and integrating them in a manner to fulfill a stated and defined need and to accomplish a specific mission or function. When describing a system, one can envision multiple parts coming together into a whole to perform a function. The description implies interactions of some form, although the connection of parts may be static. Complexity enters the picture when parts of the system have life cycles and purposes outside of one system but are still relied upon by that system for its overall purpose.[27] "When properly implemented, systems engineering will:

- Provide a structured process for integrating and linking requirements, schedule, decision milestones, and verification.
- Enable the project team to work to a single, integrated set of requirements and processes.
- Enable integration of the system at the requirements and design stages rather than waiting until hardware and software is available.
- Reduce unplanned and costly reengineering necessary to resolve omissions and integration difficulties.
- Enables the project team to sense, anticipate, and prepare for continuing change in an orderly manner."[28]

It has been shown time and again that risk to the project compounds dramatically if the appropriate level of systems engineering rigor is not implemented. It can be expected to a high degree of certainty that if the appropriate rigor is not followed, the project will take longer, cost more, and will not meet the required technical specifications, ultimately leading to a dissatisfied set of stakeholders.

The systems engineering life cycle model focuses on every phase of a project from concept development to the capability's end-of-life state. It starts at the concept development phase in which the context of the project, the concept of operations, a functional definition, and system requirements definition are formed between the systems engineer and the stakeholders. From there, a system definition evolves.

A system definition requires the elaboration of the functional view to a physical architecture, which is further decomposed into elements and sub-element definitions. For each of these, from the element and sub-element areas down to the component level, the life cycle of the elements will be reviewed, and trade space and feasibility analysis requirements will be identified for scheduling with the project manager. A definition of the system and/or its elements is generally modeled or simulated, to optimize the production, operations, and maintenance design. During this development phase, the systems engineer also performs the critical activities of arranging and managing all pertinent technical reviews and performing configuration management on the technical areas.

It is important to note that it is the system engineer's responsibility to ensure that the project manager, who has responsibility for the overall project, is kept informed and is part of the approval process for all changes, reviews, etc. The project manager performs the appropriate change management activities for cost, schedule, and scope changes for which the systems engineer will be providing input. During the production phase, the systems engineer will then ensure that with test, verification, and validation, the requirements are fully met and the stakeholders are fully satisfied. This is the phase that brings the system into service.

The system is then operated and supported, during which time the systems engineer oversees maintenance and repair activities and upgrades, which may, if not carefully planned and implemented, cause unintended consequences to other parts of the system. The systems engineer will also evaluate and complete an analysis of systemic performance issues, working with the appropriate responsible disciplines to fully resolve and restore the system to its functioning level. Finally, upon notification of closure requirements, the systems engineer will ensure that the integrated parts of the system, particularly of a complex project with independent elements, are taken down gracefully.

Within each of these overarching life cycle phases, specific detailed processes provide the structure to collect the information that is necessary to ensure that the system capability comes together as a whole and can function in a way that meets the anticipated use. These processes consider all of the different disciplines that impact the system, including all engineering disciplines, human resources, quality, acquisition and procurement, infrastructure, operations, and maintenance. These processes will be fully described in Chapters 3 and 4.

1.4.2 System-of-Systems Approach

Systems can be complicated or complex, but a system-of-systems is always complex. When a project is complex, its structure differs from traditional projects in a few predictable ways. A complex project will have some elements that have operational and/or managerial independence from the main system. In other words, if an element of a system can operate independent of the system, if it is not necessary for its existence independent of the system, but when integrated into the system provides significant value added, then it is a system-of-systems (see Figure 1.2).

A system-of-systems output can be unpredictable. Systems engineers appreciate that the combined elements of a system may not behave as a sum of the parts, but may demonstrate what is often referred to as "emergent behavior." Emergent behavior refers to the inputs which, when applied together, do not equal the expected outputs (actual outputs are different than expected) due to the effect of some unknown force. It reflects the fact that just the simple aspect of adding multiple parts together does not in fact predict what the additive performance is anticipated to be. There are many reasons that this occurs. There are performance impacts from the interconnections between elements. There may also be impacts from overlapping or duplicative associated processes. Learning algorithms embedded in subelements can affect performance in unpredictable ways as well. As technology continues to evolve and become more interrelated and complex, heightening risks and system unpredictability, research continues in the area of understanding emergent behavior in an effort to try to mitigate those risks and better predict system behavior.[29]

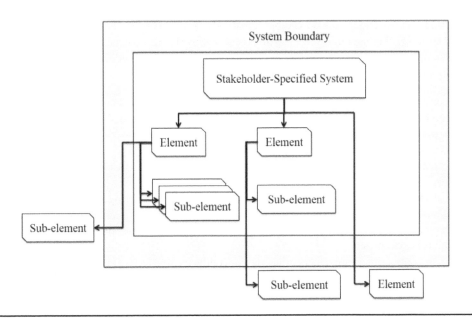

Figure 1.2   System-of-Systems Representation

A complex system's behavior, as a whole, does not equal the sum of its parts. It will exhibit emergent behavior. This means that the system, its elements, and its sub-elements can be drivers of, as well as exposed to, unpredictable outcomes as a collective. It is the relationship between the elements that provide the expected and desired, as well as undesirable, outcomes. This relationship between the elements drives the risk as well. Through the current understanding of complex systems, it is clear that when dealing with a system-of-systems project, the risk is greater, and therefore the rigor and contingencies applied are more significant.

Systems science, complexity theory, chaos theory, and operations research all try to explain the behavior of systems. Systems science research is a field that appeals to the systems engineer, as it matches the cross-discipline, across-life–cycle, holistic approach that systems engineering favors. Both systems science and systems engineering aim to resolve complexity and enable better predictions.[30] Systems thinking, on the other hand, describes a method of thinking through a problem and then resolving it effectively through analysis. It involves observing a situation and then carefully assessing all of the inputs that are effecting the situation in an attempt to determine system performance.[31]

From systems science it is understood that the system behavior can be explained by its elements. The premise in systems science is that one can understand the behavior of systems by understanding the individual parts that make up that system.[32] It proposes predictability in behavior of the system as a whole based on the behavior at the elemental level. Chaos theory focuses on nonlinearity and the appearance of seemingly unrelated consequences to even the smallest change. Complexity theory emerged from systems and chaos theory and is focused on the application of structure in organizational projects. It recognized emergent behavior but did not provide a method for addressing the emergence. Operations research methods attempt to provide a quantitatively rigorous prediction that can be applied to minimize risks of complex behavior. These theoretical concepts and tools provide ideas about how to think about complex projects. However, they do not provide structures to reduce the risks of the complexity. This is where systems engineering adds its value.

Systems engineering did not evolve from scientific systems thinking. As systems engineering evolved and projects became more complex, systems engineers began looking for ways to understand and predict the behavior of the systems. Systems engineering organizations and practitioners look to the sciences to find ways to explain the observable, apparently nonlinear behavior and to use that knowledge to add structure that may minimize the threats caused by the unpredictable outcomes.

Project complexity challenges that systems engineers are attempting to address are the areas of high risk that can often make a project unmanageable. These high-risk areas are:

- the constantly evolving introduction of new technologies at an ever-increasing pace
- dynamic conditions in the global economic and political environments that affect suppliers and other partnering and funding situations
- changes in strategies and priorities within partnering organizations
- the extension of expected product life cycles through capability expansion or revisions that keep it alive
- radical changes in technology that cause the early demise of existing capability that may be relied upon for overall success
- the level of involvement of partnering organizations within the same country and organizations within other countries
- different contracting arrangements that are not conducive to effective systems engineering (e.g., letters of agreement, in-kind or contributed effort, etc., which are extremely difficult to control, particularly if the requirements being provided to those organizations are not refined, or if they are early in the product life cycle [research and development area—a high risk project phase])
- cultural differences, communications challenges, and clarity in requirements and understanding from both the stakeholders and the project as to what exactly is to be delivered
- the level of concurrent evolution in the elements and sub-elements, such as redesign and modification, revisions to address errors and problems, and adaptations to increase capabilities

The incredible moving parts of these complex projects must all be coordinated and synchronized, organized, and synthesized if there is any hope to complete a working system. As a design is maturing, resilience and robustness need to be built in at every level, so unforeseen, unplanned (unknown–unknowns) disruptions to the performance specifications of the individual parts do not catastrophically impact the evolving system as a whole. This can be accomplished by clearly understanding the interoperability model in which the system operates, to build in as much compatibility and modularity as possible and to minimize the interfaces and interoperability that will be impacted by the above actions (version changes, heavy use, design modifications, etc.). The approach used in systems engineering to address complex projects is to use processes that can help predict, from a synergistic, integrated perspective, any potential emergent behavior early enough in the project to mitigate the risks. It is far too common in system-of-systems projects to miss schedule, have significantly increased costs, and to underperform on scope. In addition, it is common to place blame for failing systems elements onto another part of the system without actively trying to identify the true cause of the failure and then take responsibility for fixing that failure. It is therefore an imperative for this type of project to actively manage with systems engineering rigor. These processes will be fully described in Chapters 3 and 4.

The phenomenon of system-of-systems project underperformance can be attributed to several actions. First, the organizations involved and responsible for the performance of activities in support of this type of project needs to be motivated to the purpose, especially as the purpose of the project is sometimes at cross-purposes with the organization's strategic goals or vision, which may support several complex projects or programs. A system-of-systems project requires that every organizational unit responsible for the performance of its part of the system is in agreement with the larger system owners as to what their responsibilities are and how they will go about performing according to those requirements. In addition, the organization's leaders need to accept the responsibilities of the interfaces and any breakdown that might be attributed to the interfaces for which they are engaged. There are some risks that cannot be mitigated from within the project. They must be addressed in a way to set the project up for success. These include managerial and leadership constructs.

Element and sub-element systems of a defined system-of-systems project can be bound manageri-ally throughout a lengthy continuum from loosely coupled to tightly coupled. If an element and sub-elements of a defined system-of-systems project is tightly coupled, then it is centrally managed, and the operations modes of the element and sub-elements favor the system-of-systems project's central managed purpose. In other words, if a choice must be made within an independently operating element or sub-element organization as to priorities, the priorities of the system-of-system project will be the highest priority. If an element and sub-elements are loosely coupled, there will be a lack of that central management authority, no enforcement of compliance, and standards are employed at the whim of the external organizational leadership. This type of arrangement drives the risk on the project to the point where the probability of successfully executing the project with the original cost, schedule, and scope is almost nonexistent. This is because the less tightly coupled a system-of-systems project is, the higher the level of 'volunteerism' in performing the objective is required, and the limited control that can be effected on the project, and therefore, the higher the risk to the project because other higher priority requirements will inevitably and always usurp the volunteer efforts. Uncoupled or loosely coupled ele-ment and sub-element systems generally prescribe lower priority to the needs of the system-of-systems project and higher priority to the needs associated with performance as an independent unit.

The fact that organizational units often do not have leadership in place that can make decisions for the entire (cross-organizational) system-of-systems project drives the need to have formal agreements in place, with agreed-upon measures of performance (MOP); to have an inclusive decision-making Board in place that includes all project organizations; and, most importantly, to have agreed-upon conse-quences, ramifications, and alternative plans in the event of element and sub-element nonperformance.

Having structures in place will increase the probability that performance and cross-organizational issues are resolved. However, it will not entirely remove risk. This is due to the fact that in addition to the already-existing emergent behavior risk, organizations that are performing additional work outside of the specific project will inevitably evolve and make impactful changes that will trigger unintended consequences for the system. Key considerations with any system-of-systems project is the autonomy of the constituent parts, where continued evolution continues within the unique organization, unrelated to the system-of-systems project, but impacting it because of known or unknown interfaces. Concepts for managing a system-of-systems project are not any different from a complex project; however, more rigor must be implemented to decrease the risk and increase the probability of success.

1.5 Case Study: Trans-Greenland Expedition

1.5.1 Structure and Purpose

The purpose of each case study in this book is to demonstrate that systems engineering processes and practices are a natural controlling mechanism associated with projects across all fields and disciplines, and that consciously being aware of the natural alignment of activities to systems engineering processes when appropriately applied can increase the probability of a successful outcome. Systems engineering processes can be used in most every scenario to develop and implement a roadmap that will lead to successful outcomes and that meet the desires of the key stakeholders. In this next case study, although specific systems engineering activities were not consciously planned into the project, it demonstrates the natural inclination to categorize and organize in a methodical way in order to increase the potential of achieving success.

The focus of this case provides an opportunity to review an activity that would not typically be cho-sen to represent systems engineering such as is found in other texts; however, it well represents standard systems engineering process use. This provides the reader with a greater understanding of the applica-bility of these processes to their own situations, which may fall outside of the typical organizational examples. All case study quotes are from interviews held with the individual presenting the background. In this next case, the interview was held with Sue Stockdale, and all quotes are attributed to her.

1.5.2 Project Charter and Plan

In 1999, a team of four explorers, including British polar explorer, Sue Stockdale, embarked on an expedition to ski coast to coast across the Greenland Ice Cap in less than one month. Stockdale was highly qualified for the expedition. She was the "first British woman to ski to the Magnetic North Pole in 1996 during an expedition led by David Hempleman Adams."[33] Also in 1996, she served as Raleigh International Deputy Expedition Leader for three months in Chile and then as a member of the staff on the ship Professor Khromov on a trip to Antarctica in support of a UNESCO expedition. In 1998, Stockdale joined an international expedition to the Geographical North Pole.[34] Her experiences on these extreme adventures served as a solid knowledge base from which to prepare for this new expedition, which would be at least as physically and mentally demanding as her previous experiences. "The Arctic is an extreme environment and the price for failure can be death."[35]

They started on April 27th from Kangerlussuaq, Greenland, in the west coast fjords. The terrain that would be crossed included a badly crevassed section of glacial terrain starting at the ice sheet edge (Point 660), then a relatively flat expanse of ice cap, an ascent to the summit in the center of the ice cap at a height of around 2600 meters (~8,530 ft.), and then a descent from the ice cap back to sea level to Isortoq, a remote Inuit village on the east coast. From there they would fly by helicopter to Tasiilaq, and then to Kulusuk before returning home.

Each adventurer's physical fitness level was critical. They would be required to hike in the snow and ski in both downhill and Nordic styles, all while pulling a sled, commonly referred to as a sledge or pulk, which weighed in at around ~60 kg (132 lbs) at the start. The sledge contained all their necessary food, fuel, clothing, and equipment. They would also require skills in extreme Arctic camping in up to −45 degrees Celsius (−49 degrees Fahrenheit). This expedition would require them to have the strength and mental stamina to continue under any circumstances for up to 30 days without outside assistance.

1.5.2.1 Infrastructure

The expedition team was surrounded by the one element that they could not survive without: water. However, they needed to boil the snow/ice in order to make it drinkable. The fuel was the limiting factor as to how much water they would actually be able to consume.

They did not have to carry out human waste but were allowed to bury it. That allowed a reduction in their carry weight of materials. Carrying out human waste would have ultimately offset the weight they reduced through food and fuel consumption, to some degree.

1.5.2.2 Human Resources

The team members of this Trans-Greenland expedition included two males and two females from different countries (Norway/Germany/UK). They had not met face to face as a group prior to the expedition, although everyone knew at least one other person.

During the three-month planning phase, almost all communications (e.g., conference calls, faxes, and emails) were preparatory and factual, focusing on the logistics and plans of the expedition. To optimize their communications, they decided to use a common language, English. This was a helpful strategy that facilitated the planning process; however, the practice of only using English did not proliferate throughout all communications, which was discovered during the expedition when it became apparent that all of the medical supplies were identified in Norwegian—a mistake that could have cost someone's life.

Human factors can often make or break an expedition. One person can drive everyone to achieve great things, or he or she can bring everyone down to a point of despair. Group cohesiveness is important, and respect for one another's contributions and gifts is paramount. With this expedition, there was not one specific individual who held the role of leader. The expedition was more of a collaboration of experts, which made it slightly more complex and challenging. The team did not have enough time to develop the knowledge of who brought what skill sets and expertise to the experience.

While everyone was trying to establish their own place in the team, conflicts emerged. This made for some uneasy times during the first week when each individual was pulling the maximum weight on his or her sledge, over the most treacherous terrain, and while trying to find his or her sustainable group daily speed/distance.

Stockdale remembers, "To know what capabilities a team has, each individual must understand their own strengths and weaknesses. This point was reinforced to me very clearly on the expedition in Greenland. After a few days it became clear that there were varying skiing abilities in the team. Terje Fardal, the Norwegian, was extremely strong and fit and was always out in front. He was closely followed by the Norwegian woman and the German photographer. That left me, the Brit, at the back struggling to keep up! I initially felt annoyed because having completed three previous polar expeditions, I knew the sort of pace that one must keep to. After a while my annoyance turned to despair, as the others seemed able to cope, so I began to focus on failure rather than success and naturally my skiing performance suffered. At the end of the day, inside the tent, nothing was said. The next day Fardal did raise the issue of our skiing speed and as a team we had to come up with a solution. He suggested that he could take some of the weight from my sledge so that I could ski more quickly. I had to face up to my weakness but in doing so I would be helping the entire team to succeed. This was a tough lesson. It takes courage to recognize a weakness in yourself, admit it, and then do something about it."

Over the course of the expedition, the team's natural skills did emerge. Fardal was technical minded and practical with his hands. He was able to solve issues with broken and/or malfunctioning equipment. The German, who was the photographer of the group, also became the dominant navigator. Janne Sogn, the female Norwegian, was a physical therapist and therefore naturally focused on the well-being and health of the expedition participants. Stockdale's role was motivation and teamwork. Everyone had a unique role that was based on his or her own natural abilities. However, that did not mean that everything always went smoothly. Stockdale explains, "As the stress and exhaustion of skiing up to twelve hours a day began to take hold, the behavior of the team members changed. On one particularly tough day, after six hours of skiing in −40 degrees Celsius, my teammate suggested that we should stop because I was looking tired. I was incensed because I felt fine, but he would not admit it was actually him. Everyone else knew what the real issue was and so we decided to stop. Later inside the tent when he felt relaxed, he admitted he had not been entirely honest and that he was feeling vulnerable because he could not keep up."

One of the ideas they incorporated right from the beginning was to swap tents every night. They thought that this one action would help to mitigate any risk associated with social challenges brought on by the expeditions, such as a small group like this being together for an entire month under very challenging physical conditions. This approach would help them cope with small irritations associated with diverse personalities and ensure that no socially detrimental behaviors, such as forming cliques or ostracizing other team members, were able to take hold. An agreed-upon exit criteria for the expedition was that they would all leave as friends! So, they could not let the journey destroy their camaraderie.

1.5.2.3 Closure Phase

As a closure phase to this expedition, Stockdale set up a motivational talk less than a week after her return to assist in her assimilation back into normalcy. Having a goal set she felt was an important step. "Everything seems different. After spending a month in the stark white Arctic environment, the 'greenness' of the plant life around you is just overwhelming," says Stockdale. "There is such an unbelievable difference. You have a new appreciation for nature."

1.5.3 Stakeholder Engagement and Communications Management

The primary stakeholders in this expedition were the expedition participants. The success or failure of the project to cross Greenland would not have major implications on anyone outside of the team. Other than Fardal, the other participants did not have spouses or children anxiously awaiting their return. They had sponsors, and those sponsors did have requirements. They wanted high-quality photos, preferably of extreme conditions, and they wanted comments on the dehydrated food that was provided. The manufacturer of the skis wanted feedback on the performance of the equipment. Because of the isolation of Greenland from the wider world, where internet availability is not common, any contact with sponsors and other stakeholders (such as family and friends) required waiting until the expedition was complete.

1.5.3.1 Decision Management

Decision management is an important systems engineering process. This expedition team developed an effective way to manage their decision making: following Tuckman's "storming, forming, and norming" performance model of group performance evolution. According to Stockdale, they first experienced some personality clashing, which then evolved into a functioning team, and then ultimately exhibiting group norms. The team's subject matter expertise became evident over time. However, the team then looked to the subject matter expert to make a recommendation, which they would then agree on or debate. Safety was more important than getting there and was always the primary consideration. Stockdale explains, "Around day 26, the ice we were about to cross looked iffy. We were heading downhill and it looked nondescript but very smooth and almost too easy. The waypoints took us on a route that looked very old and more difficult and out of the way. Upon discussion, the team agreed that although the potential route looked sensible, it seemed safer to follow the waypoints, just as we had done up to that point. There was no way to tell what was under that ice and it just didn't make sense to risk it."

1.5.4 Scope Management

As this expedition was unsupported, all necessities were required to be transportable on the sledge. Obviously, weight was a considerable issue. "When you have to drag a sledge for a month weighing around 60 kg, you think carefully about what is inside it. Every item is selected to be adaptable—socks become gloves, empty food bags become rubbish sacks, even the pages of a book can have other uses!" Stockdale exclaims.

Their basic equipment included their sledge, tent, snow pegs, sleeping bags appropriate for the harsh negative-Celsius conditions; inflatable and foam sleeping mattresses and repair kit; stoves, cookware, and eating/drinking cup/bowl/utensils; fuel; maps/global positioning system (GPS)/compass and extra batteries, snow shovels, alpine ski touring or Nordic skis and boots, ski poles, lightweight harness, jumar clip (ascending device), climbing carabiners, lightweight ice axe, ropes and knowledge to self-rescue from a crevasse; full winter clothing, including water and windproof jacket and trousers, insulated down jacket, face mask, expedition gloves/mitts/hat/balaclava, lightweight crampons, 100% UV sunglasses and/or snow goggles, insulated water bottles, toiletries, knife, and food.

Food was the most variable element to consider for weight. It needed to be nutritious, dense in calories, and filling, but be as light as possible. Ensuring that the food, most of which was dehydrated, did not get wet or succumb to the environment was an imperative. To break the monotony, the team allowed themselves the luxury once a week of sharing a dehydrated chocolate mousse dessert, which they would look forward to for days and then argue as to "who got to lick the spoon and the bowl!"

Fuel requirements were calculated by the number of days of the expedition plus the number of times they would cook, along with how long they would burn the fuel at a time, considering what was needed to cook food, melt snow for water, and for psychological comfort and warmth.

The team also needed to have a way to cope with needed repairs as well as communicate with the outside world, in the event of an emergency. For this they had a satellite telephone, medical kit and emergency flares, and some basic repair materials. In addition to this basic equipment, sponsors requested that certain untested ski equipment be incorporated and field tested during the expedition. One expedition participant, a Norwegian ski race manager, had expertise in this area and agreed to carry out a comparative test of ski bases. The team also brought along kites to experiment with, which added weight to the sledges.

1.5.4.1 Architectural Design

The architectural design of this expedition was quite simple. The Trans-Greenland Expedition was designed as a transit of the Kangerlussuaq to Isortoq route—a well-traveled, well-marked, and way-pointed path. There was no official trail over the ice, but previous expeditions had supplied explicit GPS coordinates to follow.

The team would start on the west coast because it provided a port of easier access, flights were less costly and more readily available, and gear could be shipped less expensively. The east coast exit would require helicopter transits; the cost and schedule risk would be minimized because the weight of their equipment would be less at the end of the expedition, once food and fuel was used. The start was planned at the time of year when the risk of weather and traveling over the snow and ice would be minimized.

The plan was to ski approximately eight hours a day, taking a five-minute break every hour. The lead skier would track the time on a watch. Strict adherence to the schedule was kept in order to keep up the progress, keep from becoming chilled, and ensure excessive food was not consumed. All food and fuel had been carefully measured to last up to 30 days. It was expected that during the beginning of the expedition, when the sledges were the heaviest, the speed and distance of travel would be reduced, and that toward the end of the expedition this time and distance could be made up.

From an overarching design-synthesis perspective, each individual activity that led up to the whole of the experience was important. And as in most complex systems, the outcomes can be different from what would be expected. As Stockdale explains, "The biggest thing you must remember is that regardless of how well everything is working that you have planned, every day you don't know what you are going to face, not even every hour. Accept that every day is likely to be different. Things that will be static—an hour of skiing, etc.—give a certain stability to a chaotic experience. What you will encounter is unknown, but you can control that you will ski for an hour. Control the controllables. Be in the moment; only control what you can control. Only focus on yourself, because you cannot control others, or their experiences."

1.5.4.2 Integration and Interface Management

Because the expedition was self-contained, and everyone was responsible for their own gear and progress, they had few integration and interface requirements. At the end of the expedition at Isortoq, they needed to coordinate a flight by helicopter to Tasiilaq, and then to Kulusuk, before returning home. There were only two flights per week out of Isortoq. So the integration and interface requirement had to do with timing associated with their arrival into the town. If they arrived too early, they would have to wait, and then food and fuel became an issue. If they arrived too late, room for the team and gear on the helicopter might already have been claimed by other expeditions, and they might be forced to wait for the next flight. When they ended, they had two days to wait and finally concluded the expedition on day 30, when they were transported to Tasiilaq.

1.5.5 Schedule Management

The Trans-Greenland four-person expedition took place in the April/May time frame when the weather was more predictable and warmer but not too warm, which would cause snow melt issues along the west coast. The team would cross the Greenland Ice Cap, a distance of approximately 540 km (335.54 miles) in a time frame of 22 to 35 days under 24-hour daylight.

1.5.6 Acquisition Management

Supplies for the expedition were purchased by the team members and either brought with them or shipped to the start location. Some supplies, such as fuel, required purchase at the starting location in Kangerlussuaq because of restrictions from shipping carriers. One team member, based in Greenland, took responsibility for obtaining the materials required in Kangerlussuaq.

1.5.7 Risks and Opportunity Management

Risks associated with this expedition were discussed and fairly well understood based on the team members' previous experience in arctic adventures. They monitored these risks carefully throughout the expedition, realized some of them, and experienced some that they had not considered as serious threats. The risks they identified fell into the categories of weather, terrain, wildlife, gear, overall health, and speed/distance.

The 1999 April–May weather proved to be very challenging, with a major storm that hit them with whiteouts, blizzards, and gales, in temperatures that went down to –35 degrees Celsius. Although the expedition team was appropriately equipped, the ferocity of the storm required them to take extreme measures to survive. A number of other expeditions attempting to cross the Inland Ice were not so lucky, suffering from the extreme conditions and requiring evacuation.[36] As Stockdale describes, "The weather changes very fast in the Arctic, which can mean the ice opening up in front of you, or temperatures rapidly dropping. Sometimes we had to make decisions quickly about whether to change route, stop, or carry on, and almost always with limited information. Being able to act on imperfect information requires an acceptance of risk and sometimes getting it wrong. The question we always asked was, 'What is the worst that can happen?' and if we could accept the consequence, we would take the action."

On May 9th, day 13 of the expedition, the risk tolerance would be tested. The team by chance met up with "a Denmark/Norway two-person team who were crossing from the east coast with dogs to collect a party on the west coast. On the highest point of the Inland Ice they had encountered extreme weather conditions including −38 degrees Celsius temperatures and wind. They were frostbitten and in indication of the severity of the weather, this experienced team lost several dogs due to the low temperatures on each leg of the journey and also ran low on food and fuel."[37] During this fortuitous encounter, they were advised to secure their tent with extra poles to guard against collapse from the strong winds because it was not survivable if they were exposed directly to the weather, from the lack of ability to breathe in the cold and wind. This was disheartening news as they were heading up to the high point, at 2600 meters (~8,530 ft.). When the storm hit with a vengeance, however, they were prepared. They took their extra set of tent poles, putting two sets up for strength, had everyone get into that single tent, and prepared for the worst. A full 36 hours were spent inside the tent, with little food and water, until the storm subsided. Stockdale explained that they "just prayed that the wind would die and the tent would remain upright."[38]

The first few days of the trip took the expedition team through a section with glacial terrain and crevasses or deep cracks and fissures in the ice. These crevasses are sometimes several meters long and must be carefully navigated. They can be extremely deep (sometimes up to 45 meters), wide (up to 20 meters), with walls that are vertical sheer cliffs. The four teammates were roped together, and the rope had to be kept taut. This was in the event that one of them fell into a crevasse. The expectation was that if one fell, then the others would arrest the fall and hold fast while the victim used his or her personal knowledge and equipment to pull himself or herself back out. None of the team members had experience in crevasse self-rescue, and they were only provided instruction cards on how to take steps to survive. Stockdale's position was, "Don't rely on the process alone! Melting snow—underwater streams—whatever it is, use previous expedition's experiences to avoid risk."

The risks from wildlife encounters were low, although there was a potential to encounter polar bears along the west and east coasts. The team felt that the risk was low enough that no unique mitigation was considered necessary. In the event of an encounter, they anticipated using their flare gun to scare the animal away. In this way, they would just use their standard safety equipment, if necessary. In reality, the only wildlife encounters the expedition team members had were with birds and an Arctic Fox.

Carrying minimal supplies to repair or replace and limited duplication, with the weight restrictions, meant that gear risk was generally high. Stockdale mentions that she was unable to obtain her ski boots, a critical piece of equipment, in sufficient time to try them on for fit and had to ship them directly to the starting location, which was due to the supplier having a limited supply of the boots readily available. This vastly increased the risk to the project if they did not fit, because it would be exceptionally difficult to find a replacement. This was a high risk, which ultimately was not realized. In addition, her sledge was borrowed from another adventurer. His only request was that she send him a postcard upon completion of the expedition. This suited her well; however, the sledge was not in the condition it needed to be in for the expedition, and she was required to commission some changes through a local saddler company, who was willing to sponsor her on the trip in exchange for the work.

Fuel was purchased at the expedition start location and was carefully measured for each day. The team had factored in enough fuel to cook their meals and melt their daily water, with some left over for general heating purposes. On the first day, one of the tops to the fuel can came loose and they lost about a half-day's worth of fuel. The discovery that the fuel had been lost was sobering and alarming. It was necessary for the team to make the decision to carefully monitor and use the remaining fuel in the most conservative way so as not to run out of this critical resource before they finished the expedition.

On the first day, two hours after analyzing the ramifications and new requirements associated with the fuel incident, one of the shafts on a sledge broke. Both of these individual, isolated issues actually affected the ability of the entire team to move forward. The team worked together, and their emerging subject matter expert on equipment, Fardal, executed a fix that would allow the team to continue.

Another piece of critical gear was the GPS. As this device required batteries, they would only switch it on in the morning and in the evening to determine their position. The rest of the time they would ski on the compass bearing. The compass was worn on the lead skier and would always be visible to that person. This was an accurate and reasonable approach, and Stockdale confirmed that they "pretty much always knew where they were." However, they did need to keep the spare sets of batteries in their clothing next to their bodies to keep them warm so they would not discharge.

A piece of equipment that they chose to bring but did not end up using (and which actually added weight [and risk]) was a kite. Several past expeditions employed these kites to cover significant distances using the wind, similar to the sport of windsurfing. However, the use of kites would require experience and expertise in packing, unpacking, and sailing, with the kite in an environment where the wind speeds and direction can be unpredictable. They were unable to use the kite throughout the expedition so, in effect, it was excess baggage.

Finally, the introduction of a significant equipment risk was made when a sponsor requested testing of ski equipment during the expedition. The skis are critical equipment, and the level of risk was high, even though Fardal was a professional Norweigian ski race manager and highly qualified to carry out a comparative test of ski bases. "Between them the party had two pairs of conventional waxable skis, one pair of non-wax fish-scale and one pair which had a section of synthetic 'skin' let into the base."[39] The tests were performed as planned, and no risk was realized.

Without a doubt, injury, illness, or any other inability to be able to perform physically was a major risk to the overall expedition. Even if one of the team fell victim, the others would also be put in danger as they adapted and addressed the situation. They managed this risk by carefully determining their ski distances, taking a break every hour, and monitoring their health and fitness. Stockdale explained, "At the beginning, we were skiing 22–25 km (13.67–15.53 miles), however after 11 days, we realized that at that pace we were going to run out of food! We needed to move faster. This is difficult to do when you already feel that you are performing at your absolute maximum. So in order to make this change, we needed to mentally reframe our approach. We all knew that we could ski for eight hours with five-minute breaks because we had been moving at that pace. So we agreed to take our five-minute breaks after an hour and fifteen minutes rather than every hour. This would extend our distance (and hours) per day in an imperceptible way." This helped tremendously, and they were able to achieve much longer distances, including the longest day—40 km (24.85 miles)—near the end.

Speed and distance were a key success criteria of this expedition, but everyone was not in agreement. Fardal expected to finish in 22 to 25 days. Everyone else felt that the crossing would likely take 32 to 35 days. It ultimately took them 28 days, which they achieved mostly due to, as Stockdale expressed, "their desire for fresh food!"

1.5.8 Quality Management

The expedition team agreed on two criteria from the outset: (1) that the experience needed to be about the journey, and (2) that they needed to end as friends. According to Stockdale, "It was the hardest expedition she had ever done—the sledge was heavy, the weather challenging. Being in the Arctic reminds you how insignificant you are."

For this expedition, the measurements that mattered came in daily quantities. They included the fuel levels, the battery levels, and the distance completed, including the number of hours that it took to do so. It also included the confirmation of their GPS coordinates, and the amount of food they had remaining at any given time.

1.5.9 Test, Verification, and Validation

Two test/verification/validation activities were required during the expedition. These systems engineering processes provide the structure to assure that performance of a project or system meets the specified purpose. In this case study, there were two products that would be brought along. One was on the critical ski equipment. The results of the ski performance testing were provided to the sponsor. One test with the kites was performed during the expedition, when the weather was good; however, the team actually ended up going slower, so they abandoned the idea of using the kite.

1.5.10 Governance

The primary changes that required agreement and implementation during the expedition were the restriction of fuel, the increased speed/length of daily skiing, and the actions required to survive the storm. Changes were discussed and agreed upon by all team members, although as Stockdale described, "Who shouted the loudest, might have at first 'won,' but over time individual expertise evolved and emerged and then those team members became the subject matter experts that others would defer to."

1.5.11 Outcomes and Lessons Learned

The Trans-Greenland Expedition was a success and met all of its outcomes. Some of the major lessons learned include:

- "As we got more and more exhausted, it would have been easy to criticize one another's weaknesses. Instead we decided to focus on strengths and make sure we used the best person for the task, regardless of their role in the team."
- "We learned to let go of the expectation that you had to be good at everything, which made us really value strengths in others that were not in ourselves and made for a far better working environment."
- "Not to get too upset about why things happen or what people say. Don't take things personally. There will be days when people will annoy you. Keep in mind you have to make it work for a month, not forever. In our case, we were all there to get this expedition to work. We didn't have to make it work forever."
- "There is value in blind naiveté. Sometimes, you may not have any real clue as to what all the risks are; it is worth it to move ahead with conviction and adapt along the way. Don't get fearful. If you want to make it happen, you will find a way."
- "When it is unclear how you can make an overwhelming change, try 'reframing' the problem. Look for ways you can make it work."
- "Deliver on your commitments. The expedition sponsors were very pleased because they got some great photographs to use for their publications and valid results from the gear tests."

1.5.12 Case Analysis

This case is an excellent example of the inclination we have as humans to describe the activities that come naturally to us as a concise set of processes that, if we spend time thinking through before we embark on the activity, may help secure optimal outcomes. In this case, a project was well defined, the architecture was straightforward and well understood, and the requirements were well established and virtually unchangeable once the project (expedition) began. Change came in the form of decision making based on the unknown–unknowns that were encountered during the expedition. The actual changes that could be made were restricted to the level of remaining resources and their own human ability and ingenuity. They adapted within their team to take advantage of the emerging talents of the individuals. They managed their risks and their stakeholders and delivered as promised. Most importantly, they finished safely and within the parameters of their overall anticipated outcomes, on time, and as friends.

1.6 Key Point Summary

The focus of Chapter 1 is to describe how the discipline of systems engineering originated and evolved into a methodology that impacts organizational and project effectiveness. The positive effects it can have on a project, particularly complex projects, is profound. The following are the key concepts from this chapter. Key terms are compiled for quick reference in the Glossary.

Key Concepts

- Systems engineering is a discipline that provides the structured processes necessary to manage projects that require performance from many disciplines in order to achieve their objectives.
- Systems engineering processes can be used in most every scenario to develop and implement a roadmap that will lead to successful outcomes and that meet the desires of the key stakeholders.
- The top-down systems engineering approach includes a need to understand and correctly identify what stakeholders want, to convert that vision into a straightforward functional architecture, and then to be able to decompose that architecture into a physical architecture, which can be modeled and tested, in order to create the best design to ensure that the stakeholders' expectations are met.
- Standards, which can be used as established methods for accomplishing tasks within a discipline and for which quality (compliance) can be measured, is important in promoting effectiveness, managing risks, and increasing the probability of success in projects of all types. Implementation of sound systems engineering practices is meant to minimize undesirable emergent behaviors and reduce complexity.
- Organizational structure and culture can impact the implementation of systems engineering and the resulting successes. The implementation of systems engineering into organizations requires commitment to the discipline, as a whole, to be successful.
- Although the systems engineering discipline has been established to be applicable across all industries and includes processes written independent of any domain, its application in different specialized domains (e.g., information technology firms, financial industry, medical industry, etc.) might require tailoring of the processes to address specialized vocabulary or terminology.
- Taking the time and effort necessary to install a systems engineering discipline into an organization has been shown, with clear and compelling evidence, to add significant value.
- Systems engineering views a project from the life cycle perspective—that is, the end-to-end life of the project from conception through to closure, disposal, or restoration to its natural or original form.
- The discipline of systems engineering, when applied to a complex project, requires additional discipline and rigor. A complex system's behavior, as a whole, does not equal the sum of its parts. It will exhibit emergent behavior. This means that the system, its elements, and its sub-elements can be drivers of, as well as exposed to, unpredictable outcomes as a collective. A system-of-systems approach provides the strongest defense against undesirable emergent behavior.
- Having structures in place will increase the probability that performance and cross-organizational issues are resolved. However, it will not entirely remove risk. This is due to the fact that in addition to the already-existing emergent behavior risk, people in organizations that are performing

additional work outside of the specific project will inevitably make impactful changes that will trigger unintended consequences for the system.

- Key considerations with any system-of-systems project is the autonomy of the constituent parts, wherein evolution continues within the unique organization, unrelated to the system-of-systems project, but impacting it from known or unknown interfaces.

1.7 Apply Now

The most effective strategy in learning about systems engineering is to apply it to a personal example. Each chapter will build on the material from the previous chapter. In order to solidify the key material in this chapter, the reader should review the following summary points and answer the questions that have been posed in regard to his or her own knowledge and experience.

1. *Systems engineering originally emerged from a need to achieve better project outcomes.*
   Describe a project that you have completed where systems engineering processes were not followed. What were the outcomes? How could the outcomes have been improved with the implementation of systems engineering?
2. *Interact with various engineering disciplines as they are related to a single project or product.*
   Describe the disciplines that were involved in the above-stated project. What made the involvement of other disciplines challenging?
3. *Take into formal consideration the stakeholders' vision.*
   Describe the activities you participated in to obtain an understanding of the stakeholders of your project and their vision. Did the outcomes match the stakeholders' vision?
4. *Form a holistic and synergistic perspective of the product.*
   Describe the activities that you used to obtain a holistic view of the project. What were the challenges? Where there any synergies that could have been taken advantage of that were not?
5. *The discipline of systems engineering, when applied to a complex project, requires additional discipline and rigor.*
   Describe a project that appeared to be complex. Identify any emergent behavior. Describe what systems engineering processes you would apply to minimize the risk if you were to complete the project today.

References

1. Liu, D. (2016). *Systems Engineering: Design Principles and Models.* Boca Raton, FL: CRC Press/Taylor & Francis Group.
2. Ibid.
3. SEBoK contributors. "Guide to the Systems Engineering Body of Knowledge (SEBoK)." *SEBoK.* http:// sebokwiki.org/w/index.php?title=Guide_to_the_Systems_Engineering_Body_of_Knowledge_(SEBoK)& oldid=53122 (accessed March 17, 2018).
4. Liu, D. (2016). *Systems Engineering: Design Principles and Models.* Boca Raton, FL: CRC Press/Taylor & Francis Group.
5. Wasson, C. S. (2016). *System Engineering: Analysis, Design, and Development,* 2nd Edition. Hoboken, NJ: John Wiley & Sons, Inc.
6. Ibid.
7. Kossiakoff, A., Sweet, W. N., Seymour, S. J., and Biemer, S. M. (2011). *Systems Engineering: Principles and Practice,* 2nd Edition. Hoboken, NJ: John Wiley & Sons, Inc.

8. de Weck, O. L., Roos, D., and Magee, C. L. (2011). *Engineering Systems: Meeting Human Needs in a Complex Technological World.* Cambridge, MA: The MIT Press.

9. Liu, D. (2016). *Systems Engineering: Design Principles and Models.* Boca Raton, FL: CRC Press/Taylor & Francis Group.

10. Wasson, C. S. (2016). *System Engineering: Analysis, Design, and Development,* 2nd Edition. Hoboken, NJ: John Wiley & Sons, Inc.

11. NASA. (2007). *NASA Systems Engineering Handbook.* Washington, DC: NASA. Accessed April 24, 2013, from http://www.acq. osd.mil/se/docs/NASA-SP-2007-6105-Rev-1-Final-31Dec2007.pdf

12. Wasson, C. S. (2016). *System Engineering: Analysis, Design, and Development,* 2nd Edition. Hoboken, NJ: John Wiley & Sons, Inc.

13. Wingate, L. M. (2014). *Project Management for Research and Development: Guiding Innovation for Positive R&D Outcomes.* Boca Raton, FL: CRC Press/Taylor & Francis.

14. INCOSE. (n.d.) "INCOSE Principles." Accessed April 9, 2017, from http://www.incose.org/about/principles

15. IAPM. (n.d.) International Association of Project Managers. Accessed January 8, 2018, from https://www.iapm.net/en/start/

16. APM. (n.d.) Association for Project Management. Accessed January 8, 2018, from https://www.apm.org.uk/about-us/

17. IPMA. (n.d.) International Project Management Association. Accessed January 8, 2018, from http://www.ipma-usa.org

18. PMI. (n.d.) Project Management Institute. Accessed January 8, 2018, from https://www.pmi.org

19. SEBoK contributors. "Guide to the Systems Engineering Body of Knowledge (SEBoK)." SEBoK. http://sebokwiki.org/w/index.php?title=Guide_to_the_Systems_Engineering_Body_of_Knowledge_(SEBoK)&oldid=53122 (accessed March 17, 2018).

20. Walden, D. D., Roedler, G. J., Forsberg, K. J., Hamelin, R. D., and Shortell, T. M. (eds.). (2015). *Systems Engineering Handbook: A Guide for System Life Cycle Processes and Activities,* 4th Edition. INCOSE-TP-2003-002-04. Hoboken, NJ: John Wiley & Sons, Inc.

21. SEBoK contributors. "Guide to the Systems Engineering Body of Knowledge (SEBoK)." SEBoK. http://sebokwiki.org/w/index.php?title=Guide_to_the_Systems_Engineering_Body_of_Knowledge_(SEBoK)&oldid=53122 (accessed March 17, 2018).

22. Ibid.

23. Ibid.

24. Kossiakoff, A., Sweet, W. N., Seymour, S. J., and Biemer, S. M. (2011). *Systems Engineering: Principles and Practice,* 2nd Edition. Hoboken, NJ: John Wiley & Sons, Inc.

25. Walden, D. D., Roedler, G. J., Forsberg, K. J., Hamelin, R. D., and Shortell, T. M. (eds.). (2015). *Systems Engineering Handbook: A Guide for System Life Cycle Processes and Activities,* 4th Edition. INCOSE-TP-2003-002-04. Hoboken, NJ: John Wiley & Sons, Inc.

26. Ibid.

27. Kossiakoff, A., Sweet, W. N., Seymour, S. J., and Biemer, S. M. (2011). *Systems Engineering: Principles and Practice,* 2nd Edition. Hoboken, NJ: John Wiley & Sons, Inc.

28. Harwell, Richard (ed.). (1997, August). "Systems Engineering: A Way of Thinking, A Way of Doing Business, Enabling Organized Transition from Need to Product." Brochure prepared as a joint project of the American Institute of Aeronautics and Astronautics (AIAA) Systems Engineering Technical Committee and the International Council on Systems Engineering (INCOSE) Systems Engineering Management Methods Working Group.

29. Walden, D. D., Roedler, G. J., Forsberg, K. J., Hamelin, R. D., and Shortell, T. M. (eds.). (2015). *Systems Engineering Handbook: A Guide for System Life Cycle Processes and Activities,* 4th Edition. INCOSE-TP-2003-002-04. Hoboken, NJ: John Wiley & Sons, Inc.

30. Ibid.

31. Crawley, E., Cameron, B., and Selva, D. (2015). *System Architecture: Strategy and Product Development for Complex Systems.* London, UK: Pearson Publishing Company.

32. Liu, D. (2016). *Systems Engineering: Design Principles and Models.* Boca Raton, FL: CRC Press/Taylor & Francis Group.

33. Smith, A. (n.d.) "Scot of the Arctic; Sue Conquers the North Pole." *Daily Record.* Accessed January 29, 2014.

34. "Surprise! Cool Sue Bumps into Fellow Explorer in Arctic Wastes." (n.d.) *Oxford Mail.* May 12, 1998.

35. Stockdale, S. (n.d.) "What Really Counts." Accessed January 8, 2018, from https://www.suestockdale.com/what-really-counts-leader-striving-achieve-tough-goals
36. Fordham, D. (2000). Greenland 1999. *Alpine Journal, 105,* 230–234. Accessed March 2, 2017, from https://www.alpinejournal.org.uk/Contents/Contents_2000_files/AJ%202000%20230-234%20Greenland.pdf
37. Ibid.
38. The Scotsman. (n.d.) "No Business Like Snow Business for Motivation." Accessed January 8, 2018, from https://www.scotsman.com/lifestyle/no-business-like-snow-business-for-motivation-1-938670
39. Fordham, D. (2000). Greenland 1999. *Alpine Journal, 105,* 230–234. Accessed March 2, 2017, from https://www.alpinejournal.org.uk/Contents/Contents_2000_files/AJ%202000%20230-234%20Greenland.pdf

Chapter 2

Systems Engineering as a Project Enabler

The elements of systems engineering that will be explored in this chapter include:

- project structure
- process complementarity
- driving outcomes

In Section 2.1, this chapter first explores the structure of projects, which includes the basic definition and types of projects, and the processes that are used in the project management domain. In addition, the specific project management knowledge areas are described in order to set the stage for the next section, which aligns and overlays the project management discipline with the systems engineering discipline.

In Section 2.2, the complementarity of systems engineering processes to project management processes is discussed. Both disciplines have defined processes and knowledge areas and a flow that moves the project from conception through to the end product or service. These processes and knowledge areas both complement and enhance the overall project outcomes when they are aligned correctly. Conversely, an organization that does not take advantage of the synergy and natural alignment between the two disciplines will find that there is resulting conflict, particularly in the areas of overlapping responsibilities. Section 2.2 also explores the process alignment between the two disciplines that add value to the project, enabling a successful outcome.

Section 2.3 introduces the unified approach called the Complex Systems Methodology™ SM (CSM™ SM). The Complex Systems Methodology provides the framework for actively managing complexity using selected project management and systems engineering processes that have been shown to optimize outcomes, particularly for complex projects.

A case study demonstrating the application of systems engineering processes is provided in Section 2.4. In this chapter's case study, systems engineering activities were specifically planned into the project and were directly related to the project outcomes that were experienced.

A summary checklist of the key concepts discussed in this chapter is available in Section 2.5. Apply Now exercises, which allow for the immediate application of the information in this chapter, are included in Section 2.6.

Chapter Roadmap

Chapter 2 focuses on systems engineering as a project enabler. It specifically:

- describes the overarching discipline of project management
- provides a high-level description of the systems engineering discipline
- explores the process alignment between these two key disciplines
- identifies the value-added, project enabling systems engineering processes
- introduces the structure provided by Complex Systems Methodology
- uses a case study to demonstrate the application of systems engineering processes
- provides a summary checklist of the key concepts to assist learning
- provides Apply Now exercises to assist in the application of the concepts presented in the chapter to a real situation

2.1 The Structure of Projects

This section introduces the definition of projects and describes how they are typically structured.

- Project management holds overall authority and accountability over the success, or outcomes, of the project.
- A project provides an overarching structure for controlling cost, schedule, and scope.
- Processes support the overall mission of the project, including those processes that come from other disciplines, such as systems engineering, quality engineering, etc.
- Systems engineering holds overall authority and accountability for the technical scope elements of the project.

The project management discipline emerged in the 1950s as a useful way to manage and control project cost, schedules, and scope. Project management is a widely accepted and proven methodology that can be traced to the achievement of effective outcomes for projects around the world. There are many project management standards, "best practice-setting," and certification-granting institutions currently in existence, such as the International Association of Project Managers (IAPM),[1] the Association for Project Management (APM),[2] the International Project Management Association (IPMA),[3] and the Project Management Institute (PMI).[4] This book points to the types of standards that serve as the important processes for successful implementation of the project management discipline in all these organizations. This discipline can be applied to all activities, regardless of the industry or discipline, for example, those including science and technology, arts and music, adventure and manufacturing, because "following the discipline of project management provides the ability to categorize activities, bound them, and then assess progress along a defined course of action."[5]

As mentioned in Chapter 1, systems engineering began to emerge in the 1940s. It is an activity that engages in all areas of a system life cycle. However, when it is applied within a project, systems engineering's responsibility is to develop an optimal technical solution for the stakeholder requirements that have been agreed upon by the stakeholders. There is considerable overlap with the activities in which both the project managers and systems engineers participate; however, they have different roles and responsibilities. And many of the activities that systems engineers engage in directly affect project management activities and decision making. As will be shown in the following sections, optimal project outcomes can be achieved by the appropriate interweaving of these two critical disciplines in a way that enhances the roles of each.

2.1.1 Basic Definition of a Project

Projects are defined by all of the project management standards-setting organizations as temporary activities designed to create something unique. "ISO 21500 differentiates projects from operations by stating that projects are performed by temporary teams, are non-repetitive and provide unique deliverables."[6] They are typically activities that can be uniquely time-bounded—that is, completed within a specified time period. Projects can range from formal to semiformal or even informal.

Formal projects can be such as those found in large and complex projects. These formal projects typically require the most rigorous application of project management processes. Informal projects often undergo significantly less rigorous scrutiny and oversight because the cost of applying all of the project management processes can be more expensive than warranted for the project need. A semiformal project would fall somewhere in the middle.

The amount of formality required in the application of these processes is derived through analysis of risk, scale, scope, and complexity. All projects do not require or benefit from the same level of application of the processes. Many factors are in play, and a knowledgeable and experienced project manager will be able to tailor the project processes appropriately to enable optimal results. The level of process rigor applied should be in direct relation to the flexibility in the cost, schedule, and scope. Projects that have a great deal of leeway in performance to these attributes do not need a significant amount of applied rigor. The less flexibility there is in when the project can be completed, how much it can cost, or what the end results need to look like, the more rigor must be applied in order to have the best potential for fulfilling the project within the parameters established for scope, schedule, and budget. Even if only one of the attributes is flexible, for example, if the schedule is flexible but the cost and scope are firm, then the rigor must be appropriate to a fixed scope, schedule, and budget because each one of these attributes drives the other in a zero-sum manner. If the schedule can slip, the cost of staffing the project will go up. If the scope can slip, additional requirements will drive up execution costs. And if the budget is flexible, more features will undoubtedly be added, which means schedule and scope increases.

Projects have funding dedicated to completing the tasks as defined by the stakeholders. The funding is usually applied in the form of a project budget. A contingency may also be allocated as a subset of the overall project budget. A contingency is a sum of money that is set aside to cover unexpected expenses within the project.

In addition, projects include activities that must be completed within specific time allocations. These schedules may also have a form of contingency associated with them. This schedule contingency is called *float*. Float is assigned within the lower-level tasks of a project schedule. If the task schedule can lengthen without negatively affecting the overall project schedule and thereby affecting its critical path, then that amount of flexible time is the float. However, this float must be managed because if the schedule is allowed to slip beyond the point where the critical path is affected, it will cause the overall project to be late. Both the budget and schedule contingency is used by the project manager to optimize the project outcomes through cost and schedule control.

Projects also have a scope. The scope is an agreed-upon description (between all the key stakeholders) of what will be completed within the project budget and schedule. The scope definition is generally textual, described in functional (not physical) form, and provides the basis for the systems engineer to further define and refine the scope into stakeholder requirements. Stakeholder requirements drive systems engineering activities in developing the function and physical architecture designed to meet the stakeholder's requirements.

2.1.2 The Project Management Discipline

Each of the many project management standards-generating organizations offers perspectives on the life cycle and processes of project management that generally echo the same intent, although these perspectives may be phrased somewhat differently. They can be generally categorized as:

- understanding the stakeholder's needs and having an agreed-upon interpretation of that need
- communicating the activities in the appropriate way to the respective stakeholders
- documenting (for ongoing assessment of progress) and managing the project activities
  - cost management
  - schedule management
  - scope management
  - risk management
  - quality management
  - contracts and procurement management
  - integrative management
- realizing the project outcomes, measuring them against the plan, and obtaining confirmation from the stakeholders that their needs were met
- gracefully closing down all related project activities

They have a life cycle that generally includes these types of activities:

- understanding of the need
- planning to perform the activities to address the need
- performing the activities to address the need
- ensuring that the activities being performed are of sufficient quality and are controlled so that the anticipated outcomes are met
- completing and closing all of the related project activities

Project management processes "ensure the effective flow of the project throughout its life cycle."[7] They generally are grouped by life cycle phase to show the linear transition from one set of activities, such as understanding the need, to the next, such as planning to perform. This grouping is described in Section 2.1.2.1. The life cycle of a project may be shorter than the life cycle of a system that a systems engineer might be involved in developing and completing. However, the system life cycle phases overlap throughout the associated project life cycle phases. The systems engineering life cycle phases will be described in Section 2.1.4.1.

Within the overall life cycle of the project, project management processes describe activities that should be completed in order to assure that what the stakeholder's expect is what is delivered. The project management processes described in Section 2.1.2.2 also overlap certain systems engineering processes, although the project management and systems engineering disciplines approach the tasks within the processes in different ways. The systems engineering processes will be reviewed in Section 2.1.4.2.

2.1.2.1 Project Management Life Cycle

The project management life cycle can be described as a project's period of existence that can be divided into logical and progressive "life" events. These events move the project into different phases of activities, appropriate for the next cycle of life. These phases can be described as:

- project discovery and establishment
- preparation and planning

- realization
- governing
- completing

The life cycle of a project may be shorter than the life cycle of a system, and the life cycle phases overlap throughout the associated project life cycle phases—for example, project planning and technical definition and concept development can overlap as can execution, monitor/control, and systems design and development.

2.1.2.1.1 Project Discovery and Establishment

Project discovery reflects the earliest state of the project, when a concept is just being defined. Once the idea is formalized and a project will be initiated, project establishment begins. An example of this is when a new advancement in technology has been identified that might be considered for a project, a new service is being investigated, or an upgrade to an existing product in its operations phase is being discussed. Conversations with the stakeholders will occur to determine if there is interest in pursuing the project.

Project establishment occurs when it is clear that a project will be done, and includes the definition and authorization phase of the project that is reflected in a project charter. A project charter provides a summary-level description of the desired outcomes of the project. Funding for the project is typically allocated at this time, and a time frame for project completion is identified. The list of stakeholders associated with the project will be identified. More information about these stakeholders and their requirements will be identified in the next phase of the project. The project manager is officially named, and his or her authority to apply resources to the project is described.

2.1.2.1.2 Preparation and Planning

During the preparation and planning phase of a project, all associated project plans will be developed to document all the planned activities. These project plans will document requirements associated with leadership; teaming; collaboration; stakeholders; communications; the delivery of capability, including scope; schedule; cost; quality; resources; risk; procurement; measurements and metrics; and many other areas, depending on the project management methods and standards being applied. This is the phase in which the stakeholders' requirements are fully identified and an associated description of the activity that will fulfill stakeholder expectations is documented. The plans that address stakeholder engagement and communications management will be developed and documented as well in this phase.

A governance model is developed, even in the smallest of projects. The use of the term *governance* here is different than in some of the project management standards. In preparing to align the project management and systems engineering disciplines, the term governance is used to describe activities associated with coordination and control of the project's resources and actions. It is an all-inclusive term, which addresses policies, guidance, and processes for decision making, oversight and accountability, change, and configuration-control processes. This governance model ensures that change, normal in all projects, is effectively managed to minimize the negative effects of uncontrolled change and resulting unintended consequences. The only time governance is not of concern is when the project is not bound by limitations in cost, schedule, or scope. In general, project change will affect at least one of those.

The preparation and planning is complete when all the documentation is developed and accepted as the official version. Not all projects prepare the documentation in full detail. Although large, complex projects might fully document all aspects of the project, small simple projects may have a simple governance model, with basic documentation.

2.1.2.1.3 Realization, Governing, and Completing

It is at this time that the project is ready to be realized, governed, and ultimately completed. During this phase, active management, progress assessment, and continuous governing of the entire project plans' areas of performance are important. Also within this phase, information about the performance of the activities will be communicated to the stakeholders as per the plan. This phase of the project's life cycle will continue until all activities associated with the project, documented in the project plan, and all approved change requests have been executed.

A project can be designed to end at the entrance to the production/construction phase, which can be a separate project with a separate project manager in place. This is often done to take advantage of the leadership abilities of the project manager—one who has production/construction experience versus one who has experience in the design and development of a capability. Also, a closure/decommissioning phase of a project's life cycle can be implemented as a separate project. These separate projects would follow the same processes of discovery, establishment, preparation, planning, realization, governing, and completing as the design and development project in the first phases of the life cycle.

Completing and closing the project will include the disposition of any remaining contingency funding, closing all procurement activity, and may include transferring human resources, infrastructure, or other resources off the project or onto another phase of the project. The project manager will document the project results as realized against the plan and will validate with stakeholders their satisfaction with the project outcomes.

2.1.2.2 Project Management Processes

The project management processes that are universally identified throughout project management standards-generating organizations include:

- stakeholder management
- collaboration and communications management
- total project scope management
- resource/people/team management
- total project schedule management
- procurement management
- cost/budget/forecasting management
- risk management
- quality management
- integrative management

Depending on the project complexity, size, funding level, criticality, or impacts of nonperformance in cost, schedule, or scope, the application of rigor associated with the processes may change. This process is referred to as tailoring. Tailoring is discussed in Chapter 6.

2.1.2.2.1 Stakeholder Engagement

Stakeholder engagement refers to the processes that are used in the identification of people or groups with an interest in, or influence over, the project and its deliverables. During the process of implementing stakeholder engagement, the project manager analyzes their expectations and assesses their needs for participation in the project and in obtaining communications. The customer is a key stakeholder. For ease, when the book refers to stakeholders, the customer is included in that category.

In each life cycle phase, the stakeholders play an important role. During project discovery and establishment, they must agree to the project charter and indeed may be responsible for writing the project charter. In the preparation and planning phase, interfacing with stakeholders to obtain their view of their requirements is an imperative because that is what will drive the ultimate design of the solution. Throughout the realization phase, they play an important role in confirming that the design, development, production, and construction are on track to fulfill their needs. And ultimately, the stakeholders must validate that the system meets their needs. Stakeholder engagement is a fundamental role of any project manager.

2.1.2.2.2 Collaboration and Communications Management

Collaboration and communications management refers to the activities associated with carefully planning and then attentively managing activities associated with project collaborators, while ensuring that project communications are effective throughout the project. The more complex a project is, and the more collaborations that exist within the project, the more complicated communications become. For example, a project structure that is reliant on a number of external organizations to perform part of the project scope under collaborative agreements has a critical need to have clearly defined boundaries, interfaces, communications channels, and binding performance agreements.

Communications can be one of the most important and difficult processes to employ in a project. In order to be effective, a project manager must have a clear understanding of the roles and responsibilities that each participating organization, department, division, or work team is bringing to the project. The project manager must also define the communications methods and mechanisms, the expected reports, the process for escalation, and any terms and conditions associated with nonperformance. Without these basic constructs, the project manager will find that the project risk increases significantly. The project manager must also set the expectations on communications mechanisms throughout the project, defining when, where, how, what, and why specifics about the project are to be communicated. The processes for assessing performance and addressing change within the project must take into consideration the tempo that must be employed in communications to ensure that project leadership and management can engage and make value-added changes in a timely manner.

All aspects of communications should be carefully planned and implemented to minimize misunderstanding and maximize understanding. Communication considerations should include typical barriers and enablers such as languages and terminology (acronyms and multiuse words), level of detail, tempo or frequency for updates, restrictions (such as International Traffic in Arms Regulation [ITAR], intellectual property [IP], or classification restrictions), technical methods and technologies (email, web, wiki, formal letters, etc.), escalation processes, and authorization requirements.

2.1.2.2.3 Total Project Scope Management

The project manager's responsibility with scope management is to ensure that all work associated with achieving the outcomes of the project is included in the project scope definition. It is also of critical importance to determine what activities are outside the boundaries of the project and to ensure that they are not inadvertently included within the project scope. The project scope is determined through interactions with the stakeholders in the initiation and planning phases of the project, during the development of the business case, project charter, project scope management plan, and project plan development.

A useful process associated with scope development is in the development of a requirements traceability matrix. Each requirement is documented, along with any specifications and the activities that will be used to verify the performance. A valid requirements traceability matrix always has a sign-off column for the stakeholder's formal validation. Because the development of a requirements traceability matrix is a standard process within systems engineering, this process will be used to identify and track the technical performance of the project. However, an overarching project scope matrix can also be used to ensure that the optimal scope is delivered to the stakeholders. When an agreed-upon scope is in place, this will form the scope baseline upon which performance is measured and for which all change is assessed.

The scope must then be decomposed into a work breakdown structure (WBS) and associated lower-level work packages. Work packages represent the smallest unit of work that will be tracked and will usually have a product as a deliverable. This decomposition of the work breakdown structure into the work packages is an imperative activity in order to effectively schedule a project, as will be discussed later in this section. The tasks associated with each work package lead to the activities in the schedule and must be understood along with the resources, both human and infrastructure. At the project level, the work breakdown structure must capture all of the activities that the project manager will use to assess how costs have accrued for the project. Figure 2.1 shows a basic project-level work breakdown structure.

1.0 Wind Farm Project		(Chargeable Control Account)	
	1.1	Project Office	(Control Account (Summary Costs)
	1.2	Project Management	(Control Account (Summary Costs)
	1.3	Systems Engineering	(Control Account (Summary Costs)
	1.4	Design and Development	(Control Account (Summary Costs)
	1.5	Construction	(Control Account (Summary Costs)
	1.6	Systems Test and Evaluation	(Work Package)
	1.7	Site Activation	(Work Package)
	1.8	Support Equipment	(Work Package)
	1.9	Spares and Repair Parts	(Work Package)

Figure 2.1   Basic Project-Level Work Breakdown Structure

2.1.2.2.4 Resources/People/Team Management

Every project requires some level of resources in order to perform. All organizational elements that facilitate the project, from the humans who lead and are engaged in the project, to the buildings, office space, materials and supplies, and furniture that are associated with the provision of a project workspace. Project Resource Management includes processes to identify and bring in the resources that are appropriate to the needs of the project, and then to manage them in an effective way.

There are many different organizational models associated with resource management, and the subject matter expertise associated with the disciplines is often held at the organizational level. The project manager will work within the organizational construct to secure the appropriate talent and resources necessary to facilitate the project.

2.1.2.2.5 Total Project Schedule Management

Projects, by definition, operate within a constrained schedule. To effectively schedule a project, the scope that has been decomposed into a work breakdown structure is further analyzed to understand and document each of the activities in the schedule. This action of identifying the activities that will be tracked is often not straightforward. An experienced project manager will be more easily able to identify the appropriate level of activity tracking for the schedule; however, there is no one right way to complete this activity. The goal is to be able to clearly explain, when necessary, how the project costs are accruing (in relation to the budget for that activity) at any given time during its life cycle in a way that will allow risk-mitigating decisions to be made in a timely manner. If the breakdown is insufficient and not detailed enough, it will not provide the project manager with enough fidelity to make informed decisions. If the level of activities is too detailed, then time will be wasted throughout the project tracking information that is not useful.

Next, the tasks must be sequenced, resources assigned (labor, materials, services, and infrastructure), and the critical path determined. This provides an overarching view of the full project schedule aligned with the staffing, materials, equipment, and services that are needed to support the project. The simplest description of a project's critical path is that it is the compilation of the longest tasks throughout the schedule. Defining the critical path provides the details in which to see float. As mentioned in Section 2.1.1, float is the additional time that can be allocated to a task without impacting the critical path. When an agreed-upon schedule is in place, this will form the schedule baseline upon which performance is measured and for which all change is vetted.

Schedule management is a combination of art and science. There are different methods to approaching schedule development. However, schedule management is consistent and can be summarized in two actions: (1) construct the baseline schedule; and (2) actively manage the execution of the schedule (includes resource leveling to address resource overallocations, crashing the schedule to shorten some activities, addressing changes requests, etc.).

2.1.2.2.6 Procurement Management

The knowledge area of project procurement management covers activities associated with any purchase or acquisition of products or services that are needed by the project from others outside of the project team. Because various contracts and other legal agreements, such as memoranda of agreement, are typically used to secure goods and services, a subject matter expert, with knowledge and experience of these external requirements, is the authority and has the responsibility to carry out the planning, execution, and closeout of these project activities. The contracts manager works closely with the project manager to execute these legal agreements on behalf of the project. They may also be involved in assurance, verifying that the as-received materials and services meet the requirements as specified in the original agreements.

2.1.2.2.7 Financial Management

The processes that are used for project cost, budgeting, and forecasting reach into other domains outside the project in ways that require subject matter expertise. There are many legal and regulatory laws and rules, individual organizational directives on proper treatment of costs, and stakeholder preferences associated with costs that must be considered. For example, there may be a requirement to perform earned value management (EVM) on a project. A project may also be required to use specific techniques for developing cost estimates used in forecasting, or to use predefined terms and conditions for the acquisition of materials needed for the project. The project manager is responsible for working with the appropriate subject matter experts, and to address all organizational requirements when developing and managing the project budget. When the final budget is in place, this will form the cost baseline from which performance will be measured and all changes will be assessed.

2.1.2.2.8 Risk Management

As mentioned earlier, project managers will develop cost and schedule contingency to address risk and opportunities during the life of the project. Risk management processes provide the structure to plan for and execute risk management. The process must be actively applied throughout the life of the project through the identification of risks and opportunities, qualitative and quantitative analysis of those risks and opportunities, taking actions to resolve the risks (called *mitigations*), or to seize opportunities. This is an ongoing, nonstop activity. As risks are mitigated, or if an opportunity is taken, the scope, schedule, and budget will most likely be impacted. This impact is managed within the existing project through the use of cost and schedule contingency. Risk is an important component of systems engineering as well and will be discussed in Section 2.1.4.2.6.

2.1.2.2.9 Quality Management

Within project management, quality is planned, executed within the scope, and ultimately assessed as the "degree to which a set of inherent characteristics fulfill requirements (ISO 9000)."[8] From a project perspective, the activity associated with quality ensures that the project scope is first verified by the appropriate technical disciplines under which the technical specifications fall and then is validated by the stakeholders. Quality engineering is a discipline external to the project; however, the project manager is responsible for the overall quality associated with the performance of the project scope.

2.1.2.2.10 Integrative Management

This overarching process is associated with activities that are central, overarching, unifying and synthesizing in structure. Activities that rely on multiple disciplines or multiple organizations to perform necessary tasks within the project—areas where there are interfaces that must be addressed—are all considered integrative. Integrative Management includes stakeholder interfacing activities, such as those used to develop the project charter or to understand the needs of the stakeholders; processes used to assess the progress of the project; those used to assure verification of the technical solutions and validate the outcomes with the stakeholders; and, most importantly, to manage change against the cost, schedule, and scope baselines. Integrative management is an incredibly important process that, if neglected or implemented without the appropriate level of rigor considering the risk of the project, will impact the final outcomes of the project.

2.1.3 Types of Project Management Methods

There are many different types of project management methods that can be used to manage projects. These can be broken down into either the "traditional" method of project management or "flexible" methods, which encompass a wide and ever-evolving variety of project management methods. It is important to understand the method that is being employed for the project so that the systems engineering processes can be tailored to the project management method appropriately. Appropriately tailored systems engineering processes will increase the chance of the project achieving optimal outcomes.

2.1.3.1 Traditional Method

In the traditional project management method, typically referred to as the Waterfall Method (as it resembles the flow of a waterfall), the project is expected to progress sequentially through a series of developmental steps. Generally, the project management processes that are defined in any of the standards organizations provide an accounting of, and descriptions of, the full complement of processes that may be applied to a project. The project management processes described in Section 2.1.2 reflect the steps in the traditional method of project management. Tailoring of these processes is described in Chapter 6.

2.1.3.2 Flexible Methods

Alternatively, flexible project management methods use cyclical and iterative approaches to achieve a succession of outcomes that are defined over time and as earlier outcomes are achieved and their outcomes become apparent. Earlier successes drive new targets. Change is an integral part of the process and is encouraged. This flexibility is seen as valuable to the overall project success. Some of the flexible project management methods include Spiral Development, Agile, Rapid Application Development (RAD), Feature-Driven Development (FDD), and Evolutionary Project Management (EVO), although there are many others as the flexible development realm continues to evolve. In general, the project management processes described in Section 2.1.2 are all appropriate to the management of flexible projects.

The tempo, flexibility, incorporation of change, and expected evolution does not insinuate that a structured process is not followed. Indeed, the required processes can be more rigorous than what is found in a project that is being managed with the sequential method. Baselines are established, change is carefully monitored, and stakeholders are typically heavily involved throughout the entire process. In addition, risk reduction is a hallmark of the flexible methods. The intent is to understand what part of the overall scope will be done in short time frames, then to identify the highest risk items associated with the development activity (most difficult to achieve; highest potential cost impact), and to resolve them first. "Identifying the riskiest parts of a plan and resolving and testing these first leads to one of two possible outcomes: (1) if the testing validates the assumptions, behavior, performance, hypothesis, or other stated desired outcomes, then risk is reduced because performance was achieved, and one can move on to the next activity; or (2) if the testing does not validate the hypothesis or assumption, or if the behavior or performance does not match the stated desired outcomes, then risk is not reduced and it is necessary to take an action. One might end the exercise, restate the outcome, or perhaps redesign to try to achieve the original outcomes."[9]

This type of recursive and incremental behavior inherent in the flexible project management processes is well matched to the systems engineering processes that will be described in the next section.

2.1.4 The Systems Engineering Discipline

Systems engineering, as a discipline, provides a unique multidiscipline approach that, when properly applied and tailored to project needs, significantly increases the potential for success. The life-cycle perspective of systems engineering—from ideation to system life end—provides a powerful set of processes designed to enable the realization of projects, from the simple to the complex.[10] The systems engineering discipline consists of an evolutionary progression of a project across a full life cycle, and then is operated, maintained, and ultimately decommissioned or closed. There is a recognized need for a structure such as is provided by systems engineering to address system life-cycle requirements; whether it be to conceptualize the system, design it, produce it, operate and maintain it, or close it.[11]

As an important discipline in the execution of a project, systems engineering adds the technical processes to scope management. In performing these duties, it will add fidelity to the project stakeholder register, communications plan, the work breakdown structure and the schedule, the procurement plan, risk register, quality management plan, and the governance model. These interfaces with project management will be discussed in Section 2.2.

The objective of the systems engineer is to obtain enough fidelity from the stakeholders to clearly document their requirements and then to design and build a technical solution that meets their needs.

It is the project manager's responsibility to ensure that the technical solution fits within the budget and schedule of the overall project. However, the project manager must do this in close collaboration with the system engineer, ensuring that the systems engineer takes the lead on determining how the system design can be optimized within the budget and schedule, and in assessing any technical change recommendations driven by cost and/or schedule constraints. In best practice, the systems engineer designs the system within the cost and schedule constraints of the project, but in general practice, this is an iterative process that requires extensive analysis and dialog.

Designing the scope with cost and schedule in mind at the beginning assures that the solution that is proposed optimizes the use of project time, effort, and money. Designing scope independent of these factors leads to rework for the engineers and may also lead to a perceived or real reduction in the ability to meet the stakeholder requirements. In some cases, a cost reduction or schedule compression exercise, necessary if the design was done independent of cost and schedule considerations, most likely will lead to a significant reduction of the project scope.

2.1.4.1 Systems Engineering Life Cycle Phases

The systems engineering life cycle phases can be categorized in the following way:

- new idea exploration
- definition
- concept development
- design and development
- production/construction
- operational maintenance and support
- deactivation/closure

As was described in Section 2.1.2.1, the life cycle of a project may be shorter than the life cycle of a system. The life cycle phases overlap throughout the associated project life cycle phases. This is particularly true for systems engineering concept, design, and development phases, which are highly iterative.

2.1.4.1.1 New Idea Exploration

During the new idea exploration phase, the systems engineer may be working with stakeholders or individuals in other disciplines such as engineering or technical staff to identify new ideas that will become projects in the future. This type of exploratory research and development can occur with regularity as organizations investigate new business ideas that will attract the interest of their stakeholders. The level of effort and costs devoted to this type of exploration is usually low, and depending on the organizational desires, tailoring of systems engineering may or may not be applied.

2.1.4.1.2 Definition

During the definition phase, the systems engineer works closely with the stakeholders to define the project scope associated with the system concept. Inherent in this phase are activities designed to understand the stakeholders' perspective on what the system is intended to do through the process of stakeholder needs analysis. In complex cases in which the stakeholders may not be fully aware of their desired outcome, a concept of operations or use cases can be developed that clearly articulate how the stakeholders intend to use the system. The optimal system definition will distinctly describe the state of the system in its operational environment but should not describe any solutions that would facilitate the performance of that mission because the technical solution is defined in the next phase in response to the stakeholders' definition of the desired system. Measures of Effectiveness (MOEs) should also be collected. These are quantitative measures that are aligned to the stakeholders needs and that provide a way to assess if the capability provided is acceptable to the stakeholders. Having these effectiveness measures at this phase helps the technical system designers to develop a system that can be assessed against the agreed-upon measurement criteria.

In addition, a series of technical management plans is developed during this phase to address the approach to communications, decision support, configuration control, procurement, risk, and quality. Decomposition to a work breakdown structure at the system, product, subsystem, assembly, sub-assembly, and part levels will be completed, and a technical schedule developed. An interface charter will be drafted, which outlines the boundaries of the system. All of these systems engineering plans will be developed in collaboration with the project manager, and then they will be provided to the project management team to be incorporated into the formal project management documents. These processes are discussed in Section 2.1.4.2.

2.1.4.1.3 Concept Development

In the concept-development phase, the systems engineer is responsible for collaborating with the project manager and the stakeholders in developing a register of stakeholder requirements, along with validation criteria for those requirements. Key performance parameters (KPPs) need to be identified and established. Key performance parameters identify those specific quality ranges most important for system performance, and that, when monitored, provide visibility for decision making. These (performance) parameters should bound activities (parameters) and also be tractable.[12]

This performance information derives from the effectiveness measures developed in the definition phase and is required to scope, or constrain, the technical solutions that will be developed in response to the desired system performance outlined by the stakeholders in the previous phase. This is the phase in which the solution set for the system concept is formalized. Both functional and physical architectures describing how the system, as allocated (decomposed), will meet the specifications and requirements of the stakeholders are developed during this phase.

Additional plans will be developed during this phase, including interface, systems integration, and test and evaluation. Technical milestones and effort assessment will be finalized and provided to the project manager for incorporation into the project management plan and schedule. In addition, statements of work for procurement purposes will be developed and provided to project procurement management for execution. During the concept phase, depending on the complexity of the project, other activities may be occurring simultaneously such as informal or formal reviews and technical risk reduction in the form of prototyping, trade studies, and/or modeling and simulation. Reporting at this stage will be initiated, and focused on technical progress.

2.1.4.1.4 Design and Development

During design and development, reporting increases to cover more activity. This may include technical effort and progress, quality reviews, interface control, engineering analysis, systems and subsystem performance analysis, evaluation and integration, and test and verification reports. From the measures of effectiveness and key performance parameters identified in the previous phases, and built from the system specifications, measures of performance and Technical Performance Measures (TPMs) may be developed at the system and subsystem levels. Measures of performance are used to identify the acceptable performance parameters and specifications. Technical performance measures provide insight

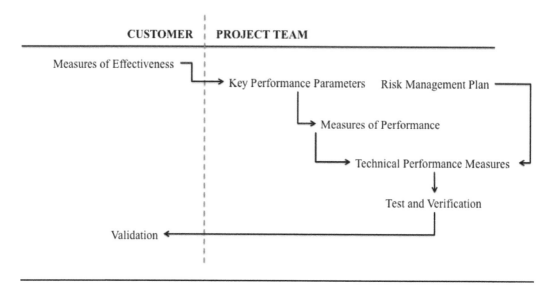

Figure 2.2 Measuring Technical Performance (Reproduced with permission from Wingate, L. M. [2014]. *Project Management for Research and Development: Guiding Innovation for Positive R&D Outcomes*. Boca Raton, FL: CRC Press/Taylor & Francis Group.)

into technical risk reduction activities associated with the most vital technical goals. These measures will ultimately be used to validate the performance of the system with the stakeholders, as shown in Figure 2.2.

A maintenance concept, including specific requirements and budgets, specialty engineering design analysis and compatibility requirements, and subsystem specification requirements will all inform and impact the as-designed configuration of the system. Interface control documents (ICDs) will be compiled where identifiable impact to the system, subsystem, or system integration might occur. Interface control documents are documents that describe how one object, process, or other item is connected to another.

During this phase of the project, the design evolves from the as-allocated (in the concept phase) to the as-designed configuration. Risk reduction and test and verification activities continue to drive design changes and evolution. This process is carefully controlled and documented through a configuration management process and quality checked through a series of reviews, such as subsystem preliminary design, system preliminary design, system integration, critical design, and production/construction readiness. From the maintenance concept, support equipment procurement requirements and statements of work are provided to project procurement for execution at this time.

2.1.4.1.5 Production/Construction

After the critical design review (CDR) is successfully completed, the design and development phase is officially over. The final design is now at a stage where first articles, low-rate initial production/construction, production/construction procurement, and ultimately full production/construction can begin. A production/construction baseline is in place. The system will have been verified and validated by the stakeholders. A commissioning or preoperational deployment plan will be developed by systems engineering to ensure a smooth transition from production/construction into operations. An as-built configuration will be maintained and configuration controlled throughout the production/construction phase. Preplanned production improvements will be built into the formal schedule.

Systems engineering will receive reports associated with test and verification, quality, discrepancy/latent defects, first article and initial low-rate production/construction performance, functional configuration audits, and production/construction status updates throughout the build phase. Systems engineering may also plan for and be involved in managing warranty claims.

2.1.4.1.6 Operational Maintenance and Support

Once the transition into the operations phase has been completed and formal acceptance of the produced/constructed system has occurred, systems engineering will ensure that the operations and system upgrade plans are in place. For future use, they might also facilitate the development of a new set of documents covering system closure/deactivation requirements, a risk management plan for the closure/deactivation activity, and a closure/deactivation review plan for future use. Preventative and corrective actions reports and operations and maintenance reports will require ongoing review. System upgrade requests from stakeholders, as identified by the systems engineer, will be passed along to the project manager or responsible management for disposition.

2.1.4.1.7 Closure/Deactivation

For the final phase of the system life cycle, systems engineering may be required to submit a formal closure/deactivation plan and undergo a formal review to move forward with the actions. Notification must be made to the stakeholders and a complete system closure/deactivation plan and schedule must be developed. Directives to close any open procurement activities are made, and the final formal review must then be held. At the culmination of these actions, the system life cycle has concluded.

2.1.4.2 Systems Engineering Process Areas

These systems engineering process areas include:

- system stakeholder engagement
- communications and decision support
- technical scope management
- technical schedule management
- acquisition management
- risk management
- quality and measurements
- test, verification, and validation
- configuration management

2.1.4.2.1 System Stakeholder Engagement

All stakeholder engagement should include the project manager to ensure that decisions and commitments are captured in the project management documentation and to assure that decisions outside of the scope of the project are not included. Identification of all stakeholders in the system, whether they view the system's introduction positively, negatively, or neutrally, is an essential process. Understanding their technical viewpoint, questions, and concerns is necessary.[13] Once it is clear who the stakeholders are and the role that they play, the systems engineer can then perform the stakeholder needs analysis to constrain the design space.

The role of systems engineering in system stakeholder engagement is to work with the stakeholders to gain fidelity on the technical scope of the project. "The purpose of the stakeholder needs and requirements definition activities is to elicit a set of clear and concise needs related to a new or changed mission for an enterprise and to transform these stakeholder needs into verifiable stakeholder requirements."[14] They do this in several ways. Upon receipt of the project charter and project management plan from the project manager, the system engineer, in collaboration with the project manager who has developed the stakeholder register, begins eliciting additional technical details from the stakeholders.

The stakeholder needs analysis can be done in many different forums and using various techniques. For example, use cases, or user stories, are an effective way to elicit stakeholder requirements, particularly when the stakeholders are vague or unsure of their true requirements or if they are mixing requirements with wishes and wants. The development of user stories, a typical flexible project management process, can help the stakeholders explain their needs in text-based, conversational language. Users can often articulate what they want, or how they anticipate the system working in its operations phase. Development of use cases requires a methodical approach to help the user identify the future functional vision of the system while avoiding specific technology in their descriptions.[15] If this methodology is used by the systems engineer, the project manager should be involved to assure best practice implementation.

Other methods that can be used to elicit stakeholder requirements include interviews, focus groups, or other face-to-face activities, which can be used to document a concept of operations. The development of this concept document is text based and, again, uses common language to clarify the ideas that the stakeholders have as to what they will use the system to accomplish. When considering what a day would look like once the system is fully functioning in the operations environment, all activities are considered. It is literally as if the stakeholder explained step-by-step what actions the system will complete in the course of operations. There may be many gaps in the concept as the systems engineer works through the exercise with the stakeholders. However, that is also meaningful because the systems engineer can help the stakeholders work through these gaps to solidify their ideas.

A key activity that stakeholders must engage in is the development of measures of effectiveness and key performance parameters associated with their stakeholder requirements. These measures and parameters set the expectations of what is tolerable within their operations model. They identify how effectively, and under what conditions, the system must perform to achieve its intended outcomes. These are the values that are deemed as minimally acceptable to the stakeholder.

Once a full list of stakeholder requirements, including their effectiveness measures and performance parameters, are known and documented, the technical solution is designed to this baseline. Validation measures are also identified for every requirement. The stakeholder-written validation measures, which describe how they will assess the system as it evolves throughout its life cycle phases, are used as a confirmation from the stakeholders that the system is achieving the anticipated outcomes. The action of validating the progress helps manage stakeholder expectations, confirms that the development is on the right track, and if done correctly, ensures that once the system is through design and development, the stakeholder is satisfied with the system and accepts it without issue.

In the operational maintenance and support phase, stakeholder involvement is associated with acceptance of the produced/constructed system and in providing upgrade requests to the systems engineer.

2.1.4.2.2 Communications and Decision Support

The project communications plan informs the technical communications plan. Technical communications to the stakeholders, as well as to the project team and management, must conform to the project communications plan so as not to overlap or conflict. Ideally, during the definition phase of the project, the systems engineer should write a technical communications plan that is aligned with the project communications plan. A description of the decision-making processes should be included so that it is clear what each discipline is responsible for as it relates to project and technical communications.

As the project is progressing through its life cycle, the focus of communications is on conveying information that has been collected and analyzed. Decision-making activities typically are in response to information related to analysis and performance, technical effort and progress, quality, interface control, test and verification, and stakeholder notifications, as well as production/construction first article/low-rate and full-rate status updates, discrepancy and latent defects reports, operations reports, maintenance reports, and if necessary, closure/deactivation actions.

2.1.4.2.3 Technical Scope Management

The area where a significant amount of systems engineering effort is applied is in the area of technical scope management. This process area is informed by the project charter and project management plan. Within the technical scope are the activities associated with defining the technical system that will fulfill the stakeholders' requirements. The systems engineer is responsible for decomposing the project technical work breakdown structure to the system, product, subsystem, assembly, subassembly, and part levels.

A Systems Engineering Management Plan (SEMP) is generally developed to document how the technical elements of the project will be managed. It often includes information about the systems engineering technical schedule, test and verification plans, approaches for risk reduction, quality and technical review plans, and any other descriptions associated with the standard systems engineering activities appropriate to the project. Figure 2.3 shows a technical work breakdown structure.

Within technical scope management is the identification of required enabling systems. During this phase, the systems engineer determines if any enabling systems requirements exist and provides the information to the project manager for resolution. These enabling systems are generally organizational capabilities, which sit outside the project but are critical to the successful implementation of the project. For example, computer-aided design tools, modeling and simulation tools, or other infrastructure, operations, or subject matter expertise may be needed by the systems engineer to complete the project scope. If these capabilities are owned by other organizational entities, access to them must be secured early in the project.

The stakeholder requirements inform the architectural design. Although the term architecture can have many different meanings, for the purposes of this book, it can be described as the highest-level description or diagram of the overarching system structure.[16] In architectural design, one focuses on the form, fit, and function of the design, as well as the tradeoffs that optimize the overall design. Form refers to the physical, corporeal, or tangible structure or results of the design. In software, for example, this could refer to the documented code. In hardware, this could refer to the reference diagrams or schematics of the physical form that will be built. Form defines the outline, contour, and shape.[17] Fit refers to how the parts of the architecture interface and "fit" into the overall architecture. This is generally associated with physical interconnections associated with placing parts. It is important to note "that form and function are related to product specification, and that fit is related to drawing dimensions and tolerances."[18] Function refers to the activities that the system engages in and that directly lead to the system performance outcomes. The way function is defined is associated with activity or action. It is in the operation of a process that moves something forward through the system. To diagram a function, for example, would be to document each starting and ending point and the flow between the two.[19]

It is common to approach architecture with limited information regarding the desired total scope. The process is nonlinear and creative in nature. However, it is an important, even critical, process that allows "reasoned" choices to be made proactively about the overarching and subordinate designs, rather than reactionary choices that would be made further down the design life cycle. The development of an architecture for a system is as important as developing an architecture for a house. Without the building architecture, the result could be a livable domicile; however, chances are it will not be elegant, efficient, or as attractive as it could have been.[20]

Architecture is reflected in several standard ways. The information gathered on stakeholder specifications, including the performance parameters and measures, is used to develop a specification tree

1.0 Wind Farm Project					(Chargeable Control Account)
	1.1			Structure	(Control Account (Summary Costs))
		1.1.1		Main Shaft	(Chargeable Task Code)
		1.1.2		Main Frame	(Chargeable Task Code)
		1.1.3		Tower	(Chargeable Task Code)
		1.1.4		Nacelle Housing	(Chargeable Task Code)
	1.2.			Rotor Blades	(Control Account (Summary Costs))
		1.2.1		Blades	(Chargeable Task Code)
		1.2.2		Rotor Hub	(Chargeable Task Code)
		1.2.3		Rotor Bearings	(Chargeable Task Code)
	1.3			Electronics	(Control Account (Summary Costs))
		1.3.1		Generator	(Chargeable Task Code)
			1.3.1.1	Magnets	(Work Package)
			1.3.1.2	Conductor	(Work Package)
		1.3.2		Power Converter	(Work Package)
		1.3.3		Transformer	(Work Package)
		1.3.4		Brake System	(Work Package)

Figure 2.3 Sample Technical Work Breakdown Structure

and requirements allocation flow down. Systems engineering focuses on providing and maintaining traceability between the top-level stakeholder specifications through to the lowest-level decomposition at all times and on analyzing change recommendations against impacts to the overall system through the use of the traceability. When done correctly, the traceability can be effectively used to assess the impacts to the greater system from optimization opportunities, performance trade-offs, and other risk reduction options.

From the specification tree and requirements allocation flow down, a system context and functional architecture can be developed. These take the form of a more detailed drawing that demonstrates all the major elements of the system and their relationships to one another. They do not denote any physical attributes, nor do they propose any specific technical solutions. For functional diagrams, they specify just that—the function that will be carried out within the system.

These diagrams provide visualizations that can be used to ensure that project boundaries and interfaces, as well as flow, are well known and defined. This is a critical step before embarking on the design of the physical architecture. "The aim of the approach is to progress from system requirements (representing the problem from a supplier/designer point of view as independent of technology as possible) through an intermediate model of logical architecture, to allocate the elements of the logical architecture model to system elements of candidate physical architecture models."[21] Ultimately, the system's logical or functional architectural diagram should provide a level of confidence to both the stakeholders and the project team that the stakeholder requirements are well understood and that the system concept will be well designed and developed in a manner that meets stakeholder expectations.

The physical architecture can be designed from the functional architecture and interface requirements. In fact, as the physical architecture is designed, and interface requirements emerge, interface control documents can be developed and communicated to the stakeholders so that during design and development, the required interfaces will not be overlooked. Developing interface control documents is an ongoing activity throughout the design and development phase of the program. However, a robust functional and physical architecture completed during the concept phase will provide more confidence to the project team and stakeholders that the overall system, including the interfaces, are captured, understood, and planned.

The physical architecture, which is the solution set meeting the stakeholder requirements, describes all major elements of the system and identifies the physical solution that is expected to support the function. Physical connections, hardware, and software should all be specified within the physical architecture. One should be able to review the physical architecture diagram and have a clear understanding of the basic technology that is supporting the function of the system. Network capabilities, hardware requirements, server specification, software specifications, etc., should all be clearly defined.

As part of technical scope management, measures of performance and technical performance measures are derived from measures of effectiveness and key performance parameters, as described in Section 2.1.4.1. Technical performance measures are used to identify the anticipated performance of the technologies that are key to project success, and then to track actual performance and risk reduction results throughout the design and development phase. These important measures of technical scope management, provide a level of comparison between planned activities versus actual activities, which the project manager can use to make decisions.[22] It is important to identify technical performance measures that are the most profoundly important to the success of the project and should be directly traceable to critical, high-risk, project technical requirements.

In addition to measures of effectiveness, specialty engineering design analysis compatibility requirements are assessed for both subsystems and the overall system. The maintenance concept, requirements, and technical budgets, as well as the interoperability requirements, are identified and factored into the design, as appropriate, to ensure the most efficient and effective system capability. The use of "technical budget" in this case refers to a technical solution boundary that the technical solution must not exceed. For example, a power budget is the total amount of power available, for which the power requirements of the project are compared and are not allowed to exceed.

In the production/construction phase of the project, technical scope management includes managing the production/construction baseline, developing a warranty plan, commissioning or predeployment plans, and the operations transition plan so that there is a smooth transition from production/construction into operations. Systems engineering often is not responsible for the development of the operations plan but may facilitate the development of that plan. Once in operational maintenance and support, the systems engineer is responsible for the processes and plans associated with systems upgrades and the development of closure or deactivation requirements.

2.1.4.2.4 Technical Schedule Management

Technical schedule management is informed by the project's integrated master schedule and overarching project work breakdown structure. The systems engineer is responsible for developing the technical work breakdown structure to the system, product, subsystem, assembly, subassembly, and part levels, as described in the technical scope section. A technical schedule is then developed from the technical work breakdown structure. The technical schedule, often referred to as a systems engineering master schedule (SEMS), provides visibility into the activities within the systems engineering areas of responsibility. A systems engineering master schedule typically include a time-phased and sequenced description of the technical event, reviews, significant tests, verification events, validation activities, and any other pertinent milestones. In other words, each technical task is shown in terms of a standard timeline. The technical labor effort that is associated with each task will also be identified within the schedule and will then be provided to the project manager so that it can be merged into the project integrated master schedule and managed as part of the overall project.

Throughout the project design and development phases, approved changes to the systems engineering master schedule will be coordinated between the systems engineer and project manager so that they are always in sync. Once the project enters the production/construction phase, a related schedule will be implemented and managed by the production/construction project manager. The next time a technical schedule is required is in the closure/deactivation phase of a project.

2.1.4.2.5 Acquisition Management

A diverse set of activities fits within this process category. The project procurement management plan serves as the overarching governance document associated with procuring resources that are not produced by the project and yet are required by the project. There may also be specific technical requirements that must be acquired through various legal and organizational contracts, subcontracts, memoranda of agreement, vendor management, supplier agreements, etc. These could include the purchase of commercial-off-the-shelf items, development contracts, or other products or services required.

In the definition phase of the project, the systems engineer develops the acquisition plan and an acquisition integration plan that specifically identifies the approach that will be used in order to integrate external-to-the-project goods or services. Information that must be provided by systems engineering to the organization responsible for acquiring the goods and services include technical specifications, tolerances and other measures, schedule and need date, acceptance criteria, processes associated with reporting, problem resolution, and warranty. Impacts for noncompliance of performance should be clearly defined, along with any other constraint or requirement that is appropriate to the project. As laws and/or other governance drive many of these activities, a procurement and acquisitions subject matter expert should be the final authority on these types of negotiations. Any acquisition requirements are provided to the project manager for inclusion in the project procurement management plan and to the procurement team for execution.

During the design and development phase of the project, new requirements for acquisition of configuration items, support equipment, subsystems, component, or parts might emerge. These will follow the process as outlined previously. In preparation for each new phase of the project, for example, production/construction, operational maintenance and support, and closure/deactivation, an acquisition plan should be developed, and directives given to the project manager and project procurement management to close the previous phase's acquisitions.

2.1.4.2.6 Risk Management

Risk management in the technical area of systems engineering is approached in a slightly different manner than in the project management discipline; however, because of the potential schedule and budget impact that technical risk and opportunities present to the overall project budget, it is an imperative that technical risk management is tightly integrated within the project's risk management process. The project manager holds the cost and schedule contingency to address impacts from risk mitigation and opportunity seizure and is, therefore, a key collaborator in this process area.

Throughout all life cycle phases, the systems engineer assesses risks and opportunities and raises them in priority to the project manager for consideration. In addition, the systems engineer develops a technical risk assessment and management plan that is specifically focused on reducing technical risk within the scope of the project. Technical risk reduction is associated with the technical solution space that is under development. Once the stakeholders requirements, measures of effectiveness, measures of performance, and technical performance measures are defined, and the architecture is developed to the point where it is understood, then a technical risk reduction strategy and plan can be developed.

As part of the analysis supporting the development of the technical risk reduction strategy, the physical architecture will be further decomposed to determine which sections of the architecture can be fulfilled by standard off-the-shelf items, and which of the sections require the design and development of subsystems, components, or parts. For those that require design and development, it is necessary to determine which phase of the life cycle they are in, such as early research all the way through late development. The earlier they fall in the life cycle (early research), the higher the risk, and the more aggressive the risk reduction strategy will need to be for that capability if the technical schedule is to be met. This is due to the fact that research and development tends to take much longer and cost significantly more than anticipated. Once the technical risk has been identified, a technical risk reduction strategy and plan will be formulated to determine the appropriate approaches necessary to meet the technical and overarching project schedules.

Technical risk reduction strategies that may be used include trade studies, modeling and simulation, and prototyping. The project manager and the systems engineer collaborate on any additional funding or schedule that may be required to perform these risk reduction strategies. Trade studies develop proposed solutions to requirements. Subject matter experts analyze these options, often modeling or simulating how they behave in the system, as a way to select between near-equal alternatives.[23] Modeling and simulation reflect the activity of developing a representation of the system and then simulating different scenarios in an effort to determine changes in performance characteristics. This is generally done to determine the optimal solution. Prototyping is a method that physically produces a representation of the system or parts of the system for testing purposes. This allows key performance parameters, measures of performance, and technical performance measures to be tested before fully committing the funds and schedule to the full development. All key performance parameters, measures of performance, and technical performance measures associated with technical risks should be retired prior to the critical design review and prior to entrance into the production/construction phase. If they are not retired, then they are accepted into the next phase as technical debt that must be resolved within the production/construction budget and schedule—a high-risk strategy.

2.1.4.2.7 Quality and Measurements

There are multiple activities associated with quality and measurements within the technical aspects of the project. Within the definition phase of the project, a Technical Quality Reviews Plan will be prepared to identify the quality reviews required for the project. Generally, reviews are called for within the broader set of project management reviews. All of the potential reviews are not noted here; however, the key standard reviews for design, development, production/construction, and operations are described below.

In the concept phase of the life cycle, they can include a system definition, conceptual design, and/or a system requirements review (SRR). In the design and development phase of the life cycle, subsystem and system preliminary design reviews may be held to assess the readiness of the system development to progress to the system integration phase. In the system integration phases, system integration, critical design, and production/construction readiness reviews are conducted. In the production/construction phase, an operational readiness review is done. Finally, a closure/deactivation review is held in the event that the system is being closed or deactivated. Additional technical reviews most likely would be executed to ensure the progress and quality of the technical solutions being developed. The number, type, attendance, and timing of the reviews are dependent on the desires of the project manager and systems engineer and will be documented in the Technical Quality Reviews Plan.

For established key performance parameter technical performance measures, technical quality reviews will be held regularly to assess progress in the risk reduction and development activities. Prior to entrance into the critical design review, all key performance parameter technical performance measures of risk should be retired, and the stakeholders' requirements that are reflected in the system-specification requirements documents should be validated by the stakeholders. All designs should be within the appropriate tolerances and parameters. The critical design review establishes that the system is ready to proceed into production/construction according to the stated performance requirements. If the risk has not been retired, then that risk reduction will be paid for in the next phase and carried as technical debt, as described in the previous section on risk.

2.1.4.2.8 Test, Verification, and Validation

Testing of all elements of the system is required to verify that the system, product, subsystem, assembly, subassembly, and part levels not only work as expected but also conform to the specifications and work in an integrated manner, as expected. The development of a comprehensive test plan, sometimes referred to as a Test and Evaluation Master Plan (TEMP), in the concept phase of the project is important because it will define how the evaluation of the system and all the subordinate elements will occur, as well as when and under what conditions. The test plan typically describes the requirements for test and evaluation, types and categories of tests, test procedures, test facilities and equipment (or other resources) needed, and the description as to the circumstances that will have occurred in order to warrant the test. When developing a test plan, the systems engineer must pay particular attention to testing not only the elements of the system, but each combination of two elements, and ultimately a test of the entire system.[24] Test outputs not only verify the expected conditions but also confirm the expected emergent behavior at the system level or the identification of an unexpected emergent behavior. Once the system and its associated sublevel elements of the system have been tested and verified, the stakeholders are brought into the process to validate that the performance meets their requirements. This is a critical step that, if left out, will inevitably lead to dissatisfied stakeholders who may not provide final system approval and therefore will affect the success of the overall project in a negative way.

During each phase of the project, there will be items to test, verify, and validate, inclusive of functional, environmental, qualification, destructive, and nondestructive testing. If any changes are driven to the design based on these activities, those changes must flow through formal change control and be captured in the design. This will also be recorded in the configuration-controlled requirements verification traceability documents.

The test, verification, and validation environment can drive specific requirements that must be factored into the overall cost and schedule of the project. Within the systems engineering domain, verification refers to the confirmation that the technical solution meets the technical specifications, whereas validation refers to the key stakeholder's confirmation that the solution meets their needs. A common omission in planning this phase is in including planned testing of both the elements and the whole system while considering each life-cycle phase. Assessing the requirements within the development, production, and anticipated operations and maintenance environments will drive design considerations and should always be considered.[25] Special test equipment, test facilities, test analysis software, and other special infrastructure and operational support are required and should be planned for by the systems engineer in collaboration with the project manager.

2.1.4.2.9 *Configuration Management*

Having knowledge at all times of the composition, history, and current version of the elements and the system as a whole is the responsibility of the systems engineer. To perform this role, a configuration management process is established and implemented throughout the full life cycle of the system.[26] In the definition phase of the project, the system engineer will develop the configuration management plan. This plan will describe the configuration control process for the life of the system, identify and establish a unique identifier for all system elements that will be managed under configuration control (typically referred to as Configuration Items or CIs), and establish the appropriate technical baselines including those products, subsystems, components, or parts to be acquired. The processes must contain the information for the broader project on how to initiate change requests, commonly known as *engineering change requests* (ECRs) and engineering change notices (ECNs), and how to document and communicate the status of these actions throughout the project so that they are appropriately integrated into the associated activities.

The system engineer is responsible for identifying, recording, documenting decisions, communications, and, most importantly, controlling the configuration of the design. During the concept phases, the as-allocated configuration is managed. During the design and development phase, the configuration is referred to as-designed. The design continually evolves because risk reduction strategies and test, verification, and validation activities affect the design, and this is a critical period for the management of the configuration. If the configuration is allowed to get out of sync, it will be extremely difficult to recover, and the project will be affected.

Once the system moves into the production/construction phase, an as-built configuration is documented and always reflects the current configuration of the product. Preplanned production/construction improvements will be recorded in the as-built configuration as well. In the operations phase, the as-maintained configuration will always provide the current view of the system, with any maintenance changes, preventative, and corrective actions reflected within.

2.2 Process Alignment Between Two Disciplines

This key section describes how the two disciplines, project management and systems engineering, are aligned, and how systems engineering provides the next level of fidelity for the technical elements of the project.

- Both disciplines are rigorous; both add value to the project.
- Both use processes that address some of the same areas of focus.
- When project management and systems engineering combine efforts on coinciding areas, they can optimize the overall project results.

There are many activities that are often executed independently of one another, although they have similar objectives. With project management and systems engineering, these activities may even be similarly titled. In many organizations, the project management and systems engineering functions are managed through separate departments or divisions. Sometimes they are unequal in stature. With the use of different project management methodologies, the activities may be implemented in inconsistent ways. Each of these situations can lead to tension and duplicative work. It is prudent to deconflict these responsibilities as much as possible and as soon as possible.

There are nine significant areas where this overlap occurs:

- stakeholders
- communications
- scope
- schedule
- procurement
- risk
- quality
- test, verification, and validation
- governance

These will be described in Sections 2.2.1–2.2.8. In some of these areas, the project manager is the process owner, even though there is a supporting requirement by the systems engineer. In some instances, the systems engineering discipline does not provide a process of its own to manage the activity. A good example of this is in the management of the budget. The project manager manages the project budget. The systems engineer's activities developing a technical solution are highly interdependent with the development of the budget. The solution must fit into the budget envelope, and any contracts for acquisition or procurement must also fit within the budget. Technical risk will also use the budget contingency. The systems engineer is responsible for acquiring any product, subsystem, assembly, subassembly, and part levels associated with the project, so he or she is obligated to ensure that the project manager is aware of the interdependencies with the budget.

Other areas of strong interdependencies involve the area of resource use—whether it be infrastructure or levels of effort of labor. These are best managed at the project level, rather than the technical level, because the project activities span all disciplines, whereas systems engineering is responsible for the technical areas of performance. Section 2.3 will describe methods for using the combined approach on those activities that benefit from collaboration in order to drive positive project outcomes.

2.2.1 Stakeholder Management

Stakeholder management is both a project management knowledge area and a systems engineering process. They are not two separate views; however, the systems engineering focus is on understanding the technical scope rather than the working with the stakeholders to address the overarching project scope, schedule, and budget. "Realizing successful systems requires reasoning with stakeholders about the relative value of alternative realizations, and about the organization of components and people into a system that satisfied the often conflicting value propositions of stakeholders. Stakeholders who are critical to the system's success include funders, owners, users, operators, maintainers, manufacturers, and safety and pollution regulators."[27] The questions systems engineers are trying to resolve include: (1) will the system as designed, as built, and as operated meet the stakeholders' needs in the way they envisioned it would? And (2) will the stakeholders be able to operate and maintain the system as planned?[28]

Each discipline focuses on a different set of relevant responsibilities within the stakeholder community. The project manager is responsible for managing stakeholder involvement at the highest level (cost, schedule, scope, change management) and to ensure that the stakeholders are satisfied with the outcomes of the project. The systems engineer is responsible for managing the stakeholder involvement in the technical area, and for providing the stakeholders with a technically sound solution to their desired system as captured and configuration controlled.

If the project manager and the systems engineer are not working in sync with the stakeholders to elicit requirements, assess change requests, and obtain stakeholder buy-in and approvals, it is likely that the outcomes of the project will not be in alignment with the stakeholders' view of what the project outcomes should be. Even if the project manager and systems engineer have captured the stakeholders' view of success, that view can radically change over time; hence the need for continual stakeholder management.[29] If technical changes are accepted without evaluation as to the schedule, budget, and overall project risk, it is likely that these changes will negatively affect the overall project performance. That is why it is imperative to utilize the project management stakeholder, communications management, and change-control activities to manage stakeholders effectively. If organizations are changing from a waterfall method, for example, to flexible methods of project management, which seems to be the trend, then the imperative to have a conjoined stakeholder management process increases.

2.2.2 Communications Management

Because both the project manager and the system engineer are communicating with the stakeholders, a joint communications plan, with clearly defined decision support responsibilities, is an imperative. Each discipline has critical communications responsibilities that are separate from the other. The project manager is responsible for communicating about the overall project to the stakeholders. The systems engineer has the responsibility for communication about the technical scope of the project. However the communications responsibilities are distributed, the status of the entire project must be communicated to the stakeholders in a timely, efficient, and effective manner.

2.2.3 Scope Management

Both the project manager and the systems engineer have responsibilities associated with ensuring that the project scope meets the needs of the key stakeholders within the cost and schedule that has been allocated for the project. However, the systems engineer also has the responsibility for ensuring that the technical solution meets the technical needs of the stakeholders. It is the systems engineer who, while working with the subject matter experts, will document the stakeholder requirements, decompose these to technical requirements, develop the technical solution, compile and assess the results of trade studies and modeling and simulation, and maintain traceability throughout the design.[30]

Technical scope management may be handled differently, depending on how the project is being managed and given the project's location in the life cycle phases. Depending on the life cycle phases of each work package, technical scope may also significantly vary between them. Impacts from these structural considerations may take the form of process iterations, evolving requirements and risks, or other activities that impact technical scope. Regardless of these factors, the systems engineer must ultimately ensure that the baseline technical scope is captured in the overarching project documentation, and that any proposed scope changes are brought to the project change-control board or configuration-control board as appropriate and, once approved, provided to the project manager for incorporation into the project documentation, as appropriate.

2.2.4 Schedule Management

The overall schedule for the project is the responsibility of the project manager. The technical schedule is the responsibility of the systems engineer. The systems engineer must ensure that the baseline technical schedule is captured in the overarching project schedule and that any proposed schedule changes are brought to the project change control board and, upon approval, provided to the project manager for incorporation into the project schedule.

2.2.5 Procurement Management

The project manager is responsible for the procurement management, as well as any acquisitions, purchases, memoranda of agreement (MOA), or other negotiation and commitments to obtain goods or services for the project. The systems engineer may identify needed technical resources from outside the project and often from outside of the organization. The project manager must have access to information associated with any commitments made on behalf of the project, agreements made, or negotiations that are initiated by systems engineering so that the project budget and schedule can be effectively managed. Ideally, all procurement activities should route through the project manager for these reasons. These two disciplines must work in close collaboration to ensure that the most efficient procurement strategy is in effect.

2.2.6 Risk Management

Technical risk reduction activities and technical opportunity seizure fits well into the overall project risk management methodology. Where project managers see risks and opportunities in the sense of the triple constraint—cost, schedule, scope—system engineers see technical risk as a natural occurrence of design, and one that can be reduced through the use of trade studies, modeling and simulation, prototyping, and testing.

   Although on the surface the overlap may not be apparent, on closer examination, risk to the project overall and technical risk are often interrelated and affect one another. If technical risk is managed outside of the project risk management process, impacts to the project from changes could impact the overall project and/or schedule. Aligning these two processes makes sense and will ensure that technical risk reduction and opportunities are prioritized within the greater project and that they are supported at the level that ensures project success.

2.2.7 Quality Management

Quality management for the overall project includes metrics and measures that demonstrate performance to the project and quality management plans. "Typical project measures, such as cost and schedule performance indicators, provide the project manager (and systems engineer) visibility into how well the project is tracking against its planned budget and schedule targets."[31] The project manager also has the responsibility to track progress against scope targets. This is particularly critical because scope could directly impact budget and schedule and must be actively managed as part of the project manger's areas of responsibility. In practicality, it is clear that measurement helps focus attention on activities that are important.

   Quality in the area of technical scope management also has additional meaning. Measures of quality come in various forms, such as the establishment of, and ongoing tracking throughout the project on, quality measures—for example, measures of effectiveness, measures of performance, technical performance measures, etc., as described in Section 2.1.4.

In addition to measurements, both the project as a whole and the technical scope activities drive review activities. Reviews serve an important role for the project. They provide the project team, the organizational management, external reviewers, and all of the stakeholders the opportunity to review the status of the project and look with "fresh eyes," to bring a perspective that is outside of the project. This is valuable to the project and to the organization because it minimizes the risk of "groupthink" and other mental models that might not be conducive to the achievement of optimal project outcomes. Project reviews typically will not dive deeply into technical details, whereas technical reviews do just that. However, there is a synergy between the project reviews and the technical reviews that should be taken advantage of. If managed efficiently, they should provide enough technical detail to satisfy the stakeholders and allow management and leadership to make sound decisions on behalf of the project.

Finally, the stakeholders will assess project and technical quality in a fundamental way to determine if the final delivered system meets the needs and the expectations of the stakeholders. To ensure this occurs, the project manager identifies the scope, while the systems engineer elicits clear technical requirements and then develops a system that efficiently and effectively meets those needs. Test, verification, and validation activities, with the stakeholders' involvement throughout the project, ultimately lead to stakeholder satisfaction.

The quality management function cannot be managed independently from either the project management or systems engineering disciplines without increasing the risk to achieving this positive outcome. Both disciplines play a significant role, and if conjoined, will present a well-constructed and well-managed project that produces a system that meets the stakeholders' needs.

2.2.8 Governance

Project governance has to do with the careful controlled management of change on a project that, if done well, will lead to optimal outcomes. Methods used to control natural change that occurs on projects are essential. Change without governance leads to chaos, overspending, schedule impacts, and unchecked scope evolution.[32] The project manager is responsible for overall project integrated change control. Any change to the baselines of scope, schedule, or budget must be vetted and approved and then implemented. Integrated change control is an interdisciplinary activity. And, it is a key part of actively managing a project to a set of project management documents.

Configuration management is a technique that a systems engineer will use to control and manage change within the technical areas of responsibility. It is a critical methodology that is implemented so that the form and function of the system and all of the products, subsystems, component, or parts are known at any given time. It is an important governance activity. As important as this is to the systems engineer, it is equally as important for the project manager to understand throughout the project life cycle the as-allocated, as-designed, as-built, and as-maintained configurations to assure that the anticipated project outcomes are on track.

2.3 Using a Combined Method to Drive Project Outcomes

It takes the appropriate application of both disciplines, in an optimal configuration, to achieve the best performance outcomes for the stakeholders. When used in a suboptimal organizational configuration, such as when systems engineering sits outside of the project, the risk to the project increases because both disciplines make independent decisions and may vie for the authority to implement actions that impact the scope, schedule, and cost of the project, which is the domain and authority of the project manager.

- Projects will be seen as successful only when they perform to the intended outcomes expected by the stakeholders.
- Technical decisions have impacts on project objectives associated with cost, schedule, and scope control and therefore must be considered within the greater project objectives.
- Using project management and systems engineering together, in the appropriate configuration, will dramatically increase the probability of achieving successful outcomes in a project.

The implementation of the common and unique project management and systems engineering processes, as described throughout this book, are blended into a unified approach called the Complex Systems Methodology. Using Complex Systems Methodology combines the most logical process activities across all life cycle phases of a system into a cohesive, inclusive, and highly adaptable and tailorable framework. Complex Systems Methodology guides the natural evolution of projects along their life cycle phases, while providing entrance and exit points along known and accepted process paths and provides the project managers and systems engineers the ability to focus their attention on creative project management instead of focusing on dealing with process-merging challenges. Using Complex Systems Methodology provides a solid framework for actively managing complexity for all technology systems across any discipline, including system-of-systems projects.

This chapter described the project management and systems engineering disciplines and how the overlaps add value to both. It also explained which processes overlap or are complementary enough to consider a joint approach to implementation methods and approaches. The minor differences in project management methods that might affect decision making associated with the application of this combined method were discussed. Complex Systems Methodology was introduced as a viable approach to merging the processes for use in addressing complex projects. In Chapter 3, details about the joint processes area will be presented in depth, as well as the anticipated outcomes from the conjoined application of these processes.

2.4 Case Study: FIRST Robotics Competition—University of Detroit Jesuit High School and Academy Robocubs Team #1701

All case study quotes are from interviews held with the individual presenting the background. In this next case, the interview was held with Michael J. Vinarcik, and all quotes are attributed to him.

In 2015, Michael Vinarcik, a Senior Lead Systems Engineer at Booz Allen Hamilton (a FIRST Strategic Partner), became a FIRST[33] mentor to the Robocubs Team at the University of Detroit Jesuit High School and Academy. FIRST is a program with a mission to "inspire young people to be science and technology leaders, by engaging them in exciting mentor-based programs that build science, engineering, and technology skills, that inspire innovation, and that foster well-rounded life capabilities including self-confidence, communication, and leadership."[34]

The FIRST nonprofit organization, founded in 1989, provides "innovative programs that motivate young people to pursue education and career opportunities in science, technology, engineering and math."[35] The FIRST organization provides opportunities for students and mentors to engage in robotics and technical challenges. On average, FIRST reaches "over 400,000 young people annually."[36] Just in the 2016–2017 time frame, over "460,000 students and 230,000 mentors, coaches, judges and volunteers in 85+ counties"[37] were involved.

The Robocubs Team[38] focuses on the FIRST Robotics Competition (for high school students). An essential part of the Robocubs mission is to "partner students with knowledgeable technical mentors in a rigorous design and build process." Vinarcik became involved after his son joined the Robocubs as a freshman. He saw his involvement as a good way to bring and share systems engineering concepts and principles into the robotics team's experience.

2.4.1 Project Charter and Plan

The Robocubs (FRC Team 1701) first competed in 2005 (winning the "Highest Rookie Seed" and "Safety Awards" in their first competition). Many dedicated parents and mentors have been involved with the Robocubs, and the design and fabrication experience they gained enabled the team to be an effective competitor. Vinarcik's 2014 entrance into the mentorship of the team led to the implementation of a more formal systems engineering approach in time for the January 2015 competition. The team has gone on to further success using these systems engineering strategies and techniques (including the team's first "blue banner," a district event win in 2016).

Vinarcik's method was to teach the team how to approach each season by designing a robot that can win competitions, not building the "best robot," but by building a robot that meets all the requirements and that is more effective than the competitor's robots in achieving the required tasks. He states that initially "you want to build a robot that can win the competition by not being disqualified (instantly losing)." He elaborates that he advised them to "pay close attention to rules that will get you disqualified." Then he focused them on "framing the problem correctly" through understanding exactly what the requirements were to achieve the highest number of points.

2.4.2 Stakeholder and Communications Management

The Robocubs Team had many stakeholders. They wanted to do well for themselves, for the school, for their families, and for the sponsors. Communications were broadly distributed through a website http://robocubs.com/.

2.4.3 Scope Management

There is one bedrock constant in FIRST every year: Robots must move, handle game pieces, and score points. The difficulty of scoring tasks (both autonomously and under operator control) varies; however, all events are timed. Therefore, the path to victory is always to convert seconds into points more efficiently than one's competitors. Vinarcik also believes that many engineers tend to explore too many possibilities or jump to a single solution. His philosophy is that the "art in systems engineering comes from seeing the tree, cutting off the dead branches (the obvious dead ends), and selecting the best of the rest." So he focused the team on the architecture and requirements.

2.4.3.1 Architectural Design

Systems architecture is one of Vinarcik's areas of expertise. He has an International Council on Systems Engineering (INCOSE) Expert Systems Engineer Professional (ESEP) certification and an Object Management Group®'s (OMG®) Certified Systems Modeling Professional™ (OCSMP™) Model Builder-Advanced certification. In addition, he recently established the Systems Architecture Guild to promote best practices in systems modeling, systems architecture, and systems engineering.[39] As he explains, "systems architecture is translating WHAT the system does into the design by selecting HOW it will do it."[40] Explaining that "1+1 = 3 because 1+1+interfaces = 3" he summarizes that "without the interactions we just have a box of parts that doesn't do anything useful!"[41] This is a foundational concept in systems design—one of several key messages that he has taught to the Robocubs in order to help them achieve their wins. Managing interfaces between components, even in the relatively small scale of an FRC robot, is critical to success—for example, mounting components can be critical to ensure stability, function, and easy serviceability.

To help the Robocubs understand the requirements and then form them into a solution space, he needed to help them perceive the requirements in a digestible form and then visualize the potential solutions while exploring the concept space fully. The architected solution had to be "good enough," and no more, because of time constraints and the limited engineering and fabrication experience of the students. It was also important to help the students understand the full context of the competition; as Vinarcik repeatedly told them, "The worst thing you can do is solve the wrong problem efficiently."

Vinarcik used standard systems engineering techniques and tools to assist the Robocubs in recording the technical requirements in a computer-based design tool (MagicDraw). Using this tool allowed the team to logically trace the requirements to the physical solution they would ultimately choose. A modeling language (SysML) provided the capability for the team to observe the systems-level behavior of the modeled robot in relation to potential changes that they might want to make. Once they settled on their desired design, they developed a physical prototype using plywood. The prototype allowed them to physically test their design and gain valuable real-life experiences prior to constructing the official aluminum competition robot. In this competition, a plywood prototype of the ball pickup/handling arm established key dimensions (to ensure robust ball pickup) and was critical to the successful fabrication of the final aluminum version. This arm's versatility (it also manipulated multiple obstacles and could deposit balls into the lower scoring goal) was key to the Robocubs' success. They also built a human-pulled chassis "sled" prototype to confirm that their wheel configuration could traverse the obstacles without becoming stuck.

2.4.3.2 Requirements Management

A typical robotics competition would have some general requirements, such as that each robot must be able to execute a series of actions, such as collecting, controlling, aligning, moving, and depositing (for a score) a self-standing article called a *game piece*. The robot has one job: to complete its actions within an established time period that applied to all applicants and turn those into points through the efficiency and effectiveness of its movements in placing the game pieces. Points are typically awarded for autonomous performance in the beginning of a match, completing various tasks during the remote control, or "teleoperating period," and additional points for meeting certain conditions at the end of the match (such as having the robot parked in a certain location or climbing a rope to a specified height).

The requirements of the robot competition are specified at the announcement of the event. In this case, the announcement came January 9, 2016. By January 23rd, the requirements had been converted to a finished prototype of the robot, and by March 3rd, a working robot was in testing at the competition site, with the competition being held March 5th.

The Robocubs' approach to interpreting the annual competition involved documenting the requirements in a spreadsheet. Each requirement had to be satisfied and compliance to the specifications verified. For example, for the FIRST STRONGHOLD competition, the strategy included "breaching the opponents' defenses (e.g., 'the outer works'), weakening their tower with boulders, and capturing it."[42] There are additional rules that must be abided by during the competition, such as being able to form alliances with other robot teams and only having one robot from an alliance in their own courtyard at a time for defense of their own tower. The win goes to the alliance with the highest points at the end of the match. As Vinarcik explains, "The robot's job is to convert time (seconds) into points. Time is the primary 'currency' that is spent; game pieces are the other. How your robot 'spends' those two commodities will determine how well you do!" There were also activities that could award opponents extra points, thereby diminishing a team's ability to win. These activities included actions built into the foundation of the game, such as perimeter defenses closing unexpectedly or points awarded for capturing an opponent's tower.

The behavioral requirements of the robot included the following:

- "The first 15 seconds, the robot must operate autonomously. Points during this phase were earned by reaching opponent's outer works, crossing defenses, and scoring boulders in goals.
- During the next 2-minute 15-second phase, the robot must be teleoperated (e.g., human drivers remotely controlling the robot). Points during this phase were earned by retrieving boulders, overcoming opponent's defenses, and scoring boulders in goals from their opponent's courtyard.
- During the final 20 seconds of the match, robots can approach the tower, extend up, and scale the tower for additional points."

The technical requirements of the robot are associated with its mass properties (physical attributes such as weight, volume, center of mass, and moments of inertia). These affect how the robot would behave within the environment of the competition field. The Robocubs robot "christened the U.S.S. Valiant in keeping with the Star Trek theme, and because NCC-1701 is the hull number of the U.S.S. Enterprise (matching (FRC Team 1701))," as Vinarcik explained, "was designed to have a low center of gravity and never tipped over, despite aggressive driving by its operator. Other robots lost matches because their designers did not understand the importance of stability, and the robots became immobilized after they flipped over."

2.4.3.3 Integration and Interface Management

Integration activities associated with the robot included the integration of the physical components of the design. Interface requirements included enabling the robot to respond to commands provided electronically on the competition field. To validate the interfaces and integration of those elements, the physical integration of the design was verified with the computer-aided design software. The interface between the hardware and software was verified through physical testing.

2.4.4 Acquisition Strategy and Integration

The parts that the Robocubs purchased included items such as hex stock and bearings, controllers, and motors. Depending on the year, purchased components may be as high as 50 percent of the robot. Purchased items typically do not have to do with the dynamic performance of the robot, such as game piece handling or navigation, which the team builds. FIRST Robotics Competition also provides certain key components (such as the controller and wireless communications system).

During the original FIRST competitions, all robots were built from scratch as component parts, such as motors and gears, which were not readily available to the teams that were competing. Over time, a thriving ecosystem of suppliers began to make common parts; components became easy to obtain and inexpensive, so the necessity to build from the ground up was drastically reduced. However, as most competing teams do not have unlimited funding, it is important to identify those parts that are readily available and cost effective to purchase versus those materials that must be developed and constructed by the team. The most efficient way to determine this is to decompose the design into its component pieces and research the availability and cost of each item, and then make the decisions based on available overall funding and the time needed to design and fabricate. Vinarcik confirmed that the team reviewed "which parts they wanted to buy and which ones they wanted to design."

2.4.5 Risks and Opportunity Management

One of the key risk management strategies that the Robocubs applied was associated with the test and verification phase in which they were making modifications "on the fly" to the model. This scenario could have left them open to unintentionally modifying something that would disqualify them. They mitigated their risk by making sure that components that were fulfilling specific requirements physically could not be modified and by routinely checking critical constraints such as weight and overall size.

An additional risk mitigation strategy that was employed for subsequent competitions was to use the previous year's model as a test bed. The Robocubs would make design enhancements and physically model and test them to obtain more accurate physical performance measurements. As they gained experience from their previous competitions, they were better poised to apply the lessons learned, to optimize their current designs.

2.4.6 Quality Management

Quality measurements for the competition are focused on the physical-performance specifications determined by the FIRST Robotics Competition. These typically include the agility, maneuverability, and responsiveness of the robot, as well as team performance associated with collaborations with other teams to achieve competition objectives, communication among team members, and physical performance (such as calling out instructions and moving pieces on the competition field). All of these activities resulted in a compiled score that would be compared across all competitors, with the highest score being awarded the winner.

2.4.7 Test, Verification, and Validation

Every FIRST team is allowed to make changes/tuning/tweaks throughout the season. There are time periods when the robots are "unbagged" and may be modified; there are also rules that permit a certain amount of weight to be "withheld" from bagging. This allows teams to learn from experience and competitors, both on provided practice fields and during matches. The "pits" at every FIRST event are full of bustling activity in each team's area and the provided machine shops. The Robocubs tried to build modularity into their robot so that they could remove, modify, and replace parts easily during this test and verification phase. Being able to remove content easily to use all rework/refit time effectively is a critical element of their strategy.

To improve the odds of success from the start, the Robocubs built surrogate field elements from the provided plans. This enabled them to test scoring, maneuverability, and other aspects of the robot's performance. "For example, in STRONGHOLD, they were able to confirm that they could cross/breach every obstacle except for the low bar (which was purposely excluded because it would have constrained the robot's height)," Vinarcik explained.

2.4.8 Governance

During the initial phases of the robot design and build, the Robocubs were able to carefully control the decisions they made about their robot. Each time a change was entertained, an elected "council of heads" consisting of student leaders and mentors with relevant expertise would meet to discuss, and changes would be reviewed by this group before being accepted and integrated into the robot. However, once the robot was constructed and moved to the competition field, change would happen dynamically. More changes would be implemented at each successive competition, as they moved from qualifiers to final competitions. This generally meant that the computer-aided design model and the actual physical robot configuration would get out of sync. This was necessary based on the need for quick action along with a low need for accurate documentation at that stage.

2.4.9 Outcomes

During the 2016 competition, the Robocubs finished the qualifying rounds in 19th place (out of 39 competitors). They learned effective collaboration techniques and were able to work with another team to advance their standing. The team then moved to the Indiana District Finals, where they moved up to 5th place.[43] They also used their lessons from the initial collaboration to place high enough to secure a position as captain of an alliance and to draft other teams to that alliance to positively impact their performance. From there, they attended the Center Line District Finals where they won the round and again placed 5th. The team continued to hone their alliance collaboration skills. On April 11th, the Robocubs team was officially qualified to move on to the state's competitions where they performed well, with three strong wins. And on April 17th, the team was ranked 60th in the state and officially qualified for the World Championship, where they competed in the Galileo Division and came in 22nd.

The robot was designed well. The early prototype ball handling system was efficient and designed to multitask. Vinarcik relayed that "it was clear that the systems engineering was an enabling activity that led to their successes. We have mentors highly skilled in designing and building robots; this helps us design and build robots with a better chance of winning because of the focus on converting time into points efficiently." He also stated that this was an exceptional learning experience for the Robocubs and that "their exposure to the processes and tools of systems engineering such as the computer-aided design and systems modeling helped them to understand that these tools require less work to be more right!"

In addition to the experience in competing, the Robocubs gained an appreciation for what is involved in running a business, allowing them to determine how best to apply available resources to maximize performance.

2.4.10 Lessons Learned

The most valuable lessons learned from the experience involved human skills. "The transfer of knowledge and expertise through mentoring of the next team was an excellent lesson for the Robocubs," Vinarcik said. "The Robocubs learned how to be competitive without being cutthroat, how to help one another, how to negotiate, and how to practice gracious professionalism." And what started as individual development evolved into passing along what was learned to underclassmen who joined the team the following year. Robocub graduates now in college are mentoring middle- and high-school teams and propagating the "convert time into points" approach.

2.4.11 Case Analysis

In this case, the project was well defined and the requirements rigorous and unchangeable. The function of the robot was also well defined because it was required to perform to exact behavioral traits. The form and fit of the robot was at the discretion of the team members, and their creativity was challenged. This case reflects an application of systems engineering processes in a way that directly led to important outcomes. It also led to emergent behavior, or outcomes that were greater and more broadly reaching than that directly associated with the robot build and competition. Learning to assess a system as a whole, as well as learning to perform as a team, were excellent experiences that will serve the Robocubs well into their future.

2.5 Key Point Summary

The focus of Chapter 2 is to describe the basic definition of a project and to outline the two critical disciplines that interact to achieve project outcomes: project management and systems engineering. Although other disciplines play a role in the project, project management is key because it provides the structure in which the overarching project activities can take place. Its role is to establish the cost, schedule, and scope baselines and to maintain control over these baselines through change management so that the end result of the project is as expected and as the stakeholders have outlined in their requirements.

Systems engineering also provides a key role as an overarching, interdisciplinary structure that formulates the technical solution that meets the stakeholder requirements. With these two disciplines intertwined effectively, projects can reach intended and anticipated outcomes successfully. Following are the key concepts from this chapter. Key terms are compiled for quick reference in the Glossary.

2.5.1 Key Concepts

- Project management holds overall authority and accountability over the success of the project or the project outcomes. Systems engineering holds overall authority and accountability for the technical scope elements of the project.
- A project provides an overarching structure for controlling cost, schedule, and scope.
- The life cycle of a project may be shorter than the life cycle of a system, and the life cycle phases overlap throughout the associated project life cycle phases. This is particularly true for systems engineering design, which is highly iterative.
- Appropriately tailored systems engineering processes will increase the chance of the project achieving optimal outcomes, particularly when utilizing flexible project management methods.
- Systems engineering adds the technical details to the scope of a project. It is an interdisciplinary approach that interfaces closely with project management and adds fidelity to activities that the project manager is responsible for, such as the project stakeholder register, communications plan, work breakdown structure, schedule, procurement plan, risk register, quality management plan, and the governance model.
- Both disciplines are rigorous, both add value to the project, and both use processes that address some of the same areas of focus.
- When project management and systems engineering combine efforts on coinciding areas, they can optimize the overall project results. It takes the appropriate application of both disciplines, in an optimal configuration, to achieve successful project outcomes, and when used in a suboptimal organizational configuration, this increases the risk to the project.

2.6 Apply Now

To feel confident in using the material in this chapter, the reader should review the following summary points and answer the questions that have been posed in regard to your own knowledge and experience. The examples you use do not need to be formal projects to be included in this exercise.

1. *The project management discipline provides a description of the processes across the project life cycle.* Describe an activity that you performed within each process area.
2. *The systems engineering discipline provides a description across life cycle phases.* Describe activities that you have performed in each of these processes.

3. *There is a complementarity between the two processes that allows alignment, which can lead to optimal project outcomes.*
   Describe a project where you used these aligned processes. What were your outcomes?
4. *Process overlaps are common across multiple disciplines associated with management activities.*
   Identify, from your own experiences, examples of overlaps in processes from each of the discipline areas. Discuss what might have been done more effectively if the process areas were conjoined.
5. *Complex Systems Methodology*™ ™ *(CSM*™ ™*) provides a structure that can be used to manage complex projects.*
   Describe a complex project that could benefit through the use of the Complex Systems Methodology.

References

1. IAPM. International Association of Project Managers. Accessed January 12, 2018, from https://www.iapm.net/en/start/
2. APM. Association for Project Management. Accessed January 8, 2018, from https://www.apm.org.uk/about-us
3. IPMA. International Project Management Association. Accessed January 8, 2018, from http://www.ipma.world
4. PMI. Project Management Institute. Accessed January 8, 2018, from https://www.pmi.org
5. Wingate, L. M. (2014). *Project Management for Research and Development: Guiding Innovation for Positive R&D Outcomes.* Boca Raton, FL: CRC Press/Taylor & Francis Group.
6. IPMA. International Project Management Association. Accessed January 8, 2018, from http://www.ipma.world
7. PMI. (2017). *A Guide to the Project Management Body of Knowledge*® (*PMBOK*® *Guide*), 6th Edition. Newtown Square, PA: Project Management Institute (PMI).
8. ISO. (2008). ISO 9000:2005. *Quality Management Systems—Fundamentals and Vocabulary.* Geneva, Switzerland: ISO.
9. Wingate, L. M. (2014). *Project Management for Research and Development: Guiding Innovation for Positive R&D Outcomes.* Boca Raton, FL: CRC Press/Taylor & Francis Group.
10. INCOSE. (2015). *Systems Engineering Handbook: A Guide for System Life Cycle Processes and Activities*, 4th Edition. D. Walden, G. Roedler, K. Forsberg, R. Hamelin, and T. Shortell (eds.). Hoboken, NJ: John Wiley & Sons, Inc.
11. Ibid.
12. Ibid.
13. Wasson, C. S. (2016). *System Engineering: Analysis, Design, and Development*, 2nd Edition. Hoboken, NJ: John Wiley & Sons, Inc.
14. SEBoK contributors. "Guide to the Systems Engineering Body of Knowledge (SEBoK)." SEBoK. http://sebokwiki.org/w/index.php?title=Guide_to_the_Systems_Engineering_Body_of_Knowledge_(SEBoK)&oldid=53122 (accessed March 17, 2018).
15. Wasson, C. S. (2016). *System Engineering: Analysis, Design, and Development*, 2nd Edition. Hoboken, NJ: John Wiley & Sons, Inc.
16. Crawley, E., Cameron, B., and Selva, D. (2015). *System Architecture: Strategy and Product Development for Complex Systems.* London, UK: Pearson Publishing Company.
17. Ibid.
18. Eisenträger, J. (n.d.). Item Interchangeability Rules—Rev. C-2013-08-26. Accessed July 2, 2017, from http://www.joergei.de/cm/interchangeability-guide_en.pdf
19. Crawley, E., Cameron, B., and Selva, D. (2015). *System Architecture: Strategy and Product Development for Complex Systems.* London, UK: Pearson Publishing Company.
20. Ibid.

21. SEBoK contributors. "Guide to the Systems Engineering Body of Knowledge (SEBoK)." SEBoK. http://sebokwiki.org/w/index.php?title=Guide_to_the_Systems_Engineering_Body_of_Knowledge_ (SEBoK)&oldid=53122 (accessed March 17, 2018).

22. Rebentisch, E. (ed.). (2017). *Integrating Program Management and Systems Engineering: Methods, Tools, and Organizational Systems for Improving Performance*. Hoboken, NJ: John Wiley & Sons, Inc.

23. Wasson, C. S. (2016). *System Engineering: Analysis, Design, and Development*, 2nd Edition. Hoboken, NJ: John Wiley & Sons, Inc.

24. Blanchard, B. S., and Blyler, J. E. (2016). *System Engineering Management,* 5th Edition. Hoboken, NJ: John Wiley & Sons, Inc.

25. Kossiakoff, A., Sweet, W. N., Seymour, S. J., and Biemer, S. M. (2011). *Systems Engineering: Principles and Practice*, 2nd Edition. Hoboken, NJ: John Wiley & Sons, Inc.

26. INCOSE. (2015). *Systems Engineering Handbook: A Guide for System Life Cycle Processes and Activities*, 4th Edition. D. Walden, G. Roedler, K. Forsberg, R. Hamelin, and T. Shortell (eds.). Hoboken, NJ: John Wiley & Sons, Inc.

27. SEBoK contributors. "Guide to the Systems Engineering Body of Knowledge (SEBoK)." SEBoK. http://sebokwiki.org/w/index.php?title=Guide_to_the_Systems_Engineering_Body_of_Knowledge_ (SEBoK)&oldid=53122 (accessed March 17, 2018).

28. Crawley, E., Cameron, B., and Selva, D. (2015). *System Architecture: Strategy and Product Development for Complex Systems*. London, UK: Pearson Publishing Company.

29. Rebentisch, E. (ed.). (2017). *Integrating Program Management and Systems Engineering: Methods, Tools, and Organizational Systems for Improving Performance*. Hoboken, NJ: John Wiley & Sons, Inc.

30. Blanchard, B. S., and Blyler, J. E. (2016). *System Engineering Management*, 5th Edition. Hoboken, NJ: John Wiley & Sons, Inc.

31. INCOSE. (2015). *Project Manager's Guide to Systems Engineering Measurement for Project Success: A Basic Introduction to Systems Engineering Measures for Use by Project Managers*, v 1.0, INCOSE-TP-2015-001-01.

32. Rebentisch, E. (ed.). (2017). *Integrating Program Management and Systems Engineering: Methods, Tools, and Organizational Systems for Improving Performance*. Hoboken, NJ: John Wiley & Sons, Inc.

33. FIRSTINSPIRES. (n.d.). "FIRST INSPIRES Organization." Accessed July 4, 2017, from https://www.firstinspires.org

34. FIRSTINSPIRES. (n.d.). "FIRST INSPIRES Vision and Mission." Accessed July 4, 2017, from https://www.firstinspires.org/about/vision-and-mission

35. Ibid.

36. Ibid.

37. Ibid.

38. ROBOCUBS. (n.d.). "Robocubs." Accessed July 4, 2017, from http://robocubs.com/

39. Systems Architecture Guild. (n.d.). "Show Me the Wow." Accessed July 4, 2017. www.showmethewow.com

40. Systems Architecture Guild. (n.d.). "FIRST Robotics Strategy Introduction." Accessed July 4, 2017, from https://www.youtube.com/watch?v=NTouZLC8tCk&list=PLDz4YEQgpXfyqUQ90ACBbFJAhi2Ow1hYU

41. Ibid.

42. FIRSTRoboticsCompetition. (n.d.). "FIRST STRONGHOLD Game Reveal." Accessed July 4, 2017, from https://youtu.be/VqOKzoHJDjA

43. Team 1701. (n.d.). "FIRST Robotics Team 1701." Accessed July 4, 2017, from http://robocubs.com/2016/03/

Chapter 3

Application of Complementary Processes

This chapter provides a thorough explanation of the concepts for applying the Complex Systems Methodology complementary processes of integrating systems engineering and project management to a project. These include:

- stakeholder-focused processes
- solution-focused processes

Section 3.1 explores the processes that enable stakeholders in a project to convey the vision of what they would like to achieve and the level of involvement that they envision for themselves throughout the project. To achieve a level of stakeholder satisfaction, the project manager must ensure that the project progress is in alignment with their expectations. It also describes the activity necessary to confirm that the outcomes match the actual needs of the stakeholders. It describes in detail how effectively to perform the processes associated with engagement of stakeholders within a project by using both disciplines of project management and systems engineering to maximize the probability of achieving successful outcomes. This section of the chapter focuses on the complementary stakeholder-focused processes of stakeholder engagement and communications management.

In Section 3.2, both project management and systems engineering disciplines have defined solution-focused processes and a disciplined flow that moves the project from conception through the entire life cycle. Certain processes are aligned to take advantage of the benefits of both, as well as the combined effects of using both processes together. These are scope, schedule, procurement, risk, quality, and governance management. In this book, governance includes activities that provide structure around change. Change control is a crucial step that must be used to address stakeholder-driven changes that, if not managed, might lead to dissatisfaction with the ultimate outcome of the project. The change process is discussed in Section 3.2.6.

It is important to point out that there are many standards-setting organizations that provide descriptions of the processes associated with both of these disciplines, and they may vary in both content and the ordering of how they are presented in comparison to how they are presented in this book. This chapter provides an overarching view of the processes most often referred to and employed on projects.

Consulting with subject matter experts or practitioners from each discipline should be considered for complicated and complex projects.

In addition to the complementary processes described in this chapter, there are also processes unique to project management and systems engineering that, when applied to technical projects, will optimize outcomes. These processes are fully described in Chapter 4. A case study demonstrating the application of these complementary processes is provided in Section 3.3. This case study describes a methodical and careful approach that follows the tenets of both project management and systems engineering methods and helped lead to the quality outcomes that were achieved. Section 3.4 includes the key concepts from this chapter, and Section 3.5 provides the Apply Now exercises that allow for the immediate application of the information described in this chapter.

Chapter Roadmap

Chapter 3 focuses on the Complex Systems Methodology complementary processes derived from standard project management and systems engineering disciplines. It specifically:

- provides methods for identifying and engaging stakeholders
- explains common approaches to communications
- describes methods for defining and managing scope
- provides guidance to develop and manage the schedule
- explores the interacting processes of procurement and acquisitions
- provides processes to manage project risk and reduce technical risk
- explores the topic of quality and measures
- describes the two primary processes used to effectively manage change
- uses a case study to demonstrate the application of complementary processes
- provides a summary checklist of key chapter concepts to facilitate learning
- provides Apply Now exercises to assist in the application of the concepts

3.1 Stakeholder-Focused Complementary Processes

This section introduces the stakeholder-focused processes or those processes that are driven by stakeholder engagement. These include the foundational processes associated with engaging stakeholders, collaboration management, and communications management. The processes associated with engaging stakeholders are necessary to develop and manage the description of what is needed and how the needs will be met. Collaboration management includes any action that involves the alignment of project team members to a common vision of how to achieve shared goals and objectives. Communications management processes define the relationship dynamics that must be developed and managed. These processes are used to ensure alignment with key stakeholder needs throughout the project and are used to assess their satisfaction with the delivered solution.

As previously stated, the complementary stakeholder-focused processes must include processes associated with engaging stakeholders and performing communications management. The key stakeholders are responsible for defining the problem, issue, or vision to be resolved through project execution. The stakeholders' responsibilities include:

- the identification of measures of effectiveness, which provide quantitative information that will be used to validate system performance

- a text description envisioned by the key stakeholders of how the system is expected to perform in each scenario and throughout its operations life cycle phase
- a method to ensure that all stakeholders, who may affect the project outcomes, are considered throughout the project life cycle
- the construct for effective and efficient communications needed to reach optimal project outcomes

Figure 3.1 shows the major stakeholder-focused complementary processes that will be discussed in this section. These include the identification and capture of the needs of the key stakeholders, including

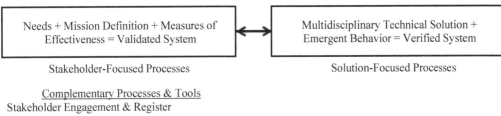

Stakeholder-Focused Processes   Solution-Focused Processes

Complementary Processes & Tools
- Stakeholder Engagement & Register
- Communications
- Technical Stakeholder Management & Requirements
- Validation Criteria
- MOE, KPP
- Document Tree
- Reports

Figure 3.1   Overarching Stakeholder Complementary Processes

an understanding of how the system is to be used within the operations environment. Project and systems engineering communications management processes will be defined and detailed in Section 3.1.2.

Measures of effectiveness describe in concrete terms how the key stakeholders will assess the system performance. This is a critical step in the validation process. Without these agreed-upon measures, the target outcomes will be difficult to design and requested changes by the key stakeholders will be difficult to methodically assess and implement. Complex Systems Methodology processes associated with engaging stakeholders, along with techniques to perform each one, will be described in detail in Section 3.1.1.

3.1.1 Stakeholder Engagement

Project processes for the engagement of stakeholders ensure that all interested stakeholders are identified and managed, and that key stakeholders are communicated with throughout the project. It is the primary method for ensuring that the outcome of the project meets the stakeholder's expectations. Project management processes provide the structure to engage stakeholders for the project as a whole, whereas systems engineering provides processes that engage stakeholders in the technical scope development of the project. These two interfacing processes, if implemented and managed correctly, can make the difference between a well-received project outcome or one that is assessed as not meeting stakeholder needs. In Sections 3.1.1.1 and 3.1.1.2, the specific techniques used to address the engagement of stakeholders are described in detail.

3.1.1.1 Project Stakeholder Engagement

For a project to be called "successful," the key stakeholders must be satisfied with the outcome of the project. This will be obvious if it has occurred, as the key stakeholders will have expressed their satisfaction at every step along the way, including the way their change requests have been processed, and will be more than willing to validate that the project is complete and that the system that has been delivered meets their needs. In addition, a "successful" project will have addressed all stakeholders throughout the project in the manner that they anticipated.

This process is of the utmost importance. The stakeholder's satisfaction with the project outcomes is the most important to measure. The perfect technical system could be delivered as per the defined technical solution, but if a system is delivered that does not meet the stakeholder's evolved needs (through careful change control), then the system could be immediately closed or used in an unintended and inefficient manner.

Fortunately, stakeholder engagement processes are fairly simple to implement. On the other hand, it takes a significant ongoing effort to manage stakeholders throughout the project. This process, similar to all of the project management processes, requires active management. If the plan to describe how stakeholder engagement will be employed is developed and then put on a shelf, the probability of achieving "successful project" status at project closure is highly unlikely.

To initiate project stakeholder engagement processes, identification of all individuals or groups with an interest in, or with influence over, the project and its outcomes is required. Making all efforts initially to recognize every stakeholder, and then continuing to add stakeholders, as identified, throughout the project life cycle, can minimize the risk of surprise requirements or negative impacts from stakeholders. A simple spreadsheet can contain the stakeholder information. This document is commonly referred to as the stakeholder register.

As stakeholders are identified, an analysis is performed to determine each stakeholder's desired level of interaction as perceived by the project manager and documented within the stakeholder register. Key stakeholders are also identified, along with their preferred method and frequency of communications. Upon completion of the stakeholder register, this document may become confidential by compilation (in its compiled version, it becomes sensitive due to the comparative nature of the information). In other words, the fully populated document may illuminate individuals who are actively resistant to the goals of the project or have conflicted or vested interests. This information should not be widely shared to protect the stakeholders' information and the perception documented by the project manager, which might lead to issues if known by the stakeholders. As such, care should be taken to assess the sensitivity of the document and protect it in an appropriate manner.

Figure 3.2 provides a simple example of a stakeholder register, including each individual's status as key or not, their role associated with the project, the level of perceived support assessed by the project manager during discussions with the stakeholder about the project, and each stakeholder's desired project communications method and frequency. The planned communications and frequency with each stakeholder will be discussed in Section 3.1.2.

In each phase of the project life cycle, stakeholders will play an important role. During project discovery and establishment, project charter buy-in from the key stakeholders is an imperative. The key stakeholders will be required to agree to the defined project charter. This agreement is generally signed by the key stakeholders and becomes the baseline for the development of all of the supporting project management and systems engineering documentation.

During the preparation and planning phase of the project, the project manager will plan the approaches that will be used to manage the project stakeholders as part of the project planning activities. Once the project has concluded, the stakeholders must validate that the project outcomes have met their needs.

Stakeholder	Key (Y/N)	Role	Level of Perceived Support	Preferred Communications	Preferred Frequency
Robert Smith	Yes	Funding Customer	Advocate	Face-to-Face Meetings	As Required
Mary Jones	Yes	Shareholder	Unaware	Activity Reports	On Occasion
Kelly Sharp	No	Activist Group	Resistant	Information Bulletins	Never
Shari Ranzano	No	Local Community Representative	Neutral	Face-to-Face Meetings	Monthly
Karen Olsen	Yes	Component Supplier	Supporter	Social Media Updates	As Required
Mariam Taylor	Yes	Information Technology Department	Unaware	Webpage Updates	Never
Maria Garcia	Yes	Vendor	Neutral	Casual Conversations	Never
James Johnson	Yes	Software Developer	Resistant	Information Bulletins	Never
Arjun Li	Yes	Environmental Advocate	Neutral	Webpage Updates	On Occasion
Lucas Hernandez	Yes	Communications Department	Neutral	Webpage Updates	On Occasion

Figure 3.2   Sample Stakeholder Matrix

3.1.1.2 Systems Engineering Stakeholder Engagement

Keep in mind that even prior to project discovery and establishment, there may be exploratory research performed to determine if the ideas are viable and should be pursued as a project. This is perfectly acceptable as long as there is some type of approval gate that deliberately transitions the informal new idea into a formal project activity, such as would occur during a quality review or a governance activity. These processes are described in Sections 3.2.5 and 3.2.6.

Once a project has been established and a stakeholder register initiated, a systems stakeholder needs analysis can begin. This is the process that helps develop the technical scope of the project. The project stakeholder register, which includes the key stakeholders, is a reasonable place to start. It would be acceptable and anticipated that the key stakeholders on the project stakeholder list would bring additional subject matter experts into detailed technical discussions. It is prudent to note on the stakeholder register which stakeholders are responsible for the development of the technical scope and related systems engineering documents.

The first project-related systems stakeholder process to be implemented falls in the life cycle phase of mission definition. Mission definition is the phase that clarifies the technical mission that the stakeholder envisions. As this is typically a fuzzy area, much care must be taken to clearly and concisely document those needs. Stakeholders will need to define what exactly they must have, the concepts of how they will use and operate the system to be developed, and how they will measure the effectiveness of the system—the measures of effectiveness.

A full set of agreed-upon and approved stakeholder requirements will be developed that describe "what" the system must do, not "how" it will perform those functions, which is part of the solution-focused processes discussed in Section 3.2. Those requirements will each have associated validation criteria or ways to prove that the requirement has been met. Without a doubt, this is the most important part of the stakeholder engagement process, defining exactly what designates completeness. Without these valuable criteria, stakeholders can hold the project from ever being completed. With these valuable criteria, the project manager can demonstrate to the stakeholder that their needs, as defined and managed throughout the project, have been sufficiently met.

Techniques used to elicit requirements from stakeholders are described in Section 2.1.4.2.1 (page 41). Stakeholder needs analysis is documented in a concept of operations and in use cases or scenario definitions. These are text-based descriptions about what the future of the system looks like. Ideally, all aspects of performance of the future system should be captured, such as how maintenance will be performed, how an operator will complete the work, and what support equipment will be required. The more inclusive the description, the more accurate the systems requirements and validation criteria will be.

Once there is a clear initial vision of how the system is expected to behave, the system concept will be developed. The system concept identifies specific stakeholder requirements, validation criteria, and key performance parameters. To document stakeholder requirements, again, a simple spreadsheet can be used to capture each requirement, along with the validation criteria, which defines acceptable performance. When the stakeholder requirements document is complete, all change will be measured against this baseline. The process of requirements management is described more fully in Section 3.2, as it is critical to managing the evolution of the requirements throughout the project life cycle, verifying that the requirements have been technically met, and validating that the requirements meet stakeholder needs.

The key performance parameters are derived from the measures of effectiveness, which provide the measures for which one knows that the system behaviors are those that directly affect the outcomes of the mission. Knowing these provides the insight to develop the key performance parameters. These key performance parameters identify the minimally acceptable thresholds, and target quantitative performance parameters, that are the key drivers of operational success. They should only point to those activities that, if not achieved, would render the system inoperable. The stakeholders must define these at a minimum, although there are many other measures available to further refine the requirements of the stakeholders. The amount of effort put into identifying and developing any additional measures will be based on both stakeholder and organizational preferences, as well as against perceived risk.

The development of the mission definition and system concept is iterative; however, that evolution needs to be carefully controlled. Evolution is encouraged at this stage as stakeholders gain fidelity and clarity about their goals. Since the system concept will drive the technical design, it is important to control change through the careful review of the change request by anyone involved in that design work. This will be discussed further in Section 3.2.6. It is important to note that the systems engineer, typically responsible for driving the system stakeholder processes, must collaborate closely with the project manager during change control to ensure that the approved changes remain within the cost, schedule, and scope baselines of the overall project.

Once the project progresses, the most effective strategy for assuring project success is to engage the stakeholders through every phase of the realization process. As appropriate, in accordance with the stakeholder engagement plan, and as part of a risk reduction strategy, the stakeholders may be involved in validation of the part, subassembly, assembly, subsystem, product, and system. Their role is to confirm that the system is being realized in alignment with their vision and baseline stakeholder requirements. Once the system is ready for operations, the key stakeholders will validate the final delivered system and typically sign acceptance documents. Only if upgrades are desired will the stakeholders reengage with the project manager.

3.1.2 Communications Management

Identifying stakeholders, understanding their needs, finalizing their scope requirements, and understanding how the performance to those requirements will be assessed is an important function of reducing risk to the project and achieving project success. In combination with having a clear understanding and agreement with what the project will deliver, having plainly defined and agreed-upon collaboration and communications plans that everyone considers stipulatory and definitive, and then managing to them, is an imperative. Many projects experience numerous hardships that are due to the challenges associated with the two interrelated activities of collaboration and communications.

As described in Chapter 2, both project management and systems engineering disciplines require communications with stakeholders. A joint communications plan must be established, with clearly defined roles and responsibilities between the project manager and systems engineer. Because these collaborations generally involve the agreement to share or provide resources important to the project, ensuring a clear understanding of what will be provided and at what time, along with an escalation path in the event that the agreed-upon actions will not be met, is critically important. Communications expectations associated with problem escalation must be agreed upon as well so that the project manager is always aware of problems before they become risks to the project. This section will describe the processes associated with collaboration and communications management, as well as the technical communications management processes, including decision support.

3.1.2.1 Project Communications Processes

High-quality communications form one of the most important success factors for projects. It is well-established that communications failures are common and lead directly to project failures. Therefore, it is important to recognize the processes that are used to minimize this risk. Communications challenges come from both internal and external interactions, and they also increase exponentially when the project is more complex and spans multiple organizations. Therefore, a special focus on project collaborations is addressed in Section 3.1.2.1.1. Specific techniques for project communications management across the entire project are then described in Section 3.1.2.1.2.

3.1.2.1.1 Collaboration Management

Collaboration refers to project interactions that leverage the strengths of individuals, organizational entities, and organizations as a whole. When communication among collaborators is not effective, the project risk increases. Collaboration with external entities requires the alignment of project team members with a common vision of how to achieve shared goals and objectives. Ineffective collaboration, such as when a collaborating organization does not have the same objectives, commitment to schedule, or scope, can cause a project to fail.

Although collaboration is required across internal and external entities and requires agreement as to common purposes, internal collaboration is more easily managed. The organizational structure provides some of the normal internal controls that provide processes for resolution or escalation that can be used in the case of nonperformance. It is collaboration with external organizations that poses the most risk as project control tends not to reach across organizational boundaries, and there are often conflicting organizational objectives that put the project at risk.

To minimize collaboration risk across organizations, a project manager must have binding collaborative agreements in place that outline the specific scope, schedule, and budget that will be committed on behalf of the project, as well as the agreed-upon cross-organization communications methods that will be employed during all relevant phases of the project. The more binding (contractually) the collaboration agreement is, the higher the probability that the expected performance of the collaborator will match the reality, particularly if the terms and conditions associated with addressing changes and for nonperformance are well-defined in the contract. If the contracts define communications methods and events that trigger communication, the project manager can reasonably expect that the contracted communications activities will be followed through to everyone's benefit.

Collaboration agreements and contracts are often negotiated between organizational contracts personnel that have expertise in the field. However, the project manager must interact closely during the negotiation of these agreements and contracts to assure that the project's needs are addressed and that the organizational standards that generally drive these agreements are in alignment to the project's needs. Once the agreements and contracts are in place, the project manager interacts directly with the collaborators per the terms of the agreement.

3.1.2.1.2 Communications Management

The more stakeholders and participants there are associated with a project, and the more diverse they are, the more challenging communications become. Typical barriers to communication often occur in these complex, diverse projects and can include language barriers, and terminology challenges such as acronym use, multiuse words, or vernacular associated with different disciplines. If not well specified, the level of detail to be communicated, the frequency of the communications, and the technology that is selected to communicate about the project, such as email, websites, social media, or others, can miss the mark of what the project stakeholders and collaborators need in the way of obtaining project information. In addition to these issues, if external controls are not well understood and planned, they can cause project failure by restricting communications that are mandatory for the success of the project. Some of these types of controls are International Traffic in Arms Regulations (ITAR), intellectual property (IP), or security classifications.

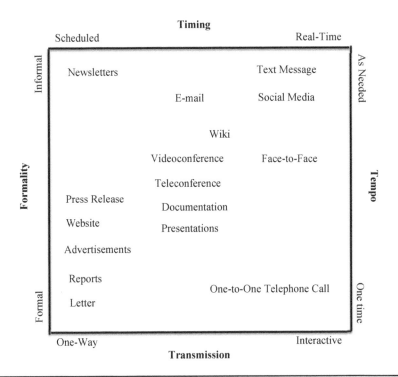

Figure 3.3   Methods of Communications

Figure 3.3 demonstrates some of the most common types of communications methods that a project may use. Along the bottom axis, the transmission ranges are identified, from one-way communications through interactive communications. The left axis shows the continuum of formality, from formal to informal. The top axis shows the timing from scheduled to real time, and the right axis shows the tempo of the communications from as needed to one time. There are many variations to these top-level categories, and some that might not be identified here. As the communications environment is constantly evolving, the most important thing to remember is to capture new methods and determine how they fit into this model in the modalities of transmission, formality, timing, and tempo and then apply them to the appropriate need within the project.

To maximize understanding within a project, communications should be planned carefully and implemented throughout all of the project phases. Of course, from a project management perspective, the most important activity associated with communications is to convey the project management plan and all associated documents, changes, and progress. Compiling a list of the documents and events that are to be communicated is an important part of this process. The most effective way to communicate those documents and propagate the knowledge contained within them is often informed by organizational norms. For example, there may be a preferred method of communicating the documents that form the baseline for the project through a kick-off meeting, or to communicate change throughout the project through a change board meeting.

A project communications plan can be developed simply through extending the stakeholder register. Information about the project team collaborators, who will have defined communications requirements, is considered as well as those of all the stakeholders. Both the binding collaborative agreements and the stakeholder register are used to assess and document the communications requirements of the overarching project, and then to add information within the stakeholder register to describe which methods, techniques, and technology will be employed for the overall project.

Receiver	Key Information to Convey	Objective/Purpose	Confidentiality/ Sensitivity Level	Intended Result	Tactic(s)	Implementation Plan	Planned Date(s) of Implementation
Key Stakeholders	Organizational Changes	Awareness	Unclassified/ Sensitive	Additional Support	Meetings and Presentations	Board Presentation	Quarterly Meeting
Collaborators	Need for Joint Project Webpage	Branding and Organizational Identity	Unclassified/ Not Sensitive	Consistent Use of Branding	Telecons and Videocons	Focus-Team Telecons	Weekly Meetings
Project Team	Project Performance	Leadership Visibility of Problems	Classified/Sensitive	Early Warning on Current Issues	Face-to-Face and meetings	Face-to-Face/ Videoconference	Weekly Meetings

Figure 3.4 Sample Project Communications Plan

The communications activities that should be captured in the communications plan should first include information about the receiver—the person(s) who will be receiving the message. This information comes from the project team, the collaborator agreements, and the stakeholder register. The key information that is to be conveyed is carefully and completely described. And the objective or purpose of the communication is documented—for example, to inform or increase awareness, gain acceptance and support, elicit a response, or incite action or reaction from the receiver. The confidentiality or sensitivity level of the message is identified next. The intended result should be described. As messages often tend to gain a life of their own and often lead to unintended consequences, it is wise to test sensitive messages before the message is released more generally.

Once the information on the communications requirements is complete, the tactics for addressing them are developed. Figure 3.4 shows a sample project communications plan. Each communication requirement can have one or more tactics. Along with the tactics, the details about the implementation and dates of delivery are documented.

3.1.2.2 Systems Engineering Communications Processes

While the project manager is focusing on the project as a whole—communicating with the overall project stakeholders and ensuring that collaborators are working in lockstep with the project to perform the project scope within the schedule and budget—the systems engineer has a somewhat different, but complementary, goal. The project communications plan is used to inform the technical communications plan. Any technical communications to the stakeholders, project team, and management need to be consistent with, and complementary to, the project communications plan in a manner that does not overlap or conflict.

Two primary areas of communications within the systems engineering processes that are important to consider include technical communications and technical decision support. Decisions made on the technical scope of the project typically span all of the life cycle phases and most often affect the schedule and cost of the project. Therefore, all technical decisions that are made by the project team (such as effort, progress, quality, interfaces, test results, production outputs, discrepancies, etc.) must be communicated effectively to the key stakeholders of the project so that the impacts of those decisions can be carefully reviewed for consequences in other project areas.

3.1.2.2.1 Technical Communications Management

The systems engineer has the responsibility for ensuring that communications about the technical scope of the work is acquired, made accessible, and distributed to the project team and key stakeholders. It is also necessary for the data and information that are obtained, created, derived, and/or developed during the project to be safeguarded and secured in the appropriate manner.

The location for data storage can span multiple information technology systems, so it is important that a clear architecture for data management be developed as early in the project as possible, and communicated to the team. Data products that are created outside of the project data management system will increase risk to the project as they will not be locatable, accessible, or follow the designated project data management controls on retention. In addition, they may not follow the appropriate data electronic exchange formats. The risk from these activities occurs further down the life cycle when the design and development, production/construction, and system integration activities are in motion. Configuration management of the design and associated interfaces are the most at risk.

To ensure clear project technical communications, an agreed-upon data management plan must be put in place. Included in this plan are the locations for all data products that will be developed and managed throughout the life cycle of the project. It should clearly identify where the data types will

Document	System/Path
Project Management Plan + supporting	Document Management System/path....
Systems Engineering Plan + supporting	Document Management System/path....
Stakeholder Requirements	Document Management System/path....
Technical Specification Requirements Traceability	Requirements Management System/path.....
Sub-systems specification Requirements Traceability	Requirements Management System/path.....
Specification Tree	Requirements Management System/path.....
Architecture Functional Diagrams	Requirements Management System/path.....
Integration Requirements	Requirements Management System/path.....
System, sub-systems, components and parts drawings	Computer Aided Design System/path.....
Materials, dimensions, tolerances, and other specifications	Computer Aided Design System/path.....

Figure 3.5   Sample Document Tree

be held; for example, presentations and documents such as the Systems Engineering Management Plan and all supporting systems engineering documents are housed in a document library or knowledgebase, and technical drawings are held in a computer-aided design system. A text file that contains a listing of all the systems engineering documents for the project and their physical locations, commonly referred to as a document tree, is developed and placed in a location accessible to all of the project team. This is a document that will inform the technical specifications tree, which will be described in Section 3.2.1.2. A sample of a document tree is shown in Figure 3.5.

Ideally, an easy-to-use knowledge repository minimizing the number of places one must go to find project documentation is preferred. This also means metadata are needed to locate items, and the quality of the information in the knowledge repository needs to be continually maintained. The guidance in the plan should also identify the technical data electronic exchange formats and interfaces, data content, and form. If this step is missed, it can cause significant cost and schedule impacts to the project that are directly related to the required conversions of products designed with noncompliant software, redesigns based on inaccurate interfaces, time spent tracking down data and information, and in configuration reviews to determine which version is the correct one.

Once the location for the technical data products has been established, codified, and accepted throughout the project, auditing must be performed to ensure compliance with the data management directive, and clear ramifications of nonperformance must be defined and enacted. With all of the project technical data and information, change records, and decision-support outcomes located in the approved locations, the project decision-making process will be effective.

3.1.2.2.2 Decision Support

Because of the iterative and recursive nature of systems engineering activities, decision making is a constant activity within projects. When decisions are needed, one must take into consideration all of the appropriate factors so that the decision is an optimal one, risk is minimized, and unintended consequences of the decision are avoided. Ongoing and consistent communications with the project team and the key stakeholders are imperative, and the most current and inclusive set of related information is needed for review before the decision can be made. Projects have limited resources. In order to minimize the impact to project resources, a structured and methodical approach to decision making is required. Thinking through decisions carefully will ensure that optimal actions are taken with the resources that are available to the project.[1]

Decision making involves the identification of the problem, the goals (specifications, targets, ranges, etc.) for comparison, research into the options, evaluation of the alternatives, evaluation of the interfaces, formulation of a best-option decision, and review of the decision by the larger stakeholder community for unintended and unanticipated impacts. The decision requires formal acceptance and implementation. Every situation that requires a decision will be unique; therefore, it is difficult to provide a single template for decision making.[2] Therefore, the decision-support processes are generally unique to each project. However, it is clear that in order to make optimal decisions, one must have access to all of the pertinent background information, backed by qualitative and quantitative evidence, and be able to discern the subset of information that is critical to make the decision. Qualitative methods include the development of diagrams, physical models, simulations, and trade studies. Quantitative methods include the use of algorithms, heuristics or axioms, multivariate analysis, and risk analysis, among others. All decisions made on behalf of the project use the documents outlined in the documentation tree (see Section 3.1.2.2.1) and must be documented through governance processes (see Section 3.2.6) and risk management processes (see Section 3.2.4).

Having access to the pertinent information required to make a decision is challenging. Design and development activities, where decision-making happens regularly, are generally high risk and uncertain. Design decisions are being made at different levels, and the selection of design concepts, choice of appropriate components, and optimization of system parameters have a direct bearing on cost, quality, and schedule. Knowing at all times where the data and information are located, which documents reflect the current configuration of the technological solution, and which information is pertinent to the problem is an imperative for effective decision making. It is often the case that too much information is presented, and it therefore becomes difficult to cull the important information from the unimportant information. Too much information can draw attention away from the real issue, and poor decision making can result. This is where information technology can help.

Decision support is generally facilitated through information systems and technology. The abilities to easily access version-controlled data and information, to locate items through search features, and to provide active links that allow access to electronic files and folders regardless of the host system, are significant risk reducers for a project. However, the features of the technology are only helpful if the data and information contained within them are of sufficient quality. The roles and responsibilities of the project staff in providing quality data and information, as well as controlling and communicating any changes that are implemented, cannot be understated. Through change control, as discussed in Section 3.2.6, the data and information used for decision making will be the most current and inclusive, and decisions can therefore be expected to be of high quality. Throughout the life cycle of the project, reports and analyses will be generated based on the documentation and processes defined in this section. These will provide a solid basis for decision making throughout the project life cycle.

Technical decisions are made using project information and the identification and analysis of alternatives. The analysis of alternatives starts with the parameters, targets, and specifications of the technical requirements. Trade studies are completed in an effort to determine which of the options best meet stakeholder needs within the project's overall cost and schedule. In an effort to reduce technical risk to the project, the most challenging and immature requirements will be identified, and actions taken to understand, advance, and mature the technology. Throughout this evolution of high-risk technology development, decisions will be made to continue, stop, or change course. This risk reduction strategy is discussed more thoroughly in Section 3.2.4.

3.2 Solution-Focused Complementary Processes

The stakeholder-focused processes provide the definition of the concept of operations and use cases or scenario definitions, measures of effectiveness, stakeholder requirements, validation criteria, and key performance parameters. These provide the basis for the project team to further refine the scope of the project and to design the technical solution. Complex Systems Methodology helps evolve the solution that will ultimately meet the stakeholders' needs. The methodology will:

- provide the solution set that will form the system expected to meet the stakeholders' needs
- ensure that the optimal technical solution fits within the project baseline scope, budget, and schedule

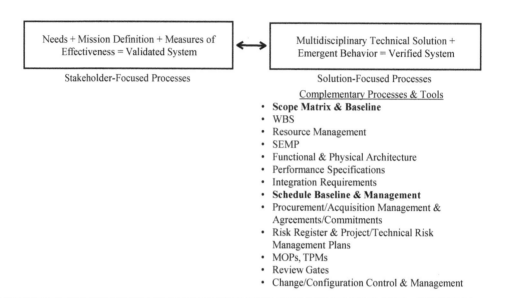

| Needs + Mission Definition + Measures of Effectiveness = Validated System | ⟷ | Multidisciplinary Technical Solution + Emergent Behavior = Verified System |

Stakeholder-Focused Processes   Solution-Focused Processes

<u>Complementary Processes & Tools</u>
- **Scope Matrix & Baseline**
- WBS
- Resource Management
- SEMP
- Functional & Physical Architecture
- Performance Specifications
- Integration Requirements
- **Schedule Baseline & Management**
- Procurement/Acquisition Management & Agreements/Commitments
- Risk Register & Project/Technical Risk Management Plans
- MOPs, TPMs
- Review Gates
- Change/Configuration Control & Management

Figure 3.6   Solution-Focused Complementary Processes

- provide the structure for technical risk reduction and quality so that the end result of the project is as needed by the stakeholders
- provide the complementary constructs to manage change and assure configuration control throughout the project life cycle so that the end result of the project is as expected

Figure 3.6 shows the major solution-focused complementary processes that will be discussed in this section. These processes include scope, schedule, procurement, risk, quality (including test, verification and validation), and governance (including change and configuration control). The project management discipline is responsible for the governance of cost management for the overall project and is therefore a unique process that will be discussed in Chapter 4. Cost management is one of the three primary activities that are interconnected and constrained—cost, schedule, and scope.

3.2.1 Scope Management

Project scope management is defined as the activities or effort identified as necessary to design, develop, realize, govern, and complete a project, as defined in the project charter and project plan. It is the complete description of the deliverables of the project for which the stakeholders will assess the success of the project. The scope of a project includes both programmatic and technical elements. As described in Chapter 2, both the project manager and systems engineer have scope management responsibilities, although the systems engineer is focused on technical scope management, which is a subset of the overall project scope. This section will outline the specifics of project scope management in Section 3.2.1.1, including the processes used to perform the technical scope of work, described in Section 3.2.1.2.

3.2.1.1 Project Scope Management

The process of project scope management focuses on identifying and describing all the work that will be performed during the project to satisfy the stakeholders' needs as assessed during the stakeholder engagement processes described in Section 3.1.1. The scope is documented, and upon agreement from the key stakeholders, is put under configuration control and becomes the baseline against which all scope change is assessed. Successful project outcomes rely on the project manager's ability to control the scope once the baseline has been established, through governance activities, in relation to cost, schedule, risk, and quality. Throughout the project life cycle, the project manager often clarifies the boundaries of what is within the project scope and what is external to the project scope. The project manager controls change to the scope through the governance processes discussed in Section 3.2.6.

As discussed in Chapter 2, project scope can be described using the project scope statement and project work breakdown structure. A spreadsheet can be used to capture the unique project requirements and to decompose them. The work breakdown structure captures elements and decomposes them into work packages that result in products within the project. Figure 2.3 in Chapter 2 (page 44) shows an example of a project work breakdown structure. These work packages, or the lowest level of the work breakdown structure, are used to schedule the project and to track costs and progress against the schedule throughout the life cycle.

3.2.1.1.1 Resource Management

During the preparations and planning phase of the technical scope development, the type of resources that are required by the project are identified. This happens as the work breakdown structure is being developed. The types of information that will flow out of this process include certain skills sets that will be needed (e.g., electrical engineer, mechanical engineer, safety officer), and the types of materials and services that are required to complete each work package. Once the project enters the scheduling process, the specific labor hours by-named individual allocations will be identified.

Often, it is not possible to determine the resource implications of the scope until the schedule is developed in full, after the technical scope management has been defined. This requires an iterative loop between the project manager and the systems engineer. The project manager may have a general idea of the staff that are required to perform the work, and may even have scheduled that work at a level that demonstrates the overarching project—for example, during the development of the project human resources plan. However, once the systems engineer starts building up the technical scope, the fidelity resulting from that activity often drives changes into the project scope, schedule, and resource requirements. Project resource management activities identify the needed resources and then are used to manage those resources throughout the project life cycle. A resource management plan outlining all needed resources and how they will be managed throughout the life cycle is developed by the project manager (informed by the systems engineer) and used to actively manage the resources needed by the project. The process associated with scheduling is defined in Section 3.2.2 and in the description of project resources activities in Chapter 4, Section 4.2.4 (page 119).

3.2.1.2 Technical Scope Management

The systems engineer uses the project charter and project management plan, as well as the project scope, to inform the development of the technical scope. The technical scope includes all activities associated with defining the system that will fulfill the stakeholder's requirements as defined in the stakeholder-focused processes. It is important to note during this process that the boundaries of the technical effort must be well understood. Specific attention needs to be given to cross-boundary constraints on the technical effort. This is discussed more in Chapter 4.

During the preparation and planning phase of the project, mission definition activities include the development of a systems engineering management plan. As described in Chapter 2, this systems engineering management plan generally includes information about the scope, along with the schedule, required reviews, risk reduction activities, and other standard systems engineering activities appropriate to the project. Defining the scope also requires developing the project technical work breakdown structure to the fidelity that the systems engineer requires to manage the work packages, which could include further decomposition to specifically address detailed tasks associated with the system, product, subsystem, assembly, subassembly, and parts of the technical solution that is being proposed, as previously shown in Chapter 2, Figure 2.3 (page 44).

The identification of enabling and supporting systems required by the project to perform that scope of work is also documented and delivered to the project manager to coordinate with organizational departments responsible for delivering those systems. These can be tools, processes, or infrastructure that are needed by the project team to perform the project scope.

As the project moves into system concept phase, efforts shift to a deep focus on the technical solution. One of the essential activities during this time, and throughout the life cycle phase of the project, is requirements management. It will be important to baseline the stakeholder's requirements and ensure that those requirements are fulfilled by the solution that is proposed. As the requirements and their specifications are converted into a functional and physical architecture, and they are allocated to lower levels of the design, it will be critical to be able to maintain traceability between each layer of the design. In other words, it should always be possible to follow the path from the lowest level of the design (e.g., a part, component, subsystem, etc.), to the top level of the design (the system level). If this is done correctly and managed effectively, then change recommendations will be effectively evaluated because one will be able to understand how much effect the change will have on the overarching system.

Specific steps in transferring the stakeholder requirements into a solution include performing systems engineering processes that define the following:

- functional architecture
- performance specifications
- physical architecture
- integration requirements

3.2.1.2.1 Functional Architecture

From the requirements, the solution design can be architected. This architecture is an abstract yet detailed graphical representation of all the parts of the systems and their relationships to one another. A functional architecture is a diagram of connected boxes that demonstrate what is believed to be a representation or model of the system as envisioned by the stakeholders. This functional representation does not identify any physical attributes or technical solutions (e.g., particular hardware or software)—only the function that they will perform. The functional architecture reflects how the system is used, whereas the stakeholders' concept of operations, use cases, and requirements reflect what they, the stakeholders, need.

The systems engineer renders a functional architecture drawing to reflect the common understanding of what is needed and where the boundaries of the system are. By sharing this vision among the project team and stakeholders, confidence that the stakeholder requirements are well understood should increase. This diagram is then reviewed and approved by the project team and stakeholders. A functional architecture can be drawn using any standard drawing program. Boxes can be used to reflect the transactional activities expected from the system in an operations environment.

The process of defining the functional architecture is iterative and often more difficult than it appears. As the project team and the stakeholders refine their visions of how things work together and what the system will be able to do based on the functional design, requirements will often change. This is expected, and when careful requirements management is maintained, the design can evolve to a significant degree. What will be important is to finalize the functional architecture before the physical architecture design is started. An example of a functional diagram is shown in Figure 3.7.

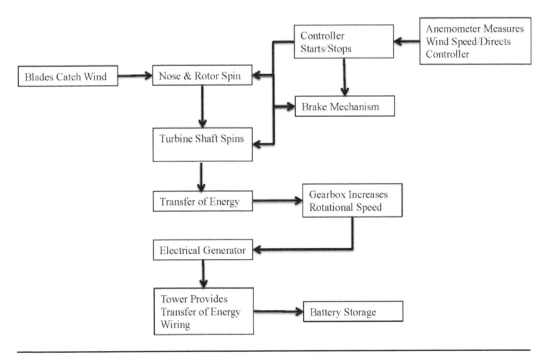

Figure 3.7   Example of a Functional Diagram

3.2.1.2.2 Performance Specifications

Once the functional architecture has been defined, it is time to allocate those activities to performance specifications and parameters. As systems engineering provides the framework to track and trace activities from the highest level of the stakeholder need through to the lowest level (component level), it is important to identify what type of activities support each functional box in the defined functional architecture. The performance specifications identify the work effort associated with each entity within the architecture. The most straightforward way to capture this information is with a spreadsheet where one can identify each block as a line item and then describe the derived performance requirements that flow from that function. This information should help the project team understand what boundaries they are working within, what the target performance is anticipated to be, and how performance must be measured by the project team during verification and by the stakeholders during validation. This specification of functional requirements in terms of expected results will be used to verify the compliance of the technical solution. In this activity, a comprehensive and all-inclusive identification of all activities and the minimum necessary performance must be specified. An example of this performance information is shown in Figure 3.8.

One output of this exercise may be a diagram called a *specification tree*. It is closely tied to both the functional architecture and the work breakdown structure and should accurately reflect both. Visually, a specification tree resembles the work breakdown structure; however, it calls out the actual specifications at each level and for each function. How the functional requirements will technically be met is not yet defined, but will be in Section 3.2.1.2.3.

Within this area of technical scope management, both measures of performance and technical performance measures are defined. As was described in Section 2.1.4.1.4 (page 39), measures of performance are measures that derive from the measures of effectiveness, which measure overall system

Function	Requirement Specification	MOP
Blades	Diameter	60 meters
	Quantity	3 pieces
	Length	40 meters
Nose and Rotor	Rotation Direction	Clockwise
Generator	Voltage	1200 volts
	Cooling System	Wind-cooled
Tower	Height	80 meters

Figure 3.8 Example of Requirement Specifications

effectiveness as perceived by the stakeholders. The measures of performance is a performance target that will be used to compare actual performance to plan. These are aligned one-on-one to the specification and then inform the technical performance range, target, and acceptable limits associated with that function. Technical performance measures are typically specific and quantitative, although sometimes qualitative as well. They include such factors as mean time between failures, personnel skills levels, system effectiveness, weight, life cycle costs, and availability. Generally, they are used to identify the performance requirements of technologies that are key to project success. The development of technical performance measures and their use in the management of technical progress during the realization phase is discussed in Section 3.2.5.

3.2.1.2.3 Physical Architecture

The physical architecture design takes form from the functional architecture and known interface requirements. As the functional architecture is a series of boxes and lines in a drawing, the physical architecture converts those boxes and lines into actual human actions, hardware, software, and technical interface requirements. It provides clarity regarding the basic technology that is supporting the function of the system, including network throughput and capacity, hardware requirements, server specifications, software requirements, and related process activities. In other words, if there is a box and a line on the functional diagram, there is a corresponding defined way of technically approaching that function within the physical architecture design. The physical architecture design focuses on form, fit, and function. Form refers to the specific tangible structure, shape, dimensions, or configuration of what is being described (e.g., square box). Fit refers to tolerance levels for which the form "fits" into the associated space of the design. Function refers to the ability of what is being described to do what is intended in a predictable and stable manner. When designing the physical architecture, trade studies may be performed to assess options that affect form, fit, and function. Ultimately though, the most feasible solution will be the one that meets the stakeholder needs and which can be achieved within the scope, cost, and schedule constraints.

As the physical architecture is designed, physical interface requirements emerge and can be captured using control documents such as interface control documents. These documents identify the connectivity that is needed in specific technical terms. An example of an interface would be an outlet and power plug to power a computer. As interface management is a specific unique systems engineering process, it is described in Chapter 4.

A physical architecture is complete once it describes all major elements of the system and is expected to support the system function. Once the design is realized and a true physical representation of the architecture is in place, the physical architecture design can act as a schematic of the actual system.

This conversion of the functional diagram elements into technical solutions that will provide the capability needed by the stakeholders is an important activity, and care must be taken that it is not done too early in the process of systems design. Any part of the physical architecture that is chosen without regard to the functional architecture or the stakeholders' requirements will be riskier as the need for the technology is then traced upward in the hierarchy in an effort to find a proper fit. This may cause unintentional changes in the solution that could impact the stakeholders' requirements and needs, ultimately affecting the project outcomes.

3.2.1.2.4 *Integration Requirements*

As these documents are developed, it becomes clear where the risk in the project scope is held. A review of the technical life cycle for each work breakdown structure work package will highlight those that are technologically mature and can be produced/constructed or otherwise built with little risk versus those that are technologically immature and require research and development to reduce risk. Design and development activities associated with technically immature products require prototypes, models and simulations, and trade studies for risk reduction.

Prior to the project moving into the realization phase, the technical solution, functional and physical architecture design, as well as the system integration requirements must be understood well enough to perform specialty engineering design analysis, such as requirements for maintainability and reliability, and to understand the implications of the maintenance concept, requirements, and budget, along with other interoperability requirements. However, this is not to say that the physical architecture is stable and unchangeable.

The iterations associated with the design and development of the system, subsystems, and components is the most complex and risky part of the overall project. This is because change becomes the norm, and keeping careful control over change becomes extremely difficult, yet essential. If the project team is not attentive and supportive of the governance requirements, or if they do not appreciate the impact of change decisions on other interfacing disciplines within the project, then loss of the configuration of the product design can occur. Not knowing which version of a design is the correct one will lead to using various versions of the same design downstream. If configuration control is lost on multiple product designs within the work breakdown structure, it will become problematic when it comes to integration and production/construction. This could have significant consequences for project success. At best, it will drive additional work efforts to determine the product configurations before they can be used in the next life cycle phase.

Once the systems integration iterations are added to the already complex cycles of system and subsystem iterative design and development, the risk is the highest. It must be clear that the products being integrated have been tested, and that they have successfully performed according to the defined technical performance measures. What is being tested in the systems integration phase is not only the individual performance of two or more products interfaced together but also the emergent behavior of the connected system of products. Part of the activity associated with scope management is interface management. This important activity, necessary to achieve the full capability of the system being developed, is described in Chapter 4.

Only after the system design and development has stabilized, and all technical risk has been reduced or mitigated, can the design progress into the production/construction phase of the life cycle. Moving the design into this phase prior to reducing all the risk first presents a situation where the technical debt of the previous phase must be reduced in the next phase.

This is typically not a cost that has been proposed in the construction/production phase and therefore is unplanned and highly impactful. It would be unusual for stakeholders to validate the system and agree that it was successfully delivered if there was a significant amount of technical debt at the acceptance review, which is usually held after the final systems integration test and verification and before it enters the production/construction phase.

Regardless of whether the technical debt can actually be reduced during the next phase depends on many factors, including the availability of the qualified staff needed to resolve the technical issues. In many cases, the most difficult problems to solve are pushed into the next phases of the project. Unfortunately, the skilled subject matter experts who were not able to solve the technically challenging problems may no longer be available to solve them in the next phase of the project. They move onto other projects or into other organizations. Without either the budget or the qualified staff, the probability that unsolved technical debt will be resolved during production/construction is unlikely and highly risky. This is significant and points to the fact that organizations need up-to-date competency profiles of all staff members in addition to an easy way to locate them, which can be maintained in the knowledge repository. This would then make it possible to find subject matter experts in an efficient manner.

In the production/construction phase of the project, the systems engineer's technical scope management focuses on ensuring that the baseline is maintained, and that there is a commissioning, pre-deployment, and operations transition plan in place. Systems engineering facilitates the development of operations, maintenance, and warranty plans as well so that as the system moves into the next life cycle phase, it does so in a smooth manner. Once the system is operating and being maintained, the systems engineer will be required to engage again if additional upgrades or modifications are needed, or the system must undergo decommissioning or closure.

3.2.2 Schedule Management

Developing a schedule (a sequenced list of tasks assigned to specific date ranges and targets), is one of the three important constraining features of a project, the others being the scope and the cost. This process also requires a significant effort during the project preparation and planning phases, with ongoing effort throughout the project life cycle to manage both the baseline schedule and any proposed schedule changes.

The overarching schedule of the project is inclusive of all interdisciplinary activities that span the project in its entirety and across the full span of the project life cycle. As a subset of that schedule, the technical tasks identify and manage the technical scope of the project across all technical disciplines and within the full life cycle of the project. The project schedule is the responsibility of the project manager. The systems engineer owns the technical tasks in the schedule and ensures that the technical tasks and all approved changes to them are captured in the project manager's project schedule. This section will describe project schedule management and technical schedule management processes.

3.2.2.1 Project Schedule Management

During the project scope definition, a project work breakdown structure is defined. This project work breakdown structure is used to develop a project schedule. This process is iterative and often complicated. It requires heavy collaboration between all of the project team members and across all team boundaries. It requires a solid comprehension of the level of activity tracking that should be scheduled and requires an accurate assessment of the time it takes to complete a task. In addition, the ability to determine the sequencing of the tasks to optimize performance of the project is something that is learned through a project manager's experience. A good rule of thumb is to track the tasks to the level that requires cost insight. If the compilation of costs at the identified task level will provide enough useful information to make risk-mitigating decisions in a timely manner, then that is the correct level. Too low a level of tracking and time is wasted in documenting and monitoring at those levels. Too high a level and the fidelity in cost accrual versus budget forecasts will not provide enough detail to make these important timely decisions, ultimately impacting the project's outcomes.

Often, the project manager will start by identifying the tasks to track within each work breakdown structure work package and document them as an activity list on a spreadsheet. Then the tasks are sequenced, or put in order, within project scheduling software. It is important to know that some tasks can only be performed in sequence, whereas others can be performed concurrently. Knowing which tasks fall into each category is critical, as aligning the tasks appropriately will provide opportunities for scheduling in contingency or float (sometimes referred to as slack). Correct estimation of activity estimates also leads to identification of the critical path, which is the assignment of the longest path of activities in the schedule, thereby defining how long the project as a whole is estimated to take.

The art and science of schedule optimization comes from the discipline of operations research and includes various elements, such as queuing theory, critical path analysis, network optimization, or routing theory. Although one need not be a mathematician to develop a robust project schedule, it helps to have experience in some of those management science methodologies.

Once the tasks have been defined to the appropriate level and sequenced, the allocation of specific resources (such as labor hours by a named individual) is then recorded in the project schedule. Depending on the organizational structure, project structure, and complexity of the project, this part of the process can be complicated, as it takes significant effort to obtain staffing commitments. However, a schedule that is not resource-loaded in this way will not be useful, as each task requires committed effort to be successful. It is an imperative to know who is working on what task and when, and to make sure that those tasks are then completed per the plan. Once the agreed-upon schedule is in place, this forms the project schedule baseline upon which project performance is measured and for which project schedule change requests are assessed for impacts.

The tracking of project schedule performance is ongoing throughout the full life cycle of the project. Depending on the complexity of the project, schedule management can be an intricate process. An integrated project schedule may be required to pull together information about different diverse elements of a project into a single overarching project schedule. Managing this integrated schedule and ensuring it is always up-to-date and reflective of the current resource allocations, charges, actual costs, and commitments is a difficult challenge. Understanding how much progress has been made on each activity or task can be challenging without automated systems that collect this information from individuals working on the project. Regardless of these challenges, it is imperative for the project manager to always have an accurate understanding of the number of charged labor hours that have been applied to the allocated schedule. This information will provide a window of visibility that allows the project manager to make effective decisions regarding the use of float, risk, and resource allocations.

3.2.2.2 Technical Schedule Management

The technical tasks in the schedule integrate into the overarching project schedule and work breakdown structure. The work breakdown structure, tasks, and schedules are generally developed using project scheduling software. Specific engineering events, milestones, and effort assessment will be tracked within a list of the technical tasks by work breakdown structure work package by the systems engineer and provided to the project manager so that the information is reflected in the higher-level project schedule as appropriate.

In collaboration with the project manager, the systems engineer may add levels of detail to the work breakdown structure to address the tracking of technical activities associated with the system, product, subsystem, assembly, subassembly, and part levels. The technical schedule development activity will typically identify tasks associated with the technical events, reviews and audits, technical effort integration activities, technical resources allocations, test, verification and validation milestones, trade studies, model and simulation, and risk reduction targets. Also included will be milestones for procurements and acquisitions necessary to perform the technical tasks.

Many of the specific activities that must be captured within the technical schedule may not be known in the preparation and planning phase of the project. For this reason, the technical schedule will undergo significant revision during the mission definition and system concept phases and must be carefully controlled through change management, discussed more in Section 3.2.6. Any proposed changes must be coordinated through the project change control to ensure that consequences associated with the revision to the technical schedule do not inadvertently impact another part of the project. Approved changes in the technical schedule will be integrated into the project schedule as appropriate.

To develop the technical schedule, the overarching project schedule is used as the starting point. The systems engineer must work with each of the project team members to:

- understand the scope tasks
- assist teams with decomposing those tasks to a lower level as is appropriate for technical decision making and schedule-performance tracking purposes
- compiling an accurate effort assessment (i.e., labor hours and skills sets)
- sequencing the technical tasks and identifying the technical critical path
- confirming that the tasks associated with the technical performance measures have sufficient schedule contingency, as they are typically the highest risk tasks

As mentioned in Chapter 2, this technical element of the overall schedule is typically referred to as the systems engineering master schedule. It provides visibility to the systems engineer of those tasks that are directly related to the discipline. As the project progresses through its life cycle, the systems engineering master schedule is used to manage the technical activities and track them to the baseline technical schedule. Once the project exits the design, development, and integration phases and it enters the construction/production phase, the systems engineering master schedule will close and the schedule associated with the build will commence. If the system that has been built requires closure or deactivation, a new technical schedule would be developed.

3.2.3 Procurement Management

Projects may not procure or acquire any additional products or services and may complete all the project work internal to the organization. However, most projects will require some level of purchasing associated with the project. Within the construct of the overall project budget, the purchase of systems, products, subsystems, assemblies, subassemblies, and parts may occur during the project life cycle. This could be due to several reasons, such as the total project scope requiring items that are outside the design, manufacturing, or the construction ability of the organization within which the project is formed. It can also be for strategic value. By bringing in other organizations to provide value in goods or services required by the project, the cost of the project can be distributed among many project team members. The specific complementary procurement processes are typically focused on procurement and acquisition management for which systems engineering typically has significant responsibility. The contractors' detailed work package schedules are integrated into the project integrated master schedule and systems engineering master schedule.

The project manager is responsible for identifying the requirements for obtaining needed project materials and services. Working in close collaboration with the organization's subject matter experts in negotiating contractual arrangements, the project manager working with the procurement department or comparable group will determine the best terms and conditions that will service the project's needs. These contractual arrangements may include acquisitions, procurements, memoranda of agreement, or other formal commitments.

Project acquisitions typically refer to the purchases of large or complex systems, companies, or other significant capabilities in order to achieve economies of scale, expand into a different market by acquiring a niche capability needed for the project, or perhaps to achieve a logistics or support pipe-line. There are many reasons that an acquisition might be required to increase the chances of reaching successful project outcomes. When acquisitions are made, prolonged and difficult processes are often required to obtain the needed products or services. Although noted as an important process, the details of acquisition management are beyond the scope of this book.

Procurements, on the other hand, typically focus on the processes associated with compiling require-ments, ensuring cost estimates are correct, identifying potential suppliers or providers, and negotiating the terms of the agreement. Throughout the performance period, these obligated services must be man-aged, payments organized and collected or paid, logistics arranged, products tracked, and contractual agreements closed as appropriate.

Memoranda of agreement, or other equity partnerships, provide a construct to operate activities in support of a project. Depending on how they are negotiated, they can be as strong as a contractual agreement or as weak as a handshake agreement. There can be any number of other arrangements made to support the project. It should be noted that the further the agreement gets from a formal contract, the higher the risk to the project as it becomes more difficult to enforce agreements that the project is relying on without contractual backing. The total of all of the arrangements for products and services that a project requires should be reviewed and managed according to the perceived risk.

Although the project manager is responsible for all contractual arrangements made on the proj-ect's behalf, the systems engineer plays a significant role in identifying needed technical products and services and in providing requirements for these to the project manager for action. The importance of having the project manager retain responsibility for the agreements and commitments that are made cannot be understated. However, the close collaboration between the project manager and the systems engineer is critical during these negotiations.

Commitments that affect or impact the scope, schedule, and costs of the project must be agreed upon with a clear understanding of the strategic impacts of each one. Since cost management is a process that is one of the three primary control mechanisms of project management, and the project manager maintains responsibility and authority for management of costs associated with the project in a unique process that the systems engineer does not typically engage in, project cost management is discussed in Chapter 4. This section will describe project procurement management and technical acquisition and procurement management processes.

3.2.3.1 Project Procurement Management

In close collaboration with a subject matter expert in negotiated contracts, such as a contracts manager, the project manager will ensure that requirements for the procurements and acquisitions are provided in a timely manner. They may also work in tandem to negotiate these agreements, with the project manager bringing the strategic view of the project and the schedule and scope details, and the contracts manager bringing full understanding of the contracts processes. Often both the project manager and the contracts manager will continue to be involved in assurance activities, including the verification that the received products and services meet the requirements as specified in the original agreements.

The project manager is responsible for project procurement management, inclusive of any acquisi-tions or other negotiated commitments made on behalf of the project. The objective of the project manager is to secure the product or service at the best price within the time needed and to meet the requirements of the project scope. The project manager is responsible for procurement planning, identi-fying potential sources, developing solicitations, reviewing the proposals or other items the contractors provide, selecting the sources, managing the performance of the providers, and managing and closing out the contractual agreements. Just as other project process areas lean on the subject matter expertise of other organizational departments, such as human resources to provide the project team members, or finance to process the payments, the project manager relies on the contract manager in the performance of project procurement management.

The contract manager is responsible for applying the organization's sourcing processes, promoting fair and open competition, minimizing risk to the organization, assuring price fluctuations are considered during pricing negotiations (i.e., international currency exchange rates), and communicating with the suppliers during the bidding process (such as requests for quotations, requests for proposals, purchase orders, and contractual agreements). The contracts manager may also be involved in ensuring the shipping and delivery, payment, installation, and training requirements are captured in the contracts. However, it is the project manager's responsibility to ensure the delivery and completion of the negotiated contracts and for alerting the contracting manager if there are any deviations that need to be addressed.

3.2.3.2 Systems Engineering Acquisition Management

With an understanding of the requirements for performing the technical scope of the project, the systems engineer will identify within the technical work breakdown structure work packages those items that will be acquired or procured. The systems engineer will develop an acquisition and procurement plan so it is clear what, how, and when these will occur, as well as the specific management processes that will be used to ensure performance. In addition, an acquisition and procurement integration plan is developed that specifically identifies the approach that will be used to integrate external-to-the-project goods and services. The acquisition and procurement and integration plans are provided to the project manager for inclusion in the project procurement management plan and to the contracts manager for execution.

Using templates typically provided by the organization's contracting subject matter expert, the requirements for each product or service that will be externally pursued is identified. All pertinent information is included, such as specifications, tolerances, dimensions, schedule and need date, acceptance criteria, processes associated with reporting, problem resolution, warranty claims, and any other descriptive information that defines the scope of work and product or service that the external party is expected to deliver.

Impacts for noncompliance of performance are also clearly defined, as well as any other constraint or requirement that is appropriate. If the organization does not have an experienced contracts manager, the systems engineer will collaborate with the project manager and others who have experience in this area to ensure that the proper information is included.

It is critical that the technical information provided in the scope is extremely clear and detailed so that there is no misunderstanding or misinterpretation that could result. Ensuring that the statement of work for the product or service is clearly written will result in lower risk associated with the purchase. When multiple parties are bidding, it is important to compare these bids with identified selection criteria in order to make the optimal selection.

Once the contracts are in place, the systems engineer will identify these as items that will be configuration controlled. This process will be discussed further in Section 3.2.6. As the project progresses through its life cycle, particularly in the design and development phase, new requirements for acquisition and procurement may emerge. As they are identified, they follow the processes as outlined, and as the contracts are successfully performed, notification is provided to the contracts manager to close the contracts.

When the project is preparing to move from one phase to another, such as from design/development/integration into production/construction, or from production/construction to operations and maintenance, a review of the acquisition and procurement needs should be assessed and, if required, should also follow the processes as outlined in this section.

3.2.4 Risk Management

Project risk management is the process associated with identification of events that can occur to harm a project, or it can provide an opportunity that can be seized to benefit a project. Risk management is a process that requires ongoing vigilance and active management throughout the life cycle of a project. Active management of project risks and opportunities, tied into well-developed cost, scope, and schedule contingencies, will increase a project's probability of achieving successful outcomes.

The process of risk management is seen somewhat differently from the perspective of the project manager versus the systems engineer. The project manager sees risks and opportunities as they are associated with the triple constraint (cost, schedule, scope). The systems engineer sees technical risk as something that can be reduced through design and development activities. The reason that they are captured here as complementary processes is that these two processes overlap. Technical risk can often affect the project cost, schedule, and scope. Systems engineers who integrate the risk reduction and opportunity pursuits with the project risk management process minimize the overall risk to the project. This provides the project manager with the opportunity to make priority-driven decisions associated with technical risk/opportunity activities before the systems engineer commits to them. This section provides information on project risk management and technical risk management processes.

3.2.4.1 Project Risk Management

The project manager will develop the overarching project risk management plan. This vital document describes how risk management will be performed throughout the project life cycle. Specifically, it describes the process, roles, and responsibilities as well as methods for identifying, assessing, responding to risks and opportunities, and monitoring and controlling the risks. It is common that the project costs, schedule, and scope will be under pressure at all times from internal and external sources. To be most effective, the risk management plan will also provide specific guidance for managing draws on cost and schedule contingency.

When the risk management plan is complete, a risk register is developed using a standard spreadsheet. The project manager surveys the project team and stakeholders to identify risks and opportunities, reviews lessons learned from other projects, and then documents them in the risk register. Each risk/opportunity will have a unique number assigned to it for easy reference and will be described through a title and a full text description of the risk.

Each risk has an owner who is a single individual responsible and accountable for managing that risk. It is a common error to assign risks to individuals who do not have the responsibility over an area and are therefore unable to implement the necessary activities to manage or resolve the risk. Once the owner is identified, the project manager and the owner perform qualitative and quantitative assessments of the impact and probability in the event that the risk or opportunity is realized. By multiplying the two factors, a composite risk index is developed. The higher the resulting number is, the higher the risk. Risks and opportunities having both high probability and significant impact will generally rise to the highest priority. These items will most likely be the first ones that are actioned; however, that is not always the case. All risks should be reviewed and prioritized based on the needs of the project and on the risk tolerance of the stakeholders, project management, and organizational leadership. Further quantitative analysis can be used to estimate the financial and schedule impacts of implementing a risk mitigation strategy to increase confidence in the decision.

Additional information will be added to the risk register, as needed, to assist in the management of the overall risk scenario. For example, triggers can be assigned to risks so that it is understood when action must be taken on a risk or a date for resolution can be assigned. Throughout the project, for each change that enters into change control, the risks for implementing that change must be reviewed. The

Risk Identification			Risk Assessment				Action	
Unique ID	Risk Title [risk or opportunity]	Description of Risk	Responsible POC	Probability (P)= 1(low) to 3(high)	Impact (I)= 1(low) to 3(high)	Composite Index (P*I)	Rating	Mitigation Strategy
1001	The base fails to meet specification	If the base does not meet specification, there are limited options for supporting the weight of the blades.	Bob Marley, Chief Engineer	2	3	6	Medium	Risk: Identify alternative base materials, or minimize weight of blades.
1002	Key interface failure	Interface between base and blades incompatible.	Bob Marley, Chief Engineer	3	3	9	High	Risk: Confirm ICD, test plan
1003	Software support unavailable	Company that provides only software package goes out of business.	Karen Smith, Software Manager	1	3	3	Low	Opportunity: Purchase the company.

Figure 3.9 Example of a Risk Register

key is to keep the risk register updated at all times and actively manage it so that the project will not be impacted. An example of a risk register is shown in Figure 3.9.

3.2.4.2 Technical Risk Management

Project managers base their risk management decisions on experienced judgment about the potential impact to the project cost, schedule, and scope baselines. The systems engineer's approach to technical risk management appears less constrained, as it is evolutionary and occurs as a natural byproduct of the design and development process. Each discipline approaches the decisions associated with risk based on what they are trying to achieve. The project manager wants to meet the objective baselines. The systems engineer has a responsibility to perform standard risk management but also looks at risk as part of the design process. The systems engineer wants to evolve the design so that it meets the technical scope requirements. For example, technical risk and opportunity are generally high whenever a new design is being attempted, and through the use of risk reduction strategies such as trade studies, modeling and simulation, prototyping, and testing, the risk is reduced. This generally occurs over multiple iterations or cycles. As the design matures, opportunities that might be unforeseen at the beginning of the design cycle may present themselves and can be taken advantage of at the beginning of the next design cycle or iteration. The process of design and development is highly iterative and has the intent of evolving the design to a stable and risk-free version that can be reproduced during the production/construction period. Chapter 4 will describe the unique processes for these important systems engineering design activities.

The way that technical and project risk intersect and interact are related to the project cost, schedule, and scope. When technical risk and opportunities are being explored, additional resources (labor, materials, and services) are often committed. Strategic choices made during the design phase on technical risk reduction can often impact a defined milestone on the schedule, and may have a cost impact if additional budget is required to perform additional tests, build more prototypes, or perform additional trade studies. It is also sometimes the case that technical scope performance parameters are exceeded unintentionally in the creative design phase. Providing more than what is needed might waste resources even if it evokes a favorable response from the customer. This will not be a problem as long as all other requirements have also been met. But when one requirement is exceeded while others are not met, a negative impact to the overall satisfaction of the stakeholders will result. Each of these situations may impact the project in unfavorable ways and therefore must be managed. These risk reduction strategies may also require draws on the schedule and cost contingencies. Using the structure of Complex Systems Methodology for risk management will help ensure that technical risk reduction and opportunities are properly prioritized within the overall project constraints, and that they are supported at the level that ensures project success.

To perform technical risk management, the systems engineer will develop a technical risk management plan that is specifically focused on reducing technical risk and seizing opportunities in an effort to optimize the design to its requirements. The plan is integrated into the project management plan and then managed throughout the life cycle of the project. As the systems engineer assesses risk and opportunities associated with the technical scope and those that require priority action with impacts to project cost and schedule, each will be raised to the project manager for project impact consideration. Most technical risks can be retired within the budgeted and scheduled constraints defined for the scope of work, which is certainly the intention of the project manager.

The measures of effectiveness, measures of performance, and technical performance measures identify the technical areas that require focused risk reduction attention. Both the systems engineer and the project manager carefully and regularly review technical performance measures (see Section 3.2.5.2). Requests for contingency draws for either cost or schedule will be evaluated through change control processes, discussed further in Section 3.2.6.

3.2.5 Quality Management

Quality has multiple dimensions and is assessed by stakeholders, the project manager, and the systems engineer in different ways. High quality is perceived as an end result that confirms acceptable business value for the investment, be it financial, labor commitment to the effort, or technical performance associated with a solid design. Stakeholders measure quality through measures of effectiveness and validation activities, and ultimately against their perception that the system, as delivered, has met their needs as stated in the requirements and as addressed through change requests throughout the project life cycle. Or, in other words, if the delivered system performs as stakeholders anticipated and expected in the current operational environment, they would assess the system as being high quality.

Project managers measure quality in ways that demonstrate performance to the project plan and the quality management plan that are developed during the planning phase of the project. Systems engineers measure quality associated with the technical design using specific designators, such as measures of performance, technical performance measures, specifications, and others, against the comparative data from tests and verification activities.

Poor performance in any area will reflect as poor quality. The project manager and systems engineer will use cost, schedule, and scope performance indicators to track performance. Both will also call and attend project and technical reviews, which serve an important project role in assuring quality and in the role of governance. They provide the stakeholders, the project team, organizational management, and internal and external reviewers with the opportunity to understand the project progress. These reviews can serve as decision-making opportunities or only for providing status information. This section presents information on project quality management; technical quality and measurements; and test, verification, and validation processes.

3.2.5.1 Project Quality Management

Throughout the life cycle of the project, the project manager will continuously assess project quality, and at the culmination of the project, will validate with the customer that the quality of the deliverable is as expected. Validation activities are the responsibility of both the project manager and the systems engineer. The project manager ensures at the beginning of a project that agreed-upon baselines and criteria to assess quality have been identified and documented. It must be clear how quality will be measured. Project quality management is typically assessed against measurements, metrics, and specifications that are defined early in the project life cycle and then managed against proposed and accepted change throughout the life cycle. Most often, quality is associated with the way that the project scope meets both the specifications (through verification), and the stakeholders' expectations (through validation). Verification activities are unique systems engineering processes and are discussed in Chapter 4.

The ability of the project to meet the stakeholder's requirements depends on how well the systems requirements are captured and evolved, and how well communications management and stakeholder engagement are applied throughout the project. If either of those activities is not done well, then the perception of quality will suffer. If they are done well, then the fitness of the scope to meet the intended purpose will serve as the primary quality measure. The scope must meet the expectations of the stakeholders, and the results need to be repeatable, consistent, and stable.

An important consideration is that stakeholders will view quality as the measure of how well the outcomes of the project meet their current need, not the need they originally specified. If, for example, the business model of the key stakeholders changes during the project, but the requisite changes were not implemented, then the final outcome of the project could result in a system that is no longer valid or required, even if all specifications were met and the quality was assessed as superior through verification. A system that is not useful to the stakeholders cannot be considered as a quality system.

To assure quality, the project manager will determine which reviews will be completed, the tempo of these reviews, and the expected outcomes during the planning phase of the project. Programmatic reviews can be completed daily, weekly, monthly, quarterly, semiannually, annually, as-needed, or any combination thereof. They can be informal, formal, or a combination of both. The purpose of these reviews is to set expectations about what the project will achieve, set the record straight about what it is not doing, and to maintain the support and trust of the key stakeholders through visibility, transparency, and clarity. They can also be used to bring in subject matter experts from outside the project to validate the status of the project and thereby assure quality. The unique processes associated with technical reviews are described in Chapter 4. All quality management planned activities are documented in the quality management plan developed by the project manager.

3.2.5.2 Technical Quality and Measurements

Both the project manager and the systems engineer own the quality management responsibilities. Neither can make decisions associated with quality without consulting the other, the result of which would be an increased risk to the project. The systems engineer will collaborate with the project manager to identify the appropriate quality information and measurements required to assess quality within the technical scope of the project. Although there are many different measures that can be employed on the technical performance of the project, the measures that are the most profoundly important to the project, and often the riskiest, are generally identified and carefully tracked through the use of technical performance measures.

Using defined technical performance measures to manage critical technical areas provides the visibility needed to ensure constant forward movement to reduce risk and meet targeted milestones. Technical performance measures provide specific quantitative measures that track expected or projected technical performance against actual achieved performance over time. They generally identify specification ranges, thresholds or limits (upper and lower), design contingency (known as *margin*), and unacceptable risk ranges. As with schedule and cost reports, technical performance that is in the unacceptable or marginal ranges will generally trigger review and risk mitigation. The systems engineer will escalate this type of issue to the project manager if the risk mitigation requires resources that cannot be accommodated in the allocated budget and/or schedule. Technical performance measures are traceable to higher-level requirements. A lack of technical performance on technical performance measures therefore results in impacts on other areas of the project, so they must be carefully and regularly assessed. Graphical representations of progress on technical performance measures are generally reviewed alongside cost and schedule graphs during project reviews. An example of a technical performance measures graph is shown in Figure 3.10.

As part of quality activities, technical reviews are a standard activity throughout the life cycle of the project. Although there is a synergy between the technical reviews and the project reviews, they are focused on different outcomes. Often, the project reviews will ask for only enough technical detail to satisfy the stakeholders, to obtain enough information for management to make sound decisions on behalf of the project, or they may only review the technical performance measures. The systems engineer is responsible for identifying and setting up the proper reviews for the project to ensure that the development proceeds at the proper pace, and risk is being sufficiently reduced. Technical reviews serve an important function on the project. They provide a forum for describing the technical activity to a broader audience, for bringing in external subject matter experts for guidance or advice, and for performing deep technical assessments. They also serve as gates that must be passed before moving into another phase of the project—a key governance activity.

There is a standard set of technical design reviews that serve as gates. Upon clearing a gate, the project is considered baselined at that configuration. The reason to baseline at each gate is to hold steady

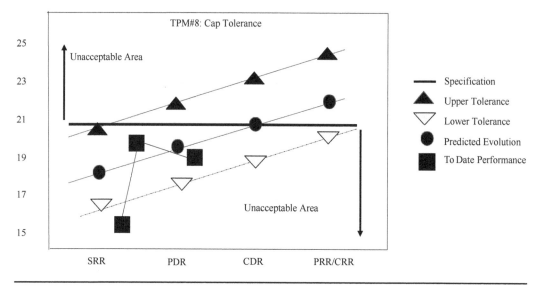

Figure 3.10 Example of Technical Performance Measurements

the design so that the next phase activities can proceed against a known configuration. If the design is not stable enough, then risk increases not only to the system design but also to the management of interfaces. One or more reviews may be held to baseline the system definition or the system requirements (SRR). A conceptual design review (CoDR) may be held during the system concept phase of the project to confirm that the design meets the stakeholder requirements before time and effort is spent in the design and development phase of the project. During the system design and development project phases, a system preliminary design review (PDR) assesses the completeness of the initial functional and physical designs, including any designs for the subsystems or components. The overall goal of this review is to confirm that the design is ready to undergo the final design activities that will lead to the critical design review (CDR). For established technical performance measures, prior to entrance into the critical design review, all technical performance measures risk should be retired, and all designs should be within the appropriate tolerances and parameters. The critical design review's purpose is to ensure that the system is ready to proceed and will meet the stated performance requirements. If the risk has not been fully retired, then the critical design review will identify the remaining risk, as well as the technical debt that will follow the design into the next phase of the project.

In the system integration phase of the project, a system integration review (SIR) may be held to assess the readiness or completeness of the integration phase of the project. A production/construction readiness review (PRR/CRR) may serve the purpose of reviewing the design to determine readiness for production/construction. Finally, an operations readiness review (ORR) may be held to determine if the design and all of the associated materials, such as training materials, maintenance manuals, etc., are ready to enter the operations phase of the project. When a closure/deactivation is required, there will be a final closure/deactivation review.

There are many additional technical reviews that can be chosen. The form of each of these reviews can be somewhat different depending on the discipline, and there are many examples available in current literature, including inputs and outputs for each one. Many best practices exist that can be used to inform and guide the systems engineer on which measurements and reviews are best for the project. Whichever measures and reviews are selected for the project, all will be identified during the preparation and planning phase and documented in a Technical Quality Reviews Plan. Additional information on tailoring these activities for different types and complexities of projects is available in Chapter 6.

3.2.6 Governance

This section will describe the overarching Complex Systems Methodology structure associated with governing the project, including both the project management and systems engineering roles in project integration, change, and configuration control processes. Governance is defined in this book as the structured, controlled management processes that are applied by the project manager and systems engineer as the project progresses through its life cycle. It is overarching and includes any activity associated with assurance, including quality activities such as performance and gate reviews, risk assessments and mitigation, procurement management, etc. If governance provides adequate, yet not restrictive, support to the project and is done well, the processes will lead to optimal outcomes. Figure 3.11 shows the overall set of complementary processes that are controlled through integration activities, change, and configuration control.

These overarching processes are associated with activities that are being performed on behalf of the project by multiple disciplines and, in many cases, in multiple organizations. These "moving parts" of the project are the riskiest and can cause overall project failure, or they can negatively impact performance variances and cause schedule slips, overspending, inability to achieve technical specification requirements, and other serious programmatic issues.

3.2.6.1 Project Integration

Project Integration is part of the standard project management activities and part of the systems engineering life cycle phases as well. These complementary processes involve active management of the activities associated with performing the project and bringing all the disparate pieces together into an elegant system design that will promote desired emergent behavior. However, this is also a high-risk set of processes that requires constant attention since this activity happens in the space between the disciplines and can often be missed, as ownership of the areas of responsibility may not be clear or obvious.

Within the integrative processes are those critical elements such as the development of the initial project charter that describes the intent of the project. Interfacing with the stakeholders and converting their needs into requirements, and then converting the requirements into a physical design while keeping careful control of the processes, is a key integrative task. Ensuring that all blockers are removed so that the project team can effectively design, develop, realize, verify, and validate the system with the stakeholders is included in integrative processes. This is an incredibly important activity. If these processes are not put in place correctly or are tailored out, the probability of bringing the project to a successful conclusion within the constraints of the cost, schedule, and defined scope, will be impossible. Improper tailoring will cause the integrative processes to fail (see Chapter 6).

3.2.6.2 Change Management

The process of controlling change allows deliberate decision making. Change will occur. It is part of the evolutionary process of projects, and it cannot, or should not, be avoided. The project manager is responsible for project change control throughout the life cycle and is responsible for reviewing each impending change from a cross-discipline, integrated perspective so that risk to the project is minimized. Any change that could possibly impact the baselines that have been set and agreed upon by the project team must be analyzed and the potential effects of implementing the change understood fully. Risk mitigation activities associated with understanding the effects of change, such as with modeling and simulation or prototyping (unique systems engineering processes) are described in Chapter 4.

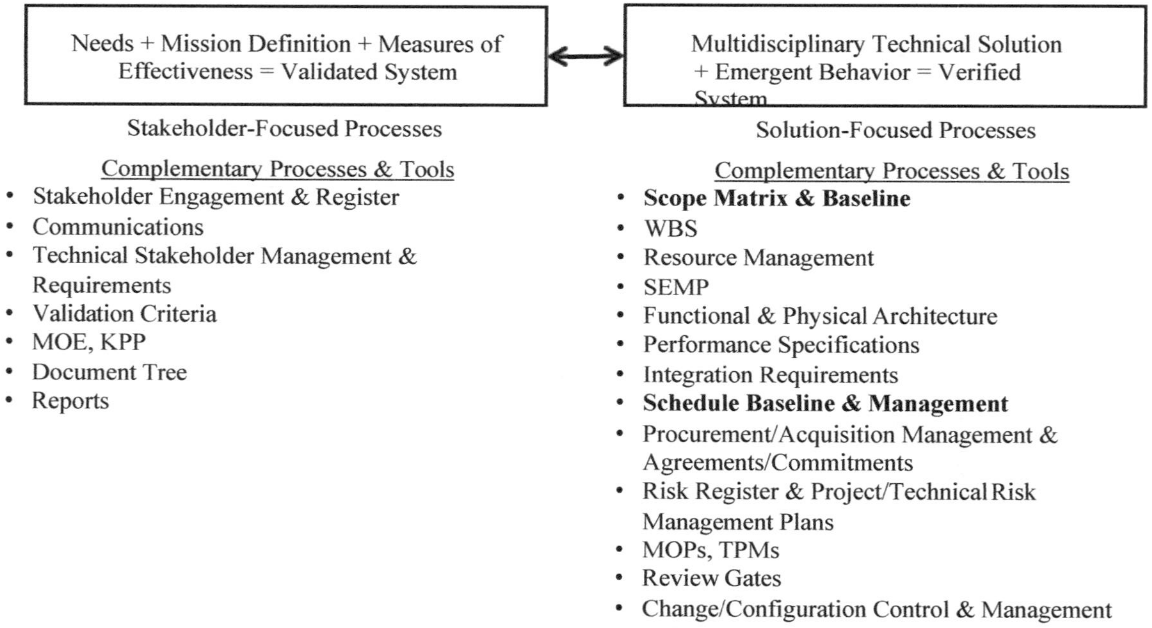

Needs + Mission Definition + Measures of Effectiveness = Validated System

Multidisciplinary Technical Solution + Emergent Behavior = Verified System

Stakeholder-Focused Processes

Complementary Processes & Tools

- Stakeholder Engagement & Register
- Communications
- Technical Stakeholder Management & Requirements
- Validation Criteria
- MOE, KPP
- Document Tree
- Reports

Solution-Focused Processes

Complementary Processes & Tools

- **Scope Matrix & Baseline**
- WBS
- Resource Management
- SEMP
- Functional & Physical Architecture
- Performance Specifications
- Integration Requirements
- **Schedule Baseline & Management**
- Procurement/Acquisition Management & Agreements/Commitments
- Risk Register & Project/Technical Risk Management Plans
- MOPs, TPMs
- Review Gates
- Change/Configuration Control & Management

Figure 3.11 Overarching Complementary Model

Change control implementation rigor depends on the project complexity. The level of rigor and formality can vary. The structure that must be implemented is one in which a change can easily be proposed regardless of the source location (across the project team and stakeholders), with impacts accurately assessed and reviewed by the appropriate decision-making body. In most effective implementations, an interdisciplinary change board is set up and staffed with individuals that oversee the key baselines under review. Subject matter experts are invited to discuss technical issues that are under consideration. The person (or persons) who brought the change request forward generally presents his or her change request and supporting evidence to the board for disposition. After deliberations, the change board makes its recommendation. In some cases, a binding decision can be made immediately; in other cases, a recommendation is provided to the project manager, who has the final decision authority.

The most effective change control process serves as a gate through which only approved changes that have been vetted by the impacted stakeholders, with all concerns resolved, may proceed into implementation. Regardless of who holds the change board decision authority, the project manager is responsible for overall project change control.

Stakeholders and the project team often drive change requests that must be analyzed and assessed, with a final determination as to the impact of that request. That analysis first focuses on determining if the request is a modification to the scope as currently defined or outside of the boundaries of the existing scope. Many changes are associated with minor modifications to the design during the realization phase of the project. These changes often can be completed without cost, schedule, or interface impacts. If they fall into this category, the project manager can deem that the change is inside of the current scope and not impactful to any other part of the project and can then modify and document the change within the scope statement. If, however, the change request impacts other parts of the project organization, it is necessary to review the impact of the potential change so that possible affected parties can weigh into the decision. In the event that a change request is deemed by the project manager to be outside of the boundaries of the existing scope, schedule, or budget, then a formal change request is required, as well as a review with the appropriate affected stakeholders.

3.2.6.3 Configuration Control

Configuration control can refer to a technique associated with version control. It can apply to any item that changes often or rapidly, and where there is a need to know which item is the most current. Losing configuration control of an item, whether it is a document version, a prototype version, or a designated configuration item, can have dire results. Time and money must be spent to resolve a loss of configuration. It is therefore an important process that must be managed throughout the entire project life cycle. The configuration control process is tightly coupled to the change management process; however, there are many items that change configuration without going through the project change board, although they may go through a configuration change board often set up to assess specific technical configuration issues.

During the definition phase of the project, the systems engineer will draft a configuration management plan. This plan will define how the configuration control process will be implemented throughout the project life cycle and how this process will interact with the change management process. Typically, engineering change requests and notices, as described in Chapter 2, impact the configuration of an item. The processes for documenting and implementing these engineering changes when they are approved must be included in the plan.

In addition, throughout the entire work breakdown structure decomposition of systems, subsystems, components, and parts, a unique identifier will be assigned to each identified item so that information about that unique item can be tracked as it evolves within the project. These unique items are often referred to as configuration items. Configuration control is an important governance activity that is implemented so that the current form and function of the system and all of the elements that make up the system are always known. The need to understand the as-allocated, as-designed, as-built, and as-maintained configurations rely on accurate configuration control.

The system engineer is responsible for managing the process, including documenting decisions and communicating the results of any pending decisions and approvals to the team members that have a vested interest in the item. As the design evolves from its as-allocated form through to its as-maintained form, there is no period where configuration control is less important than another. If the configuration is allowed to migrate out of sync, once it is discovered, effort must be put forth right away to resolve the discrepancy. The longer the mismatch is allowed to remain, the higher the risk to the project becomes.

Due to the critical nature of performance in the area of configuration control, a subject matter expert in configuration status accounting or project controls is generally responsible for the collection, verification, and validation of the data and information collected in association with configuration control activities. Configuration status accounting activities focus on management of the configuration baselines, changes, and implementation of those approved changes into the project. A project controller may oversee both configuration status accounting and project accounting and scheduling, depending on the size and complexity of the project.

3.3 Case Study: Simon Beck's Snow Art

All case study quotes are from interviews held with the individual presenting the background. In this next case, the interview was held with Simon Beck, and all quotes are attributed to him.

Artist and author Simon Beck[3] works with the unusual media of snow or sand to create his incredible geometric images. As of 2017, he had created 265 snow drawings and 87 sand drawings. His engineering education, his extensive career experience as a cartographer, and his love of the outdoors all come together in his inspired, contemporary designs.[4] As Beck explained, "After a day of skiing, I got the idea to draw a star on the small frozen lake. The day after, looking down from the ski lift, I was impressed by the result." And others were impressed as well, as he heard from the other riders, "Superb," "It's a flower!" and other expressions of delight. He opted not to say anything at that time, as he did not consider what he did to be serious art. However, after five years that opinion has drastically changed, and what "started out as a joke, is now a part of my life," Beck said.

Beck explains that as a child he enjoyed drawing geometrical forms based on Koch's snowflake, which is a fractal object that can be constructed from elementary geometry. Fractals are "infinitely complex patterns that are self-similar across different scales. They are created by repeating a simple process over and over in an ongoing feedback loop."[5] In particular, he liked making the hexagonal shape, an architectural design that is encapsulating and has clear boundaries, which are easier to replicate in that repeating process. Since he started creating his snow art, he has created hundreds of temporary pieces of art. In 2014, Beck began to switch between snow and sand as the medium. Each offers different benefits and risks, challenges, and enjoyment.

3.3.1 Project Charter and Plan

Each time Beck sets out on a new creation, he starts with a design, location, and type of medium (either snow or sand) that appeals to him. Each design that Beck creates involves walking an average of 20 kilometers (12.4 miles). Depending on the location and type, he will develop his schedule and design. Beck explains, "With snow, I have more time to complete the design. Drawing on sand on the beach has to be done more quickly before the tide comes in and washes it away." The best results, as he explains, come from completing the design in time to catch the light, as it is the shadowing that makes the design more dramatic. But it is not just the completed project that needs the right light. As Beck clarifies, "the snow drawings are difficult to see in poor light; in really poor light it is difficult to the point of not being able to see where to go when creating them."

3.3.2 Stakeholder and Communications Management

Although Beck creates his art for his own enjoyment, he has found a stakeholder community in those people that are interested in his creations and follow him on social media. Prior to having that venue to showcase his work, he struggled with how to reach his audience. Communications about Beck's artwork are included on a Facebook page titled "Simon Beck's Snow Art."

Other stakeholders include paying customers. In the past he has created designs specifically for a movie production, and he has provided created designs that were photographed for a book. As Beck mentions, "One never knows when someone will offer money for the art, but when they do I consider it."

3.3.3 Scope Management

Beck uses his past experiences to manage the scope of each new project he plans. As he describes, "I know where all the sites are that I generally work on, so I know exactly where to go to do the drawing. And I know how to get there, and sometimes I know how much time it will take to do the work. For new designs, the time to create it is rather unknown as not only is it a function of the design (which generally has not been made before) but the depth of the snow which affects the walking time, which makes it hard to predict."

Beck's location of choice is in the French Alps, the Lac Marlou near Les Arcs 2000 ski resort in Savoie, France,[6] where he has been working since he started creating his art in the snow. He explains that this is "because the Marlou has excellent vantage points for ground-based photography that can be reached using the lift system, and it is possible to ski home from it after the lifts have closed without having to go uphill."

Occasionally, he will choose a site that he has never been to before, which adds more challenge. Without prior experience at that site, it may not be known what safety issues might present themselves: avalanche danger, safety of the ice on a frozen lake. With a new site, there are other unknowns as well, such as if the required soft powder snow is at a sufficient depth, if there are cracks in the ice or other blemishes that will impact the aesthetics, if the location is level enough and if the flat field is of sufficient size, or other factors that can cause serious complications to the design. In addition, natural features may affect the size of the design. "Some drawings naturally fit the site, some not so much," Beck says. He tries to resolve all those questions before attempting a design at a new location.

3.3.3.1 Architectural Design

The process for Beck's architectural designs, as he explains, "might start with a picture or photo of a design that I then trace or plot onto a template." He enjoys using standard patterns and shapes, with the best ones being fractals where he uses a simple rule again and again at different scales. Beck says, "once you've learned the rule, you're just sort of following a very short set of instructions."

3.3.3.2 Requirements Management

To create the artwork that he desires, there are a significant number of requirements that Beck takes into consideration. The site must be flat and have sufficient area to match his vision of the size of the drawing. If on snow, a higher elevation is better as the snow is typically of better quality and has a lower probability of skiers crossing the field within his design. Also, a steep mountain will typically allow good angles for photography without employing a drone, which is costly. Beck looks for sites that are high also so that he can move downhill off the site once the drawing is completed, rather than climbing after a full day of exhausting exercise. His ideal is to "climb up and ski down," Beck says.

The snowfield would ideally be covered in about 20 centimeters (eight inches) of soft, powdery snow over a very firm surface, such as a frozen lake. But, he explains, "good results can be obtained in depths as low as five centimeters (slightly less than two inches)." The weather predictions must point to a day with clear skies, without wind, not only during the design and development of the art, but also after the design is complete so that photographs can be taken. "The key requirement is that the sun be shining on the day after for the photos. The drawing day doesn't have to be so good," Beck explains.

For the sand drawings, Beck needs a site that is not only large enough to match his vision of the size of the drawing, but also one that will not be impacted by the rising tide within the periods of design, development, and photography. Using sand locations offers a limited window of opportunity to complete the drawing—typically, no longer than nine hours.

3.3.3.3 Integration and Interfaces

Once the design is ready, Beck uses his ample knowledge, experience, and his physical fitness to execute the design on the waiting field. As he explains, "I was a competitive Orienteer, so I use orienteering techniques to start a map on the ground. And then I use distance measurement using pace counting, along with compass bearings to measure and mark the points that I will fill in with footsteps." Orienteering is a sport that uses a map and compass to find prearranged coordinates. He uses a baseplate compass, a prismatic compass, some rope, and some type of markers, often using pieces of clothing (gloves, hat, coat, etc.) that he sheds as he works through the design, the weather warms, and his level of exercise raises his body temperature.

"The prismatic compass tells you the direction you are looking at, so I can make much more accurate bearings using this," Beck says. "I always start with about two hours of measuring at a slow walking speed to accurately position the drawing. The first two stages (measuring and drawing lines) are careful procedures that require focus and concentration, just like any other technical job. Most of the thinking has been done at that point, and I've already measured and drawn out the correct points. I generally won't go wrong after that."

It is important to remember that he is drawing circles by judgment versus through precision measurements with a tool such as a mandrel, which ultimately impacts the accuracy of the curved angles and circles within the design. But as Beck says, "You won't notice if a circle is not exact." He creates his art by walking, without the benefit of an overhead perspective.

After the outlines are placed, the shading within the overall structure requires less thinking. Using his footsteps, Beck moves back and forth, rewalking the lines, and shading. Once he has the basic structure drawn in the snow or sand, he says, "I never relook at the diagram/drawing. I'm following the skeleton, the fragments of main lines. I can draw the rest of it in my mind's eye, just by walking in the snow here and there. Usually I start at the edge and work towards the center, as the largest areas to shade are usually around the edge and it is psychologically easier to get these done first, and as a general principle I do the bits I like least first." At this point, he often listens to music, or enlists others to assist. As Beck explains, "I can teach people to do this easily, but I only give them the parts where they shade an area only, they don't get to do any of the design."

3.3.4 Schedule Management

The schedule to complete one drawing is highly variable. This variation is due to many things, such as how Beck is feeling, what the snow is like, how complicated the design is, and if he is working alone or has assistance. The longest time he took to complete a design was 12 hours; however, it could take anywhere up to that amount of time to complete a drawing. This is generally preceded by up to two hours of preparatory work. Two hours is spent in the measuring stage and up to 10 hours in the execution phase. Beck moves at an average speed of approximately three kilometers per hour (two miles per hour). That speed may increase slightly if a slow jog is incorporated—not a common occurrence but he implements it on occasion.

Another consideration is to know how much time he needs to photograph his completed work. Beck needs enough time after he finishes his creation to ski to vantage points where he can take photographs. Whether this happens that day, or the next, depends on when he completes the design, as it requires the right natural lighting to properly showcase his work.

3.3.5 Procurement Management

Items that Beck needs to purchase in order to complete his art include the appropriate clothing, socks, shoes, snowshoes, compasses, rope, and markers. He currently does not own or use a drone to capture his completed artwork; however, "that is something I am considering purchasing in the future," Beck says. When he first started taking photographs of his artwork, he originally used a lower-quality camera. The purchase of a high-quality camera was needed though, in order to capture the art in high-enough fidelity to ensure that it reflected the image he was hoping to create. With the new camera and photo-editing software, he is able to get the results he desires.

3.3.6 Risks and Opportunity Management

There are many risks and opportunities associated with creating Beck's snow and sand art. Ultimately, he must be able to execute his project and "get down safely at the end of the day!" Beck says. This means that he has to mitigate the risk of running out of energy in the middle of the drawing, which he does through "eating complex carbohydrates the night before, fruit and porridge in the daytime, and taking sweets along to eat after nightfall." He also needs to conserve enough physical energy during the execution phase of the project to complete the drawing within the time frame without exerting too much energy and with enough reserve to ensure he can make his way back home. If he is not feeling fully energized, he will choose smaller designs, which "wastes a lot of the available space, or use a smaller site and keep the larger site for another day, depending on the weather forecast. Usually after a spell of poor weather, I will be well rested and recovered so the first drawing in a spell of good weather would be a big one, then the next day would be for skiing and photography and perhaps some more drawing later on, then see how I feel the day after."

Along with eating properly during the execution phase of the project, he also needs to drink plenty of water. Beck explains that he is often asked about ingesting any substances that could alter his perception, such as alcohol, to which he exclaims, "No way! This would affect your judgment and increase the risk tremendously on the mountain." Of foremost importance is safety. The risk of physical danger and the need to assess safety is a major concern, considering one is often walking on a frozen lake. The ice must be thick enough to support his weight. As Beck describes, "You really need to overcome fear of walking on a frozen lake; however, if there is the slightest doubt, then one does not take the risk." It can also be a bit disconcerting to be alone for so many hours of the day in a remote area. Beck says, "It can get terribly boring. I would never have gotten this far without these personal audio devices." Music is an important part of the experience for him, classical music being his favorite.

Beck is sometimes challenged to finish his design before some event occurs to affect the outcome in a negative way. For example, occasionally someone will ski through his design or an animal will walk through the drawing as he is creating it. Or the tide will come in higher than expected and wash away part of the design. These impacts are difficult if not impossible to rectify. And as Beck explains, "The worst is when it gets tracked before the photos are taken, or when the expected sunny day turns out to be cloudy—not being able to get photos is the most common reason for a failed drawing, although it still serves as a rehearsal for the second attempt." With sand drawings, the incoming tide will usually start to wash away the drawing within minutes of taking the photos. Opportunities can also come out of these events, and Beck will try to turn the impact of the event into something beautiful within the design. He will adapt to a new form and adjust his expectations for the outcome.

3.3.7 Quality Management

Beck approaches quality in a careful design, and through a methodical implementation. However, he does have a tolerance for minor deviations in the outcome. As he explains, "I have transferred skills from my map-making experience, a method of working that prevents small errors adding up to a noticeably degraded end result. However, sometimes something goes wrong. Mistakes cannot be undone, but sometimes an error is simply ignored. Other times the drawing has to be modified as a result."

The quality of the photographs is also important, as they are the actual final record of the drawing. Lighting has to be right in order to reflect the shadows in the most advantageous way, and the angle of the photograph must provide the optimal viewpoint to showcase the artwork. The overall quality of the photograph must be in a high-definition format if the photographs are going to be printed in a book. Beck explains, "I did not have a decent camera when I first started photographing the designs in the snow. But I learned that it is important to have a camera that can take high-resolution photographs so that they represent the design well when going to print, or for editing a photo to show details."

3.3.8 Governance

Although generally the designs transfer well to the snow or sand without issue, that is not always the case. As Beck points out, "Usually changes happen to address a mistake." Considering the complexity of some of the designs, it can be easy to make an error in arithmetic or in imaging. Realized risks, such as a person skiing through the design, or an animal walking through, may cause a redesign requirement since as Beck says, "You can't undo the mistakes, therefore you have to change the design on the fly." He will make design changes based on the evolving circumstances during the design implementation. And although he is generally satisfied with his art, as he explains, "It is inevitable that you will not get the same level of satisfaction from doing the same design over and over. You must get progressively better, so it is an ongoing evolution. It is always easy to do the same design a second time when conditions are perfect."

3.3.9 Outcomes

Beck wanted to "show people that the environment is really beautiful and is worth preserving"[7] through his art. That was the outcome he was striving for and one that he has achieved. There are few artists that work at this scale, using the environment as a canvass and as the final art product. Beck's art gradually fades away and only lasts until the next snowfall or the tide comes in.[8] Beck clarifies, "My art pieces are all created in and from nature and are essentially impermanent. The important thing is to get the photos. I plan drawing for the day before one that is forecast to be sunny. Once I have the pictures, the best thing is another big dump of snow so I can make a new drawing."

3.3.10 Lessons Learned

Several lessons learned can be captured from Beck's experiences. Of course, carefully planning the design is of paramount importance. However, organization before the work begins on site is also an imperative. As Beck explains, "I have to make sure my gear is properly organized the night before, my batteries for my headlamp must be charged in case I end up out after dark, my clothing must be laid out, as well as my food and water. I have to load everything in one rucksack." In some cases, he will store marker sticks the night before if the site he is planning on using is close to his location. He also watches the weather forecasts right up to the time he is ready to leave for the site. He has a sizeable meal of porridge and bananas for breakfast and ensures a good assortment of snacks is available for the day.

One of the key lessons that Beck learned was in how to stage food and water at points around the design, so that he can access them throughout the day without disruption to the process of implementing the design. This creates a well-distributed depot of supplies that he can access as needed.

A surprising lesson that Beck learned was that even though it seems like this type of art would have often been done before, he could not find evidence of previous art. When he started, he remembered, "I did not have internet at home. I had to go to an internet café to do research. When I did the research I could not find much on this type of art." So Beck decided to take it seriously and do as many drawings as possible so that he could ultimately complete a book of his art that he could sell. In retrospect, Beck feels that the book he put together might have been premature and that he should have waited until he had more drawings completed.[9] He continues to enjoy sharing his pictures for free on the internet, and it is clear that people enjoy his art and often offer him paid commission work from exposure to his social media.[10]

3.3.11 Case Analysis

This case provides an example of the process structure that can be instrumental in assisting in the achievement of optimal outcomes of a creative, artistic discipline applied in a natural environment. The case elaborates the systems engineering processes that are used to assist in the scope and schedule management of the design implementations. Quality is an important attribute of the outcome, and risk management is used to directly affect the quality outcomes.

3.4 Key Point Summary

The focus of Chapter 3 is to identify the Complex Systems Methodology complementary processes from the disciplines of project management and systems engineering, which, when used together, will provide positive project outcomes. These include methods for identifying and managing stakeholders, communicating well, and managing collaborations. It also describes methods for designing the solution that will meet the needs of the stakeholders. Processes to effectively manage schedule, procurements, risk, quality, and change are also identified. Following are the key concepts discussed in this chapter. Key terms are compiled for quick reference in the Glossary.

3.4.1 Key Concepts

- Complementary processes include both stakeholder- and solution-focused processes.
- Stakeholder-focused processes help to convey the vision of what is needed and the level of involvement that they want to maintain throughout the project.

- A text description of how the system is expected to perform, and the identification of measures of effectiveness that provide quantitative information that will be used to validate system performance, are required from the stakeholders.
- The construct for effective and efficient collaboration and communications must be in place in order to increase the probability of reaching optimal project outcomes.
- Solution-focused processes are used to provide the solution set that forms the system that will meet the stakeholders' needs.
- Solution-focused processes ensure that the optimal technical solution fits within the project baseline scope, budget, and schedule.
- Technical risk reduction and quality, change. and configuration control must be actively managed to ensure that the end result of the project is as needed by the stakeholders.
- The complementary solution-focused processes include scope, schedule, procurement, risk, quality (including test, verification, and validation), and governance (including change and configuration control).

3.5 Apply Now

The application of the processes that are described in this chapter to a project that is familiar to the reader will enhance the understanding of the material. The Apply Now section's summary table helps the reader visualize the chapter material (see Table 3.1). The top of the table provides an overview of all of the complementary processes and the anticipated documents that will be implemented in the project. The bottom of the table provides a template for the reader to fill out while answering the questions posed in the Apply Now section.

1. *Stakeholder-focused complementary processes include the development of a stakeholder register and the identification of stakeholder requirements.*
   Identify a project you would like to do. Complete a stakeholder register and requirements and define the quality measures, including validation criteria, for that project.
2. *Stakeholder-focused complementary processes include communications and collaborations management.*
   Using the same project from Question 1, complete a communications matrix and a basic collaboration plan.
3. *Solution-focused complementary processes include scope management.*
   Try to define the technical scope for the project idea. Identify your technical requirements. Draw a functional architecture for that idea and identify the performance specifications. Then develop a basic physical architecture with integration requirements identified. This can be a simple example.
4. *Solution-focused complementary processes include schedule management.*
   Using your example above, identify the tasks associated with your project and develop a basic schedule. Decompose the work breakdown structure.
5. *Solution-focused complementary processes include resource management.*
   Using your example above, identify the resources (including procurement) activities that would be associated with your project.
6. *Solution-focused complementary processes include risk management.*
   Using your example above, identify the risks that would be associated with your project. Build a basic risk register.
7. *Solution-focused complementary processes include quality management.*
   Using your example above, identify the quality metrics and measurements that would be associated with your project.
8. *Solution-focused complementary processes include governance.*
   Using your example above, identify the governance activities that would be associated with your project.

Table 3.1 Apply Now Checklist

COMPLEX SYSTEMS METHODOLOGY™SM (CSM™SM)
APPLICATION FOR ALL PROJECTS

Questions to Ask	Responses				
Are the complementary stakeholder-focused documents/ processes in place?	Stakeholder	Communications/ collaborations	Needs analysis	Scope baseline	Quality
	Register	Communications/ collaboration plan	Stakeholder requirements	Validation	MOEs, KPPs

Questions to Ask	Responses										
Are the complementary solution-focused documents/ processes in place?	Cost baseline	Schedule baseline	Functional architecture	Scope baseline	Quality	Resources	Physical architecture	Risk	Governance		
	Baseline budget	Work Breakdown Structure/ baseline schedule	Functional architecture diagram	Technical requirements/ baselined scope	MOP/ TPM/ reviews	Acquisition/ procurement/ labor, materials, services	Performance specs, integration requirements, ICDs	Risk register/ management	Change control	Configuration control	Integration

COMPLEX SYSTEMS METHODOLOGY
APPLICATION FOR ALL PROJECTS

Questions to Ask	Responses				
Are the complementary stakeholder-focused documents/ processes in place?	Stakeholder	Communications/ collaborations	Needs analysis	Scope baseline	Quality

Questions to Ask	Responses								
Are the complementary solution-focused documents/ processes in place?	Cost baseline	Schedule baseline	Functional architecture	Scope baseline	Quality	Resources	Physical architecture	Risk	Governance

References

1. Liu, D. (2016). *Systems Engineering: Design Principles and Models*. Boca Raton, FL: CRC Press/Taylor & Francis Group.
2. Kossiakoff, A., Sweet, W. N., Seymour, S. J., and Biemer, S. M. (2011). *Systems Engineering: Principles and Practice,* 2nd Edition. Hoboken, NJ: John Wiley & Sons, Inc.
3. Beck, S. (n.d.). "Simon Beck: Snow Art Gallery." Accessed January 12, 2018, from http://snowart.gallery/see.php
4. Beck, S. (2014, October 1). *Simon Beck: Snow Art.* S Editions.
5. Fractal Foundation. (n.d.). "Fractals are SMART: Science, Math and Art!" Accessed January 12, 2018, from http://fractalfoundation.org/resources/what-are-fractals/
6. Benjamin, E. (ed.) (2017, July 12). "This Artist Walks 20 Miles to Create Geometric Patterns in the Snow." Culture Trip. Accessed January 12, 2018, from https://theculturetrip.com/europe/articles/simon-beck-artist-snow-art-alps/
7. McCarthy, E. (2014, November 11). "11 Questions for Snow Artist Simon Beck." *Mental Floss.* Accessed January 12, 2018, from http://mentalfloss.com/article/59958/11-questions-snow-artist-simon-beck
8. Beck, S. (2014, December 19). "Snow Art." TEDxKlagenfurt. Accessed January 12, 2018, from https://youtu.be/CfPPZS4IvWs
9. Bellos, A. (2014, November 6). "Simon Beck's Astonishing Landscape and Snow Art Illustrates the Cold Beauty of Mathematics—In Pictures." *The Guardian.* Accessed January 12, 2018, from https://www.theguardian.com/science/alexs-adventures-in-numberland/gallery/2014/nov/06/simon-becks-snow-art-landscapes-mathematical-designs-drawings-alps
10. Beck, S. "Facebook Page: Simon Beck's Snow Art." Accessed January 12, 2018, from https://www.facebook.com/search/top/?q=simon%20beck's%20snow%20art

Chapter 4

Application of Unique Processes

This chapter describes in depth the application of the Complex Systems Methodology's unique systems engineering and project management processes to a project. When used in combination with the complementary processes described in Chapter 3, these unique processes will provide the structure for project success. There are two major categories of focus for these processes:

- stakeholder-focused processes
- solution-focused processes

Section 4.1 identifies those stakeholder-focused processes that are unique to each of the process disciplines of project management and systems engineering. These include the specific project management activities that start the project and keep it running smoothly until the expected outcomes are achieved. Included as well are the systems engineering processes used to clearly capture the ideas of the stakeholders in a way that will be digestible and understandable by a technical team so that a solution can be envisioned.

Section 4.2 describes the solution-focused unique processes that are associated with scope, schedule, and cost management—the three primary baselines that constrain a project. The project management processes associated with human resource management, which are imperative in projects that have team members and that utilize labor in the execution of the project, are explored. Unique quality management activities associated with testing and verifying that technical solutions meet all specifications are described. Also explored are project management activities associated with addressing risk that extends beyond the project, as well as the systems engineer's responsibility in reducing technical risk.

As was described in Chapter 3, the many standards-setting organizations in both disciplines of project management and systems engineering may present these processes in variations both in breadth (those processes that are included) and in depth (the number, type, and context for the subprocesses). This chapter provides a view of the processes in an intuitively aligned and consistent pattern to the complementary processes described in Chapter 3. The processes described in this chapter are also the most important ones for achieving expected and positive project outcomes. Subject matter experts and practitioners from both disciplines should be consulted so that they can apply the appropriate tailoring of these processes for complex projects. Tailoring is explained in Chapter 6.

In this chapter, the Section 4.3 case study demonstrates the application of project management and systems engineering unique processes. The case demonstrates the ability to apply these processes to achieve successful outcomes in a nontraditional project. Section 4.4 includes this chapter's key concepts. Section 4.5 provides the Apply Now exercises, which should be used to apply the lessons learned from the chapter.

Chapter Roadmap

Chapter 4 focuses on the Complex Systems Methodology unique processes, derived from standard project management and systems engineering disciplines, and provides the structure to help the reader understand the key concepts and to apply the learning to real examples. This chapter:

- includes processes to develop the originating documentation for the project, such as the charter and project plan
- describes the processes associated with identifying and managing stakeholder concerns
- provides methods for elucidating requirements from stakeholders
- explores the critical processes associated with interface management
- describes the activities associated with choosing and applying schedule, cost, and scope methodologies
- provides insight into human resource management for projects
- explains the necessary processes to test, verify, and reduce technical risk
- provides a case study to demonstrate the application of project management and systems engineering unique processes
- provides a summary checklist of key chapter concepts for easy reference
- provides Apply Now exercises to assist in the application of the processes described in this chapter

4.1 Stakeholder-Focused Unique Processes

This section continues to develop the stakeholder-focused processes, but from the perspective of those that are unique processes for each of the disciplines; in other words, where the project management or systems engineering processes are foundationally different but still critical for the overall project to meet its objectives. Each of these unique processes are described within its specific discipline. These unique processes include the project-management–related activities of defining exactly what will be done and by whom and then actively managing stakeholder involvement throughout the life cycle. The systems engineering stakeholder-focused unique process activities are associated with clearly defining the stakeholder's vision of what is technically needed.

Although it may seem that collaborating with the stakeholders is an activity that is performed predominantly in the early phases of the project life cycle, there is a real need to continue active engagement with stakeholders throughout the entire life cycle. If the needs of the stakeholders are not accurately captured, and if any change or evolution of those needs are not addressed and implemented as appropriate, then the ability to reach outcomes that are seen as successful is at risk. As discussed in Chapter 3, the complementary stakeholder-focused processes, along with these unique processes, will increase the probability of achieving stakeholder satisfaction with the outcomes both throughout and at the end of the project.

Both project management and systems engineering unique stakeholder engagement processes will be defined in Section 4.1.1. Figures 4.1 and 4.3 show the unique processes and tools that will be discussed in this chapter and how they align to the complementary processes and tools described in Chapter 3.

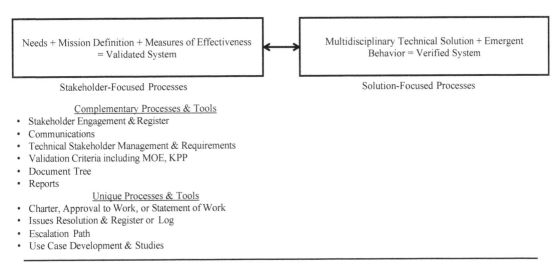

Needs + Mission Definition + Measures of Effectiveness = Validated System

⟷

Multidisciplinary Technical Solution + Emergent Behavior = Verified System

Stakeholder-Focused Processes   Solution-Focused Processes

<u>Complementary Processes & Tools</u>
- Stakeholder Engagement & Register
- Communications
- Technical Stakeholder Management & Requirements
- Validation Criteria including MOE, KPP
- Document Tree
- Reports

<u>Unique Processes & Tools</u>
- Charter, Approval to Work, or Statement of Work
- Issues Resolution & Register or Log
- Escalation Path
- Use Case Development & Studies

Figure 4.1   Stakeholder-Focused Overarching Processes and Tools

4.1.1 Stakeholder Engagement

Figure 4.1 shows the unique stakeholder-focused processes and tools. The processes for stakeholder engagement for both project management and systems engineering ensure that all of the work that goes into a project is value-added and that the outcomes are useful to the stakeholders. It is often the case that stakeholders have different views on what will create value for them. In many cases, even the key stakeholders' vision of the intended outcomes is not in alignment. This is the reason for the stakeholder-focused processes—to ensure that there is a clear and agreed-upon target for which the project can aim. The processes will provide enough structure that there is a high probability of reaching the solution that is the highest value. It is important to understand that not all projects require the same level of rigor in the implementation of these processes. Tailoring is a topic that will be discussed at length in Chapter 6. In Sections 4.1.1.1 and 4.1.1.2, the Complex Systems Methodology unique processes associated with project management and systems engineering techniques used to address stakeholder engagement are described in detail.

4.1.1.1 Project Stakeholder Engagement

Within the project management discipline is the responsibility to gain consensus as to what the scope, schedule, and costs for the project will be and to baseline each of them so that change can be effectively managed throughout the life cycle of the project. However, before any of that can occur, the project manager must obtain the approval to move forward to execute the project. The first activity that must occur is to gain an agreed-upon vision of the project with the key stakeholders. This is generally reflected in a project charter. This overarching vision is the critical first step of all processes that follow.

Although the process of documenting an agreed-upon project definition is simple and can be achieved with a basic text document, obtaining convergence and concurrence on the definition of a project can be more difficult. As difficult as it might be, it is an imperative to obtain documented approval of the project definition by the key stakeholders so that the project outcomes do not become moving targets. Most importantly, it must be conveyed to the stakeholders that concurrence across all parties is not required. Although it is important to capture the stakeholders' thoughts as to what the project should encompass, the majority position of the key stakeholders is what must be obtained to go forward on the project. This is critical, because it is almost impossible to obtain across-the-board agreement on all ideas—a problem that tends to be exacerbated the more diverse the stakeholders are. The most common approaches to soliciting ideas are from direct conversation or face-to-face meetings. There are many methods that can be used to solicit ideas and to converge on an agreed-upon path forward that are beyond the scope of this book.

Once the project is defined, agreed upon, and approved via documented commitments, such as through the project charter, then that document informs the other project management and systems engineering documentation. It is rare for this high-level document to change or be modified during the life cycle of the project, so care should be taken to ensure that the definition does not delve into too much detail. When changes are accepted, it is through formal change control so that all impacts on the related plans are understood prior to implementation. This is also the document that will be reviewed at the end of the project to validate that the project intent has been met. It is a strategic-level document that is meant to describe in broad terms the expectations of the stakeholders as to what the project will deliver.

Throughout the life cycle of the project, issues, concerns, suggestions, or requests are brought up by stakeholders and should be reviewed against the defining documents to first determine if they are truly within the project's purview to resolve and then to assess a proper path forward. Anything brought forward should be documented for future reference, along with the action taken, even if a decision was made not to make any modifications or changes to the existing path. It will also serve the project manager well to document what resolution was identified, what actions (if any) were taken and when, as well as the date the information was communicated to the stakeholders. Actively managing this process throughout the life cycle will help ensure that the outcomes match the agreed-upon and change-controlled project boundaries.

A simple spreadsheet can be maintained to track stakeholder issues or concerns and their resolutions. Figure 4.2 provides a simple example of the issues register captured on a spreadsheet. A unique identification number, a title, a full description of the request and who submitted it, the responsible point of contact, and the resolution that was provided are documented on the issues register. Both the resolution and the communication dates are included. Any amount of useful information, such as stakeholder contact information, can be included based on the needs of the project manager. However, this register should not duplicate any information already contained on the stakeholder register. Both of these documents should be used in tandem.

4.1.1.2 Systems Engineering Stakeholder Engagement

As mentioned in Section 4.1.1.1, it is often difficult to clarify the project definition. Sometimes the stakeholders have a vague idea of what they need. Often the stakeholder vision is intermingled with wants, needs, definitions of solutions, or other things that are or should be outside of the project or addressed at a later stage of the project. The stakeholders' vision is often not shared among all parties. During the project stakeholder engagement activities, these discrepancies and inconsistencies are resolved at the highest level.

Within systems engineering, a process called *stakeholder needs analysis* takes place during the mission-definition phase of the project life cycle. During this phase, scenarios, concept of operations, and use cases can be developed to resolve the vision and provide additional detail into how the stakeholders view the system that is to be developed. These text-based descriptions present information about how the system is envisioned to behave during its operational life. It will not include information about any specific solutions that might meet the needs that are identified.

Systems are often complex and, as described in Chapter 1, may behave in ways that are unexpected. It is therefore useful during the development of these stories to ensure that the knowledgeable and

Unique Identifier	Title	Issue, Concern, or Request	Submitter	Responsible Point of Contact	Resolution	Resolution Date	Communication Date
1001	Additional Review Desired	Separate critical design and production readiness reviews needed.	Karen Wilson	Bob Marley, Chief Engineer	Confirmed limited negative impact of additional review to project cost and schedule. Scheduling additional review.	12/1/17	12/5/17
1002	Stakeholder Requirement Change	Desire color be blue instead of green.	Joe Smith	Bob Marley, Chief Engineer	Confirmed major impact to plans and conflict with needs of other stakeholders. No action.	1/23/18	1/25/18
1003	Additional Communication Need	Provide website updates more frequently.	Ali Carabba	Karen Smith, Software Manager	Confirmed inability to staff within established cost. Alternative of email updates to be provided.	2/5/18	2/10/18

Figure 4.2 Example of an Issues Register

experienced systems engineer abstracts the detail into descriptions of behaviors. The systems engineer will use a method of casual, conversational-style elicitation not only to help the stakeholders identify, in their own words, how they will interact with the system, but also to define the boundaries of the system. Each story will be told in a way that incorporates the multi- and inter-disciplinary activities and identifies expected behavioral interactions between systems elements. This way, the holistic picture is captured in totality, and any characteristics that might lead to emergent behavior can be included.

This process of developing the appropriate stories that describe the stakeholder needs in clear operational dialog is difficult in the sense that most people, when describing what they want or how they see something working, can use confusing, detailed, and impressionistic language about how they envision the processes would work. Sometimes stakeholders have a tendency to focus on the solution and offer suggestions for how the needs should be filled with both processes and technology. This often causes gaps in the stories, inconsistencies, or confusing inaccuracies. These must be worked out together between the systems engineer and the stakeholders to the point that a clear and concise description of the conceptual operating model for the system is defined and where the identified needs are solution neutral.

The processes for developing a concept of operations, use cases, or other scenarios includes meeting with the stakeholders to discuss and elicit information about all anticipated activities that they can imagine associated with how the project will be employed. This will provide the appropriate context so that a solution can be designed. An experienced systems engineer will be able to parse the ideas into those that are true definitions of the intended operations, versus solutions or other out-of-scope activities.

Even though all anticipated activities must be considered, not all of them are incorporated into these stories. Limiting the number of use cases to those that truly influence the system is important. It helps to develop a useful set of system behaviors that can inform the solution design. Although it is dependent on the complexity of the system, a good rule of thumb is that between five and fifteen use cases or stories is optimal. There should be enough to capture the influencing behaviors and no more. Time bounding, by establishing a time period during which to measure the walk-through of each activity in a normal operational environment, will make things more manageable. Each story should be traceable to a system or mission objective. All assumptions, preconditions, precedents, follow-on actions, interactions, dependencies, triggers, and constraints should be documented as well.

The experienced systems engineer will also be able to assist the stakeholders with identifying needs associated with human factors, reliability, availability, maintainability, safety, security, resilience, environmental, logistics, cybersecurity, software, training, habitability, and other constraints. And the systems engineer will also be able to assist in the identification of thresholds that define the expected ability to tolerate disruptions, lack of access, wait times, repair times, and other unproductive system time frames.

A simple spreadsheet can be used to document a summary of the stories, each with its own unique identification that maps to the text documents, which capture the steps in the sequence in detail. It is often useful to apply visualizations in the form of diagrams and drawings to capture stakeholder ideas. Using figures, circles, boxes, and arrows, a clear appreciation of the activities can emerge. If this method is used, the systems engineer should be careful that each node in the diagram is an action (represented by verbs, not nouns, adjectives, or adverbs) and should denote activity interaction, direction, and sequence of events. This diagramming method is common in process definition and process reengineering in order to identify streamlining or waste-reduction opportunities.

4.2 Solution-Focused Unique Processes

A clear definition of the expectations of the users through the application of complementary and unique stakeholder-focused processes provides a solid basis for developing and creating the solution. In this section, the Complex Systems Methodology solution-focused unique processes are added that, when implemented along with the solution-focused complementary processes, will provide all the necessary structure important in a project. As described in Chapter 3, both project management and systems engineering have roles to play in the evolution of the solution that will ultimately meet the stakeholders' needs. As defined in Section 4.1, these following solution-focused unique processes will:

- provide the additional activities that will define the complete solution set expected to meet the stakeholders' needs
- ensure that the cost and schedule of the project are effectively managed using the appropriate methodologies determined by the key stakeholders
- identify and secure the human resources needed by the project and manage them effectively throughout the project life cycle
- add the processes for quality management and measures that test, verify, and validate the development, production/construction, and operations of the system
- provide the risk-management processes that obtain non-project resources to address risks and the risk-reduction processes used for demonstration and decision making

All of the complementary and unique processes and tools associated with management of the project management and systems engineering activities are shown in Figure 4.3. The baselines of scope, cost, and budget are presented in bold.

Needs + Mission Definition + Measures of Effectiveness = Validated System

Multi-Disciplinary Technical Solution + Emergent Behavior = Verified System

Stakeholder Focused Processes

Complementary Processes & Tools
- Stakeholder Engagement & Register
- Communications
- Technical Stakeholder Management & Requirements
- Validation Criteria including MOE, KPP
- Document Tree
- Reports

Unique Processes & Tools
- Charter, Approval to Work, or Statement of Work
- Issues Resolution & Register or Log
- Escalation Path
- Use Case Development & Studies

Solution Focused Processes

Complementary Processes & Tools
- **Scope Matrix & Baseline**
- WBS
- Resource Management
- SEMP
- Functional & Physical Architecture
- Performance Specifications
- Integration Requirements
- **Schedule Baseline & Management**
- Procurement/Acquisition Management & Agreements/Commitments
- Risk Register & Project/Technical Risk Management Plans
- MOPs, TPMs
- Review Gates
- Change/Configuration Control & Management

Unique Processes & Tools
- Stakeholder Needs Analysis & Stories/Drawings
- Interface Documents & Management
- Scheduling Methodology
- **Cost Baseline & Management**
- Project Staffing & Management
- TEMP
- Requirements Verification & Validation Matrix
- Technical Risk Reduction Methods & Trade Studies, Modeling & Simulation, Prototypes

Figure 4.3   Solution-Focused Overarching Processes and Tools

4.2.1 Scope Management

In Chapter 3, the complementary processes associated with project scope management were outlined, including the documentation, baseline, and configuration control activities required to control the scope against change, which is a natural byproduct of life cycle evolution. The project manager, in collaboration with the systems engineer, is typically responsible for clarifying the boundaries of those activities inside and outside of the project scope and for ensuring that the scope is controlled. This section identifies the specific activities that are unique to scope management—in this case, those of the systems engineer within the technical scope management process area.

4.2.1.1 Technical Scope Management

The clarification of boundaries associated with the technical scope of the project is a critical task. And as part of that task, understanding and documenting the interface requirements is a necessity. In many cases, projects are severely impacted by unidentified requirements that connect parts of the system together. As the systems engineer, technical scope development includes the activities associated with defining the system. That definition must include all the cross-boundary activities as well. In order to ensure that interface management is considered during the project preparation and planning phase, the systems engineer will implement an interface charter. This charter identifies the stakeholders that will be responsible for identifying, baselining, and controlling change associated with interfaces and is a good first step in reducing risks from poorly executed interfaces. In addition to the charter, the systems engineer will develop an interface management plan that describes the processes that will be used throughout the project life cycle to manage interfaces.

4.2.1.1.1 Interface Management

Throughout the complementary processes of technical scope development, such as in the development of the functional and physical architectures, special attention must be placed on interface management. Interfaces can best be described as connections between two or more parts of a defined system. Interfaces can also connect the system to other systems. Identification and definition of technical interfaces are the responsibility of the systems engineer. It is a highly iterative process that starts at a high abstract level and continues through decomposition and elaboration until all interfaces at all levels, including system, product, subsystem, assembly, subassembly, and part levels, have been identified at the physical level. In addition, performance specifications and parameters for each interface must be captured.

The text or spreadsheet documents capturing these interface requirements must be accepted and approved by each owning organizational entity in order to be useful. In addition, when defining interfaces, all owning organizations must be involved in the documentation development to assure that any standards or protocols associated with their knowledge areas are considered. If all parties responsible for the implementation of an interface are not in agreement, the risk increases that necessary connection points may not be designed into a part, causing costly rework downstream.

During functional architecture development, an abstract graphical representation of the boundaries, and all the parts of the system along with its relationships, are identified and form a type of an obligation, junction, or connection that must be considered in the physical design. This architecture can be used to identify the transfer points between one part of the system to another part. An example of an interface in a functional architecture, using the wind turbine from Chapter 3, Figure 3.7 (page 79), might be the type of connection required by which the tower is seated on the ground. Although not originally specified as a need by the customer, this example would be a critical interface to consider.

As described in Chapter 3, the physical architecture defines form (shape), fit (tolerances for associated spaces), and function (predictable and stable behavior), and this is true for the interfaces as well. During the physical architecture development, the functional architecture and known interface requirements are converted into a design solution. This design solution must fully describe and document the requirements so that the design engineers are clear on what is needed. An effective practice in the management of interfaces during design and development is in the use of interface control documents. These contain information that should include the following information:

- associated use case, story, or scenario
- expected action of the interface (what action is it to do?)
- assumptions and constraints (e.g., security requirements, access, etc.)
- interface functional requirements
- interface physical requirements
- input of the interface (or one end of the part of the system to be joined)
- output of the interface (or one end of the part of the system to be joined)
- specifications and/or parameters (including specialty engineering requirements)
- owning organization(s)
- verification technique
- validation activity

An approved and signed version of each interface control document is put under configuration control and carefully managed for change and configuration. Modifications to any interface control document may have significant design impacts that can both be costly and drive unacceptable schedule slips. These interface control documents become part of the information that progresses into the subsequent life cycle phases.

A common technique used to assure the completeness of the list of identified interfaces is to develop an N^2 chart.[1] This technique uses a basic matrix framework to tabulate functional and physical interfaces and to analyze those relationships. For functional interface analysis, the most straightforward application of this technique is to draw a grid of rows and columns and to place each known action on the chart diagonally. An input to that action, the trigger, is written into the box to the left, and an output to the action is written into the box above. Blank spaces should be left if there is not an obvious particular interface. Following a standard clockwise pattern, by analyzing every square so that all functions have been compared to all other functions, the full set of interdependencies should be captured. Each interface that is identified should be noted on a spreadsheet with a unique identifier. These interface functional definitions will be further decomposed during the physical architecture definition and then formalized in the interface control documents.

For physical interfaces, a similar approach can be used to confirm that the interface control documents capture each of the key physical interfaces. A matrix is drawn, and, depending on the level of decomposition to review (systems, products, subsystems, assemblies, subassembly, and parts), each one is given a unique identifier and placed on its own line along the left. In the exact order, each one is copied to the column header. An analysis is completed where each line is assessed against each header to determine if there is an interface requirement. If so, an interface control document unique identifier is noted in the intersecting box. Using the simple example of a wind turbine, an example of these types of N^2 charts is shown in Figure 4.4.

				Blades	Nose	Rotor	Generator	Tower
	Input			Blades		A1			
Transfer Wind to Blades	Interface: through Nose			Nose	A1		A2		
Interface: through Nose	Transfer Blade Movement to Rotor	Interface: through Rotor Wiring		Rotor		A2		A3	
Interface: Emergency Shutdown	Interface: through Rotor Wiring	Transfer Energy to Generator		Generator			A3		A4
		Output		Tower				A4	

Figure 4.4   Example of N^2 Charts

4.2.2 Schedule Management

As discussed in Chapter 3, all projects will have an interdisciplinary overarching schedule that is aligned to a work breakdown structure, includes all major tasks, has resources assigned to each activity, is time bounded, is sequenced, and has a critical path defined. This schedule will be managed throughout the project life cycle and is controlled tightly for changes to ensure that the final outcomes of the project are achieved during the expected date. This schedule is inclusive of the technical tasks and is the responsibility of the project manager to develop and maintain. The determination of which project schedule methodology is used is a unique process to project management and can be rigorous or lightweight, depending on the desires of the key stakeholders. The method to use will be decided by the project manager based on the needs of the project and the performance monitoring and reporting requirements of the stakeholders. The project manager will work closely with the systems engineer to determine the required schedule contingency for the technical areas, and he or she also has the responsibility to set the parameters for schedule contingency or float within the overall project schedule. According to the Association for Project Management (APM)[2] "Schedules are presented in many different ways in order to suit the circumstances." Four types are shown below:

- simple broad schedule with well-understood and planned outcomes
- simple broad schedule with ill-defined or sequence-driven outcomes (one output leading to the next input)
- complex or detailed schedule with well-understood and planned outcomes
- complex or detailed schedule with ill-defined or sequence-driven outcomes

A simple broad schedule might have high-level tasks that will be used to produce something that has little risk, such as a production item that has been made many times in the past. In this case, extensive levels of detail would not be needed, and the planned outcomes would be expected to be easy to attain without change to the initial schedule. In contrast, a simple high-level schedule that has ill-defined or sequential outcomes, such as a basic website software development, might not need detailed tasks identified, but it could require more scheduling activities throughout the life cycle.

A complex or detailed schedule would be needed for projects that have tasks distributed through many departments, divisions, groups, or organizations. Or, for example, where the tasks require many intricate steps that must be performed in a specific order. If the outcomes are well understood, and as long as the tasks are captured and correctly documented, the schedule can be managed without much change. If the outcomes are ill defined, both the tasks and the schedule itself may be modified throughout the project, requiring much more rigorous schedule management. The types of scheduling methods that can address the needs are described in the next section.

4.2.2.1 Project Schedule Management

Although there are different ways to approach the schedule development and control, standard practice dictates that there are two common types of scheduling methods that are deployed and utilized in projects of all sizes and complexity:

- Standard Reporting of Schedule/Duration (with a spreadsheet or a graphical representation such as a Gantt Chart)—comparisons between planned labor, materials, and services with actual labor, materials and services during planned intervals (can be full duration or rolling wave)
- Earned value management (EVM)[3]—a tool used to forecast activities, which compares planned value, based on a budget allocation and an earning rule on how to receive "credit" for work performed and budgets spent with "credited" work and spending

Each method has its positives and negatives. For example, earned value (EV) can be used to compile many different indices that are used to measure the health of the project and make projections about future performance. Each method relies on the data compiled behind it for any accuracy in assessing the current situation or in forecasting. The most important concept to consider is that rigorous approaches in schedule management, such as earned value, can work on any project but will work most effectively on projects with well-known and stable scope. There must also be a well-established and accepted reporting structure, meaning the organization is set up and functioning well in collecting accurate data on labor hours worked and materials and services spent at a detailed level. Data for earned value are collected at the control accounts of the work breakdown structure, one level above the work package level, and therefore that level of tracking must be supported in the organization. If any of these preconditions are not met, attempting to implement earned value will increase the risk to the project and will most likely negatively affect the ability to accurately assess the schedule and budget.

To provide context as to what each method looks like, Figure 4.5 shows a basic example of standard schedule reporting, and Figure 4.6 shows a basic example of earned value reporting. These can vary widely based on the project. In the example of standard schedule reporting, key pieces of information include the work breakdown structure work package being tracked, the duration of each work task, start and end dates, predecessor and successor activities (what must complete prior to and after that activity—necessary for calculating critical path), and the names of the individual(s) who are performing the work. In a standard scheduling representation, time is shown using interconnected bars. There are many variations of this type of report available for the project manager to use so that it is clear what activities are being completed during the expected time frame and by the individuals committed to complete the tasks. Details on what is included in a project schedule are discussed in Chapter 3.

Figure 4.6 shows an example of a simple earned value graph of schedule variance (SV) using Microsoft® Project® software. Time is shown along the x-axis and value along the y-axis. Planned value (PV) is the scheduled work that is expected to be complete at each time block within the budget that has been allocated. The actual value of completed work (actual cost, AC) is a calculation of the actual work completed within the allocated budget. Therefore, both the costs and the allocated work by the control account feeds into the calculation. Earned value is the percentage value actually completed as compared to the expected amount completed.

With this information, indices can be calculated that provide information in a snapshot about positive or negative project performance, allowing the project manager to make effective decisions throughout the life cycle of the project. As mentioned previously, using an earned value methodology requires accurate data to the level of the control accounts that are being tracked. Many project managers have found their projects in trouble due to inaccurate data capture feeding into the earned value process and from an overreliance on the value of information generated by the indices. For example, indications may be that the project is ahead of schedule; however, this could reflect a focus on non-critical-path activities that are being done early. It is important, therefore, to always audit and verify the supporting data with alternative methods to ensure the data integrity and to review the critical path and priority of the work being completed.

Because the schedule and cost are tightly integrated in the earned value method, an example of reporting that would be used is shown in Section 4.2.3, Figure 4.8. The use of both methods requires knowledge and experience in the art of project scheduling that, in detail, exceeds the scope of this book. A subject matter expert on scheduling should be consulted in the development of all but simple project schedules.

For projects that are using flexible methods, earned value can be used effectively as well. The key metric in these projects is to have common agreement to what is complete in the way of features finished per plan for the timeframe that has been established. It is not valid to measure velocity of the work or how much work is being completed, but must be focused on completion of a useable feature. To mitigate risk, a project manager with experience in implementing flexible project management using earned value should be engaged in incorporating the methodology. As with earned value for traditional projects, earned value, planned value, and actual cost hold the same definitions.

Task Name	Duration	Start	Finish	Predecessor	Names
Wind Turbine					
◢ Wind Turbine	313 days	Fri 9/22/	Tue 12/4/		
◢ Structure	81 days	Fri 9/22/	Fri 1/12/1		
Main Shaft	24 days	Fri 9/22/	Wed 10/2		J. Smith
Main Frame	15 days	Thu 10/2	Wed 11/ 4		C. Akba
Tower	37 days	Thu 11/1	Fri 1/5/1 5		B. Jones
Nacelle Housing	5 days	Mon 1/8	Fri 1/12/ 6		B. Wald
◢ Rotor Blades	160 days	Fri 9/22/	Thu 5/3/1 3		
Blades	22 days	Fri 9/22/	Mon 10/2		J. Jones
Rotor Hub	93 days	Tue 10/2	Thu 3/1/1 9		B. Melho
Rotor Bearings	45 days	Fri 3/2/1	Thu 5/3/1 10		K. Welle
◢ Electronics	262 days	Mon 12/	Tue 12/4/ 8		
◢ Generator	37 days	Mon 12/	Tue 1/23/		J. Billing
Magnets	22 days	Mon 12/	Tue 1/2/1		
Conductor	15 days	Wed 1/3	Tue 1/23/ 14		
Power Converte	95 days	Fri 5/4/1	Thu 9/13/ 13		P. Guen
Transformer	43 days	Fri 9/14/	Tue 11/1: 16		A. Perna
Brake System	15 days	Wed 11/	Tue 12/4/ 17		L. Mann

Figure 4.5 Example of Standard Schedule Reporting

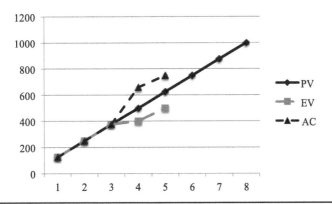

Figure 4.6   Example of Earned Value Reporting (Reproduced with permission from Wingate, L. M. [2014]. *Project Management for Research and Development: Guiding Innovation for Positive R&D Outcomes.* Boca Raton, FL: CRC Press/Taylor & Francis Group.)

4.2.3 Financial Management

The unique processes associated with the application of cost management for a project falls under the responsibility of the project manager. The project manager determines the project budget, procurement, and acquisition methodologies to be used. The project manager is also responsible for setting the cost baseline, determining the proper contingency and reserve levels, developing the overall budget, performing cost estimation and analysis, controlling change within the budget, and managing the performance to the project budget throughout the life of the project. The project manager works with the individual disciplines, including the systems engineer, to obtain cost estimates and the basis-of-estimates (BOEs) that describe how the estimates were made for the overall scope of work. These details are compiled into useful information products that are used to effectively manage the project budget and associated costs.

Standard budget performance tracking can be done using spreadsheets or, if earned value is used, a variety of tracking systems from spreadsheets, all the way to all-inclusive systems can be used. An example of budget performance tracking using a spreadsheet is shown in Figure 4.7.

Included in most budget reports will be the chargeable account codes and a side-by-side comparison between planned and actual labor, materials, and services costs by time. The costs can be as detailed as the work breakdown structure control accounts and usually roll up into summary lines. These reports may include additional information based on the needs of the project manager and those stakeholders who assess budget performance.

An example of an earned value report is shown in Figure 4.8. The numbers along the *x*- and *y*-axes represent time and value. Planned value, actual costs, and earned value, along with variances in cost and schedule, are reflected in this summary chart. In addition, indices are calculated to provide insight into the health of the project. The project manager reviews these at regular intervals so that informed decisions can be made in a timely manner.

Whichever methods are used to identify, allocate, and manage the cost management processes will depend on many factors, including stakeholder desires, legal and regulatory laws or rules, organization directives, etc., and the project manager will collaborate with subject matter experts that have the appropriate expertise to determine which method should be applied.

Project	WBS	(Chargeable Control Account)	Task code	Planned Labor (In Annual $s)	Planned Material & Services (In Annual $s)	Actual Labor $ Quarter 1	Actual Material & Services Quarter 1	Delta (% Spent versus Plan) for Quarter 1
1.0 Wind Farm Project			100000	$ 2,950.00	$ 12,497.00	$ 632.00	$ 3,497.00	94%
	1.1	Structure (Control Account (Summary Costs)	111000	$ 950.00	$ 6,118.00	$ 237.00	$ 1,507.00	101%
	1.1.1	Main Shaft	110001	$ 150.00	$ 798.00	$ 35.00	$ 250.00	83%
	1.1.2	Main Frame	110002	$ 275.00	$ 1,050.00	$ 45.00	$ 200.00	135%
	1.1.3	Tower	110003	$ 195.00	$ 4,050.00	$ 57.00	$ 1,000.00	100%
	1.1.4	Nacelle Housing	110004	$ 330.00	$ 220.00	$ 100.00	$ 57.00	88%
	1.2.	Rotor Blades (Control Account (Summary Costs)	120000	$ 800.00	$ 3,699.00	$ 165.00	$ 1,330.00	75%
	1.2.1	Blades	120001	$ 150.00	$ 2,050.00	$ 45.00	$ 680.00	76%
	1.2.2	Rotor Hub	120002	$ 200.00	$ 1,099.00	$ 20.00	$ 550.00	57%
	1.2.3	Rotor Bearings	120003	$ 450.00	$ 550.00	$ 100.00	$ 100.00	125%
	1.3	Electronics (Control Account (Summary Costs)	130000	$ 1,200.00	$ 2,680.00	$ 230.00	$ 660.00	109%
	1.3.1	Generator	130001	$ 650.00	$ 1,050.00	$ 120.00	$ 50.00	250%
	1.3.1.1	Magnets	131001	Summed Above	Summed Above	Summed Above	Summed Above	
	1.3.1.2	Conductor	131002	Summed Above	Summed Above	Summed Above	Summed Above	
	1.3.2	Power Converter	132001	$ 200.00	$ 50.00	$ 45.00	$ 10.00	114%
	1.3.3	Transformer	133001	$ 175.00	$ 680.00	$ 40.00	$ 250.00	74%
	1.3.4	Brake System	134001	$ 175.00	$ 900.00	$ 25.00	$ 350.00	72%

Figure 4.7 Example of Budget Tracking Spreadsheet

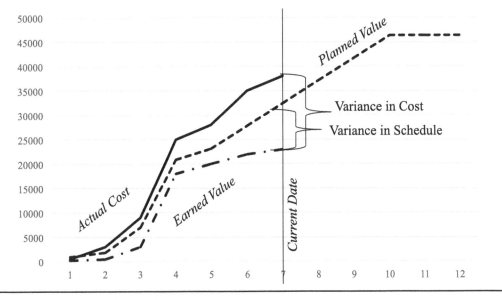

Figure 4.8   Example of Earned Value Reporting

In addition to budget implementation, cost analysis, and forecasting, the project manager is responsible for those cost-related activities that involve external entities. For example, the project manager works closely with the systems engineer to manage and fund risks from the budget contingency. In some cases, the project manager needs to obtain funding from outside of the project to fund required risk mitigations. In the areas of procurement and acquisition management, the project manager is responsible for the overall process implementation and for the performance of any external agreements. However, in many large projects, other organizational departments and divisions, such as a contracts and procurement department, may play a major role in the management of these activities as well. The project manager will form collaborative relationships with these entities to ensure that the needs of the project are fully met. A report showing the budget and schedule performance of all external entities in binding agreements to the project must be compiled regularly by the project manager and rolled into the overall project cost management reports.

4.2.4 Human Resource Management

As part of the Complex Systems Methodology complementary processes described in Chapter 3, the project manager, in collaboration with the systems engineer, identifies the required talent and aligns them to the work packages. This section describes the unique human resources management activities that are performed by the project manager. The project manager has the responsibility for identifying, recruiting, training, managing, and retaining the key project staff that will manage the project throughout the life cycle and for negotiating with other departments or divisions for all of the multidisciplinary labor and time allocations required by the project. Each organization typically has its own processes that the project manager must follow, but within those constraints is also the opportunity to organize, develop, and manage the team in a manner that optimizes the skills of the team members. A talented project manager will be able to optimally align the project's workforce skill set to the work.

4.2.4.1 Project Human Resources Management

The human resources required to successfully complete a project include not only individuals skilled in project management methods and processes, but also those with specialty skills that are integrated and part of the project management team such as systems engineers, budget managers, schedulers, and contracts and procurement managers. In addition, it includes management of the activities of project participants identified during the scope definition phase of the project. For example, all the individuals that are identified within the resource-loaded integrated master schedule need to be considered within the human resources management responsibilities of the project. It is important to consider that the staff with project responsibilities may report to various departments and divisions throughout the organization(s) involved in the project and may not report directly to the project manager.

The human resource responsibilities of the project manager depend on the organizational model within which the project resides. They also depend on the project size and complexity and the project organization itself. For example, a project may have some direct hired staff, some matrix managed full-time staff, some matrix managed part-time staff, and co-located partners and suppliers. The project may even have some staff that are remotely co-located at a partner or contractor site. Each of these scenarios may require a different method of management and provide both benefits and risks to the project.

The easiest management arrangement for the project manager is to direct hire and maintain the skills that are needed to perform all required human resources tasks. In matrix management situations, a project manager may have staff dedicated to performing activities on behalf of the project but may or may not have direct supervision responsibilities. The supervisory responsibilities for those staff are maintained in another division, department, or organization, while the day-to-day direction of the staff is the responsibility of the project manager. In another model in which staff members are identified to perform work for the project, but where both the supervisory and the day-to-day responsibilities are held by another organization, the project manager must manage by influence rather than by direct authority. This approach becomes riskier and more difficult as more non-project activities are required by fully matrixed staff members. This could take the form of functional responsibilities that are assigned or other project work that is assigned to the same staff member. Either arrangement can work well, as long as both the project manager and the manager in the other division are in agreement as to the scope of work, the timing of the work, and the performance of the staff member.

When the staff member works on various projects and is only allocated a percentage of time on the project, the complexity compounds for both the matrix and project manager who must then carefully monitor the individual's performance on the agreed-upon tasks to assure that the appropriate priority is being placed on the work and that the individual's effort is not being diverted. Regardless of the form that the project organization takes, if there are shared or matrixed resources involved, standard practice suggests that any agreement should be documented and signed by both parties such that everyone is clear on the expectations and the performance can be monitored accordingly.

Human resources activities for hiring direct project staff often include recruitment, hiring, retention, staff reductions, training, and performance management. Other activities include setting objectives for the team, organizing the work according to the project schedule, assessing the completeness of the work performed, and performing all additional supervision tasks. Within a project, it is the project manager's responsibility to employ and continually motivate a diverse and capable team that can perform appropriately across all required parts of the project. As projects evolve along the life cycle, the team requires constant rebalancing. The addition and reduction of skill sets to meet the requirements of the project in each phase of the life cycle must be considered and addressed for optimal performance. In small, simple projects that only have direct project hires, this might be quite easy. In large and/or complex projects it can be challenging, as multiple human resources models may be employed at the same time. Human resources management is a discipline in itself, and a subject matter expert should be consulted on all but the simplest of projects.

Although the project manager is responsible for management tasks, he or she also holds a position of leadership. The project manager is responsible for high-level and overarching processes that require the successful performance of many moving parts. Effective leadership is usually demonstrated by creativity, dedication, and drive and the ability to influence and compel individuals to act in ways that benefit the overall needs of the project. Effective leaders have refined communications skills and are able to operate throughout many levels of the organization. They are able to motivate a team and derive the needed project performance from all team members, regardless of their organizational alignment or physical locations.

A common and effective way to address the human resource needs of the project and assure that a solid approach is implemented is to create a workforce management plan. This plan is typically a text document that describes the processes that will be used to manage the project staff. It includes descriptions of all of the project's unique human resource processes that will be used. It should not document any currently existing organizational human resource processes, but it should reference those organizational processes whenever appropriate. The plan also identifies the needed skills over all phases of the project. The level of effort required to develop a workforce management plan depends on the size and complexity of the project. Methods to tailor processes so that only the appropriate amount of process rigor is applied is discussed in Chapter 6.

4.2.5 Risk Management

As discussed in Chapter 3, the Complex Systems Methodology complementary processes associated with risk management include risk and opportunity management performed by the project manager, as well as technical risk reduction performed by the systems engineer. In this next section, the unique systems engineering tasks associated with technical risk reduction are explored. These activities include trade studies, prototypes, modeling, and simulation. Each of these techniques can provide information on alternatives, design, and operations impacts before the full design is finalized.

4.2.5.1 Systems Engineering Technical Risk Management

A systems engineer's approach to technical risk reduction depends on the risk identified within the technical life cycle. Some risks may emerge through the natural design activities during the project. A good starting point for the systems engineer is to review each expected product that is to be produced as an outcome of the project. A well-defined work breakdown structure should align to each of the products.

Technical risk and opportunity are generally higher in early design phases and are reduced further along in the life cycle. To determine the level of technological maturity, each product can be assessed to determine where it currently resides from a life cycle perspective. If the product requires design and development, it should be expected that it is technologically immature and will require specific risk-reduction activities. Risk-reduction strategies such as prototyping, trade studies, and modeling and simulation techniques can provide useful insight, allowing good design decision making and early actions in the design process.

4.2.5.1.1 Prototyping

Prototyping is a useful method that produces a representation of a system, or one of its elements, in a form that demonstrates its form, fit, or function. It is a tangible thing that can take the form of a physical scale model or a software model that has been developed for illustrative purposes. A prototype is only meant to demonstrate what the system *could* be, not what the final form *will* be. Prototyping provides valuable information on how the system will look and behave prior to significant investment in labor or costs. The most useful prototypes for risk reduction serve as a confirmation as to the intent of the design. They may also serve as test beds for trying out alternatives designs. When stakeholders can see the design come to life, clarity around what their expectations are usually follows.

4.2.5.1.2 Trade Studies

When the solution to a stakeholder requirement can be fulfilled in multiple ways, it is important to assess the behavior of the options on the parameters and to investigate which option provides optimal performance. Optimal does not necessarily mean the best technical performance, because all standard project management constraints still apply—that is, cost, schedule, and overall scope—and the solution must still fit within those constraints. The trade study will focus on a constrained set of parameters to evaluate design alternatives, although every part of the system, including each of the sub-elements, can have trade studies performed on them, including hardware, software, operations, maintenance, integration, processes, etc. The study can be performed to reduce risk in any area of the life cycle in addition to the technical risk, such as operations, costs, reliability, maintainability, and logistical support, to name a few.

The risk reduction strategy associated with trade studies is in using a methodical analysis to provide visibility into options, allowing a solution to be chosen between two or more potentials. Trade studies that address measures of effectiveness and key performance parameters should be prioritized, because there will not be infinite funding or time to complete these studies. A trade study relies on research and statistical analysis. A subject matter expert in operations research or statistical analysis can provide valuable assistance in assuring that the setup of the trade space that will be analyzed is valid and that the results are interpreted correctly.

A trade study will be documented using a report format. It will include a summary objective and will state any constraints, criteria, assumptions that are known. Also included will be a description of the method that will be used to perform the sensitivity analysis (or what-if analysis), and the parameters to be investigated. The results will be documented, and the subject matter expert will make a recommendation. Impacts to cost, schedule, and scope will be evaluated. The recommendations from the trade studies are reviewed by stakeholders and, if the solution is preferred, will be brought into the change control processes and, with full participation by the individuals responsible for the work breakdown structure for which the trade study is done, presented for inclusion. If approved through the change management process, the change will be implemented into the project.

4.2.5.1.3 Modeling and Simulation

Modeling and simulation are methods of using both representative model and activity simulations to demonstrate the behavior of a system or elements of that system. Modeling and simulation techniques provide a way to demonstrate how the system may look or behave under different conditions. This is a significant risk reduction strategy, because how a system engages and interacts within two or more elements, all the way through the full system interactions with its operational environment, are often unknown until each element is integrated and/or the whole system is deployed. Having a model that can be used to run simulations will provide a platform upon which insights into patterns of straightforward and emergent behavior are brought to light and where the system and all its elements can be optimized.

There is also a process called *model-based systems engineering* (MBSE) that relies on the modeling and simulation concept to reduce risk in systems, particularly complex systems. In order for model-based systems engineering to work, it relies on a digital representation, or model, of the whole system in its allocated form with complete traceability. This hierarchical arrangement of data is kept in a single, well-structured database using a generic or common language across all disciplines associated with the project. The use of visualization software is an important attribute of the capability, which provides graphical representations of multiple views of the system, including the functional and physical interfaces, and process. The use of model-based systems engineering has the potential to provide great benefit to the project, in that it can provide an automated way to validate assumptions, assess change impacts, identify interface or interoperability issues, and provide confirmation as to how the fully synthesized system will perform in different environmental conditions. This acts to significantly reduce the risk of the project early on in the life cycle. In order to use model-based systems engineering, the entire project must be willing and able to provide information that must be kept updated and current, and configuration control must be exceptionally strong.

4.2.6 Quality Management

Technical quality management is a domain expertise of the systems engineer, although overall project quality is the responsibility of the project manager. For the project, quality is both quantitatively (as a specification or performance parameter) and qualitatively (as a value) assessed by the stakeholders and the project team. Measurements of quality can provide a basis from which to verify results and validate with the stakeholders that their needs are met. The measures that were identified in the complementary stakeholder-focused processes (measures of effectiveness, key performance parameters) and those identified in the complementary solution-focused processes (measures of performance, technical performance measures) are used to assess quality during a project and can be used throughout the life cycle of the system. Chapter 5 provides more information on the complete set of measures that can effectively be used to assess overall project performance.

This next section discusses the systems engineering tasks associated with identifying the testable elements of a project, technically verifying the performance, and then validating that performance with the stakeholders so that the fitness or ability of the system to meet the intended purpose meets the intended quality.

4.2.6.1 Systems Engineering Verification

Systems engineering verification is the process that confirms that every system element that is being developed, produced, purchased, and operated has process steps implemented to test and verify that the performance of that system element meets the necessary specifications, both independently of the system, and also when integrated into the system. These process steps can also be used to test for emergent behavior in the system as the capabilities from each of the independent elements are brought together. The primary activities that will provide the highest probability for successful verification of all stakeholders' requirements are the complete capture and documentation of all allocated requirements (requirements verification and validation matrix) and their position within the system (traceability).

A simple spreadsheet or a software tool can be used to document the traceability of the requirements as described in Chapter 2, plus the other information needed to document the verification and validation activities. Often referred to as a requirements traceability matrix, verification matrix, or a combination of both, this spreadsheet provides a comprehensive look at the technical requirements of the project, where they originate, how they decompose, and the performance to specifications or parameters that have been established. This is also where the project's performance is documented after tests. In complex projects, using a software tool may be the only reasonable way to demonstrate the traceability and provide additional functionality for assessing change impacts on the design.

Complicated statements using a variety of descriptors make the ability to test difficult if not impossible. Therefore, testable requirements that include definitive language without ambiguity and simple

Project	WBS		Test Case Scenario	Test ID	Verification Status	Status
1.0 Wind Farm Project						
	1.1	Structure	Structure Fatigue Testing	1.1.0001	Pass–Within Tolerance Levels	
			Bearing Axial Stress Test	1.1.0002	Fail–Outside Tolerances	Perform Mitigation, Reschedule Test
			Drivetrain Rotor Loads Stress Test	1.1.0003	Pass–Within Tolerance Levels	
	1.2.	Rotor Blades	Blade Stress Tolerances Fatigue Test	1.2.0001	Pass–Within Tolerance Levels	
	1.3	Electronics	Braking Tolerances Performance Test	1.3.0001	Pass–Within Tolerance Levels	
			Voltage Profiles Performance Test	1.3.0002	Pass–Within Tolerance Levels	

Figure 4.9   Example of Requirements Verification Matrix

statements of intent are the best. A simple example of a requirements verification matrix is shown in Figure 4.9.

For each requirement, the specifications and the intended verification activities are documented and form the basis for the tests so that the performance of the system, product, subsystem, assembly, subassembly, and part levels can be demonstrated to be repeatable and stable, both alone and in combination with the next level of integration. Testing provides the answer to four pertinent questions:

1. Does the element work as intended and expected?
2. Does the element conform to the specifications?
3. Does the element work in an integrated manner as expected?
4. Has undesirable emergent behavior been exposed?

Each part, subassembly, assembly, subsystem, product, as appropriate, and the entire system should be tested and verified before being accepted into the operations phase. As was discussed in Section 4.2.5.1.3, all of these individually and as a whole could be tested in a simulated environment. However, that step is often not implemented and may not be necessary, depending on the system.

Tests should include interfaces, effectiveness, compatibility, reliability, maintainability, producability, and any other parameter that is determined appropriate to verify form, fit, and function. Testing can be physical or simulated. Although each project has unique test, verification, and validation requirements, a basic structure for performing these activities in a manner that offers a higher probability of reaching the desired outcomes of the project is essential. This can be attained by following these steps:

- **Step 1:** Develop test cases (experiments) that can provide the explicit steps to demonstrate that the desired performance is achieved for each requirement.
- **Step 2:** Develop a test and evaluation plan that describes the processes that will be implemented to complete the information on the requirements verification and validation matrix.
- **Step 3:** Set up and document a test environment that will ensure valid and repeatable test results.
- **Step 4:** Perform and document the test and results.
- **Step 5:** Provide a process for resolving discrepancy, deficiencies, or poor test results.
- **Step 6:** Communicate results and update project documents if required.

As was described in Chapter 2, the development of a comprehensive test plan, often referred to as a test and evaluation master plan, helps describe how the evaluation of all subordinate elements will occur on what timescale and under what conditions. It also describes the system integration verification activities that demonstrate the performance of the combined system elements. The test and evaluation master plan identifies the requirements and procedures for the test and identifies any specific test equipment, information technology, infrastructure, and/or facilities that are needed.

Once the element(s) and system are verified, the project manager will bring the stakeholders together to validate that the system performance meets their needs. If communication with the stakeholders has been sufficient throughout the project life cycle, this is generally a straightforward final step.

4.3 Case Study: Jill and Julia—Singers, Songwriters, Musicians

All case study quotes are from interviews held with the individuals presenting the background. In this next case, the interview was held with Jill and Julia, songwriters, and all quotes are attributed to them.

Jill and Julia are sisters who write music and perform as a duo in musical venues.[4] Both play piano and guitar, sing, and are involved in the songwriting. Starting in 2011, when they decided to pursue a musical career, they have performed at county fairs, music festivals, sporting events, Las Vegas casinos,[5]

Nashville, West Coast, and on a promotional radio tour. Their musical genre is unique and can be generally categorized as Alternative, Alternative Country, Folk, or Indie.

In 2013, they obtained their first nonexclusive recording contract with an independent record label, Lamon Records.[6] There in Nashville, they put together their first professionally recorded extended play (EP) album, entitled *Jill and Julia*. It included four songs that showcased their songwriting and playing abilities.[7] One of their singles from the album, titled Wildfire, was the focus of the promotional radio tour. In 2015, the pair released their first full-length album, entitled *Cursed*.

4.3.1 Project Charter and Plan

When Jill and Julia decided to go into the business of songwriting and performing, they did not start out with a specific plan. They knew that they enjoyed music, which they had been exposed to throughout their lives. Music had been part of their experiences growing up, and included piano, voice, and guitar lessons for both. However, they did realize that being able to make a business out of their talents would be a desirable outcome. As Julia explained, "It is natural to look after yourself, being able to afford a roof over your head, paying bills, etc. But it would be nice to be successful doing things we enjoy."

The pursuit of music as a career began to take shape when a music instructor suggested they start playing shows, obviously identifying their unique sound and talent. They took the advice and found that they were able to achieve some success in the market. Their approach to enhancing their musical career was dependent on their own marketing ability. They would focus on writing their songs, rehearsals, and performing at venues that they could book. As Julia explains, "We were able to book shows right away, but we really didn't know much about it, and soon we were booking too many small venues and it was becoming exhausting. So we decided that we wanted to go for larger-quality venues, so started following that path."

Their music instructor then provided them with an introduction to his producer, who created a three-track demo recording for their use in pursuit of a formal contract with a known label. As Julia explained, "At that point we spent hours putting together mailings, primarily in the Nashville market, where there are many independent labels; however, we generally did not get any responses. Most labels will not accept unsolicited submissions, so we were fighting to make it in the door." However, the effort did eventually pay off when they were recognized by the Lamont record label and offered an EP deal. An EP is a compilation of four to six original songs, often created for promotional use. Jill explains, "This offer provided promotion, marketing, and manufacturing in exchange for a one-year exclusivity contract for the production side." This was the break that they were striving to attain.

Once they signed a contract with Lamont, they had to meet specific requirements and schedules, such as being available in Nashville for recording during a particular time frame and having 10 songs ready to record. This meant that they would need to write those songs within a specified time—often difficult to do with a creative process.

4.3.2 Stakeholder and Communications Management

Jill and Julia consider their satisfaction with their music as the most important factor to consider. However, they also understand that the record label and their paying customers are important stakeholders that must be considered. As Jill explains, "Early on we first tailored our music to what we thought people would be interested in. However, as we learned, we felt that we were achieving a higher level of art by following our own pattern. So in the next album we did what felt right. It has worked in our favor."

4.3.3 Scope Management

For the projects that Jill and Julia write songs for, they have a number of songs that need to be included, whether it is a demo, an EP, or a full-length album. They look for a "cohesive theme," explained Jill. "We wanted it to be transporting, unique, and to represent our sound so that people would know it was ours when they heard it." Julia elaborated, "We identify our conceptual album with a theme behind it, and then write the songs that fit in to the concept and theme." Both Jill and Julia agreed that they "knew after they finished each song if it was right for the album."

4.3.3.1 Architectural Design

When Jill and Julia are contemplating a new songwriting project, they often get inspiration from other artists and from various disciplines. As Jill explained, "We get inspiration from songs, films, or other media. We may try a new beat that we'd heard, or will model a song in the same key that we found interesting." They often have a theme in mind, as they did with their album *Cursed,* and they tailor the songs to match the theme.

It is important to Jill and Julia that they remain true to who they are. "We want our style and what we are imagining for the song to come through strongly," Julia says. They co-write all of the songs, and they play off each other's strong suits. For example, Julia will generally handle the lyrics, while Jill handles the harmonics and melodies. As they are working through the song—"If we don't like it, or if it is not 'catchy'—then we will set it aside and move on," Julia explains. If the song seems to be coming together in a positive way, they will continue working on it until they are in agreement that it is complete and ready for the album. Jill reiterates, "We will not move forward with a song if we are not in agreement."

4.3.3.2 Requirements Management

The requirements for the songs that Jill and Julia write are based on their need for the song to "authentically" reflect who they are. The songs should also fit the album theme that they have chosen. If they are working on an album for the label, then there could be requirements associated with length or style. And if they are playing at a venue, there may also be requirements associated with the number of songs, the length of time spent on the stage, or the need to bring their own amplifier equipment.

Regardless of the venue, there is always a need for high-quality musical instruments and sound equipment, which are quite expensive. Quality guitars, for example, that would be considered if a purchase was required, are generally over $1,000 USD. Lesser-quality instruments and equipment would not be considered because of the risk to the musical outcome.

4.3.3.3 Integration and Interface

In the course of managing their business of songwriting, recording, and performing, there are many interface and integration considerations. They do try to control as much as they can, because the results and outcomes of their activities have a direct impact on their reputation and name recognition as musicians and performers. Activities that most other types of business could outsource, because of the risks, cannot be accomplished by anyone except Jill and Julia. As Julia says, "We micromanage everything, because in this business you do not often get a second chance at the impression you make. And the results of that impression are long lasting."

One significant change that Jill and Julia made based on the perspective that impressions are long lasting, is that they always wear black now when they play venues. At one point a videographer suggested they both wear white. Jill exclaims, "We hated it! We knew we wanted to look like a cohesive duo, but white wasn't going to do it for us. So now we always show up in black. Experience has taught us that we want our audience to focus on our music, and not be distracted by a style or color of clothing."

For smaller venues, they often bring their own equipment, so they must be able to figure out how to connect to all the supporting electronics and equipment themselves. During performances at various larger venues, they often have to rely on production crews and sound engineers for setting up the sound systems, so they need to know enough about the equipment to provide guidance to the engineers so they get the results they need. During the professional recording studio experience, they needed to place their trust in the producer, engineers, and the supporting musicians. "If you can't articulate what you need from the engineer, you won't gain their respect, and you may not get the setup that you need to make your session successful, either at a venue or in the recording studio," Julia explained.

Although they controlled their own performance and would choose the songs that they would perform, the recording studio controlled everything else. During graphic design and photography sessions, the label had professionals who were responsible for the artwork. However, for other media exposure, Jill and Julia always have input into the processes. For example, for their *Las Vegas Weekly* cover shot titled "Cream of the Crop,"[8] Jill and Julia "recreated Herb Alpert's *Tijuana Brass* 1965 album cover," Julia explained. The 1965 album cover featured a model covered in whipped cream. Upon actually posing for their cover and being covered with whipped cream, the shear messiness and stickiness of the whole experience lead them to wonder what they were thinking, even though they were quite happy with the result.

4.3.4 Schedule Management

The most constrained schedule that Jill and Julia had to work with was in preparation for entering the record label's recording studio. "We write music all the time," Jill explained. "However, it is a different experience when you are writing to a deadline." The pressure to come up with appropriate songs is high. As Jill explains, "you are only as good as your talent and your instrument." Julia adds, "People generally won't give you a second chance." Schedules also come into play when they must prepare to play at a venue. The only time they have the luxury of writing without time boundaries is when they are investigating new concepts for songs or writing songs that they might use in the future.

4.3.5 Procurement Management

Jill and Julia, as songwriter musicians, supply a service of performance on a regular basis. This means that they are required to continually negotiate contracts for their performances or recordings. As Julia says, "We are selective on what shows we take and which ones we pass on. There are always opportunity costs. If we are covering for another well-known artist, we may not get any significant money from that, but if we perform in a small, private venue we may get a large sum but not the exposure that we need." These tradeoff decisions between obtaining exposure versus obtaining a hefty paycheck are not always easy to make.

"For the most part, we have everything that we need," Jill says. However, if they have to buy a new sound system or a new guitar because of breakage, then that is a very costly purchase. Julia explains, "Due to the fact that we need high-quality guitars, that can cost a significant amount. Since we do not currently have a guitar endorsement, that would have to be paid for by us."

4.3.6 Risks and Opportunity Management

In the business of songwriting and performing, there are a significant number of risks and a great number of opportunities. Julia explains, "A huge part of the music business is being in the right place at the right time, meeting the 'right' people that can move your career along, and jumping on opportunities when they present themselves." There is a low percentage of success in this business, and even if you build up a significant portfolio, are well regarded, and have the credibility, many talented bands do not achieve the level of success that they strive for." Jill expands on that by saying, "You may not get picked up. In so many ways, it is about luck. Even the most talented may not land the recording contract." Therefore, they take a calculated and methodical approach to their business, always making sure they keep an eye on the competition to see what works and what does not, and they always stay ready to take advantage of those opportunities that present themselves.

From the risk perspective, Jill and Julia explain that working your way through the business activities, such as negotiating contracts, writing songs, marketing and branding the business, interacting with their fan base, rehearsing, traveling, and attending meetings to discuss opportunities, all takes an inordinate amount of time. The trade-offs they make in attending to these activities could be spent doing other things such as writing more songs or performing in more venues. As Julia says, "You are risking your time—a lot of energy and time that you hope will pay off in the long run." When they signed with the record label, they were effectively taking themselves off the market for other labels. That is a conscious decision that they made with significant ramifications.

Other risks in this business that Jill and Julia try hard to mitigate include the risks to their public image, to their privacy, and to their safety. Their public image, as well as their privacy in the age of the internet, is difficult to manage. This is a "very public career, where we are putting ourselves out there where we will get critiqued and criticized," Julia says. Fans often do not respect boundaries, and there will be occasions where they cross the line. Information is easily available about where they will be and what they are doing. As Jill explains, "We always need to be on our guard and to make sure that we have some level of security around us so as not to put ourselves at risk."

Photographs often appear in the public domain without their knowledge or agreement, and there is not much that can be done if they are not complementary or do not show them in a positive light. As Jill points out, "We need to be careful with our clothes, our hair, what we say on social media." So they are always aware that they are in the public eye and try to present the image that they wish to have represented photographically. But as Julia affirms, "good or bad . . . the pictures are all published, and that can affect our image."

4.3.7 Quality Management

When it comes to their music and their performances, Jill and Julia know that the quality of the work is critical. Unless they are both comfortable with the material, they do not share it or show it. They are the only checkpoint, so if it gets through their scrutiny, they are comfortable with the results. As Jill and Julia both confirm, "music is subjective. If we are happy with our work, yet others don't like it, then that is just the way it is. We can live with that."

They do however, appreciate that if no one buys their music, comes to their shows, or enjoys their work, then their livelihood is at risk. "We appreciate our fans and understand the critical role they play in our success or failure," Julia says. They have taken a position that they are not interested in focusing their attention on internet success but will continue working with the record label. In order to build a lasting career built on being a respected musician, this appears to be the best path to achieve that goal.

That translates into a strategy that is not built on a transitory fan base, even though that might bring quicker initial returns.

When Jill and Julia are recording in a studio for a record label, quality is assessed by that organization as well. Without experience, it is easy to give up your own views about the quality of the output because there are many other people involved who have a vested interest in the success of the project. As Julia explains, "With our first EP, we had not been exposed to the studio environment, and we did not really know our way around, so in some ways we accidently forfeited our opinions. As we gained experience and understood better exactly how things worked in the studio, we gained confidence and became more comfortable stating our needs and desires. The more experience you have, the more power you have."

4.3.8 Test, Verification, and Validation

Jill and Julia will personally test, verify, and validate that their music conforms to their expectations and do not generally go to others for this confirmation of their music. Because songwriting and performing are creative endeavors, and the music itself represents the artists' style, their own personal satisfaction drives their perspectives of successful outcomes. As Jill says, "We are our own worst critics!" Julia further elaborates, "We feel that we can tell which songs will do well and represent our style. We want to be authentic to our vision. We are very picky about the music we want to produce, and we know exactly how we want to go about that to achieve our vision."

4.3.9 Governance

Dealing with changes to songs, musical scores, venues, and other business decisions, Jill and Julia feel that they are on the same page and have little debate when faced with these types of decisions. They might be at odds with the record label, other promoters, or venues as to what songs to promote or play, but they seem to have an easy way to resolve any controversy between themselves. As Julia explains, "If we cannot come to an agreement about an issue after serious debate, we will just scrap the idea. This is something that seldom happens; however, when it does, we have no concerns about walking away from it."

4.3.10 Outcomes

For Jill and Julia, their outcomes have matched their expectations. They have two albums with a respected record label and a significant fan base with a large social media following. Their music is selling, and they are able to choose the venues that they feel best expose their music to the public, and, based on these past successes, new opportunities continue to present themselves.

4.3.11 Lessons Learned

Everything that Jill and Julia have learned during this initial experience in producing for a record label and performing in some major venues is coming together in a way that is helping them strategize about their future. They have gained significant experience in songwriting, recording, performing, marketing, and selling the outputs of their talents. As Julia relates, "Our future goal is to obtain a recording contract with a major record label. We now have better insight into how the game is played, and we are a bit more savvy. This will help us as we pursue our dream of having a long-lasting career in the music industry."

4.3.12 Case Analysis

This case describes how a creative endeavor, such as occurs within the music industry, may follow standard project management and systems engineering processes. These processes can help guide activities in ways that can lead to optimal outcomes, regardless of the industry. Having visibility into how these activities align into the specific processes for something as evolutionary as songwriting and performing provides a solid background for thinking about these types of projects. This case reflects a strong focus on scope management. And opportunity management plays a large role—one that is often not demonstrated in other industries. As this business continues to evolve, these processes can be considered and applied to assist in the achievement of optimal outcomes.

4.4 Key Point Summary

Chapter 4 focuses on the Complex Systems Methodology unique processes that, when used in combination with the complementary processes, will provide the highest probable positive project outcomes. In the area of stakeholder-focused processes, these include the processes used to initiate work, to resolve and/or escalate issues, and to develop the most robust description of the anticipated needs of the stakeholders, either through use case development or other studies used to define the highest-level operational needs. From the solution-focused view, unique processes focus on developing the complete solution set that will meet the stakeholders' needs. The processes for effectively resourcing and managing the project, assessing quality, and reducing risk complete the overall process for performing project management and systems engineering on a project. Key concepts are outlined below. Key terms are provided in the Glossary.

4.4.1 Key Concepts

- Complex Systems Methodology provides an overarching structure to ensure that the most impactful processes are considered for each project.
- Unique processes from the disciplines of project management and systems engineering include both stakeholder- and solution-focused processes.
- Unique processes join the complementary processes to demonstrate the full spectrum of processes that when applied, greatly increase the probability of achieving project objectives.
- The unique stakeholder-focused processes provide insight into the needs of the stakeholder and how to address their issues throughout the project.
- Unique solution-focused processes ensure that the cost and schedule of the project are effectively managed using the appropriate methodologies determined by the key stakeholders and that the human resources needed by the project are secured and effectively managed. They also identify critical processes associated with quality and interface management, test, verification, and reduction of technical risk.

4.5 Apply Now

Complex Systems Methodology unique processes that have been presented in this chapter can be directly applied to a project that is of interest to the reader. This will result in a greater and deeper understanding of the material. The Apply Now section provides a summary table (see Table 4.1) to help visualize the chapter material, along with questions that can be applied to any project.

Table 4.1 Apply Now Checklist

COMPLEX SYSTEMS METHODOLOGY™ SM (CSM™ SM)
APPLICATION FOR ALL PROJECTS

Questions to Ask	Responses								Governance		
	Stakeholder	Communications/ collaborations	Needs analysis	Scope baseline	Quality	Resources	Physical architecture	Risk	Change control	Configuration control	Integration/ test, verification, validation
Are complementary stakeholder-focused documents/processes in place?	Register/ charter	Communications/ collaboration plan/ issues register	Stakeholder requirements /use cases, concept of operations, scenarios	Validation	MOEs, KPPs						
	Cost baseline	Schedule baseline	Functional architecture	Scope baseline	Quality	Resources	Physical architecture	Risk	Governance		
Are complementary solution-focused documents/processes in place?	Baseline budget/budget methodology	WBS/Baseline schedule/ schedule methodology	Functional architecture diagram/ interface definition	Technical requirements/ baselined scope/interface definition	MOP/ TPM/ reviews/ RVTM	Acquisition/ procurement/ labor, materials, services/ workforce management plan	Performance specifications, integration requirements, icds/design solution	Risk register/ management/ prototypes, modeling & simulation, trade studies	Change control	Configuration control	Integration/ test, verification, validation

COMPLEX SYSTEMS METHODOLOGY
APPLICATION FOR ALL PROJECTS

Questions to Ask	Responses								Governance
	Stakeholder	Communications/ collaborations	Needs analysis	Scope baseline	Quality	Resources	Physical architecture	Risk	
Are complementary stakeholder-focused documents/processes in place?									
	Cost baseline	Schedule baseline	Functional architecture	Scope baseline	Quality	Resources	Physical architecture	Risk	Governance
Are complementary solution-focused documents/processes in place?									

1. *Stakeholder-focused unique processes include the establishment of a project charter.*
   Think of a project that you have completed in the past that could have benefited by a better charter. Write a charter for your project.
2. *Stakeholder-focused unique processes include issues management.*
   Using the same project from Question 1, complete an issues register for that project.
3. *Stakeholder-focused unique processes include the development of use cases, conops, and scenarios to clearly define what the stakeholders need from the project.*
   Thinking of the operations environment in which your project would be employed, write a use case. Do not forget to include the cross-boundary activities as well.
4. *Solution-focused unique processes include the selection of a budget and schedule methodology.*
   Identify the budget and schedule methodology that you would find most helpful for the project you have selected. Develop a basic workforce management plan.
5. *Solution-focused unique processes include activities associated with development of a functional and physical architecture.*
   Using your example above, identify your design solution, including interfaces. Develop a basic N^2 diagram. Describe any prototypes, modeling and simulation, or trade studies that you would use to reduce technical risk on your project.
6. *Solution-focused unique processes include quality and governance activities.*
   Using your example above, build a basic requirements verification matrix, and identify the test and verification activities that would be associated with verifying specifications on the requirements verification matrix. Build a basic test and verification matrix.

References

1. Lano, R. J. (1977). The N^2 Chart. *TRW document #TRW-SS-77-04.* Redondo Beach, CA: TRW, Systems Engineering and Integration Division.
2. APM. (n.d.). "Schedule Management." Accessed January 12, 2018, from https://www.apm.org.uk/body-of-knowledge/delivery/schedule-management
3. Defense Systems Management College. (1997). *Earned Value Management Textbook.* Fort Belvoir, VA: Defense Systems Management College.
4. Jill and Julia. "Jill and Julia." (n.d.). Accessed January 12, 2018, from www.jillandjulia.com.
5. Dauphin, C. (2014, May 29). "615 Spotlight: Jill & Julia Talk 'Wildfire' Single, Vegas Life." *Billboard.* Accessed January 12, 2018, from http://www.billboard.com/articles/columns/the-615/6106201/jill-julia-wildfire-single-las-vegas
6. "Jill and Julia." (n.d.). Accessed January 12, 2018, from https://www.linkedin.com/in/jill-and-julia-6b45ba111/?trk=public-profile-join-page
7. Lamon Records, LRC©. (n.d.). "Sister Duo Jill and Julia Bring Unique Country Sound to Lamon Records." Accessed January 12, 2018, from https://lamonrecords.com/sister-duo-jill-and-julia-bring-unique-country-sound-to-lamon-records
8. *Las Vegas Weekly* Staff. (2015, January 22). "The Next Wave of Vegas Music: 10 Acts to Hear this Year." *Las Vegas Weekly.* Accessed January 12, 2018, from https://lasvegasweekly.com/ae/music/2015/jan/22/next-wave-vegas-music-10-acts-local-band-hear-year/?framing=home-news

Chapter 5

Success Measurements

Chapters 1 through 4 described the measurement processes and tools that are used to actively manage a project with the project management and systems engineering disciplines. Measurements are typically performed incrementally throughout a project, meaning they are assessed regularly at pre-chosen time periods. This chapter focuses on measurements that assist in the active management of the project by using repetitive assessments that evaluate the progress against the baselines and technical parameters that have been set. The processes described in this chapter are associated with identifying and using measurements to ensure successful outcomes as defined by the project manager, the systems engineer, and the key stakeholders. They include:

- reasons for measuring
- ways to measure
- methods to communicate results

Section 5.1 explores the reasoning on why measurements are important and how success is defined. It describes the way that measuring provides evidence throughout a system's life cycle about the planned and actual status and outcomes.

In Section 5.2, the supporting foundation that must be in place in an organization in order to perform accurate and useful measurements is described. Consistent assessment processes must be employed throughout the life cycle. Identifying and implementing useful measurements ensures that a system performs as predicted in each life cycle stage and provides robust decision-making information. Dangers associated with the use of measurements are also described.

There are a significant variety of measurements that can be used, and Section 5.3 focuses on which measurements are generally used for different activities throughout the system's life cycle, from the early development stage, through operations, to closure. Section 5.4 describes the essential measurements that should always be applied to a project and explains the ways that measurements are implemented. Also described are Complex Systems Methodology measurement processes associated with the combination of activities of both project management and systems engineering.

The effective communication of measurements and their results is an imperative. Useful approaches to communicate information so that emerging risks within the project can be dealt with in a timely manner are discussed in Section 5.5. A case study demonstrating the use of measurements is provided in Section 5.6. In this case study, project management and systems engineering processes, including definitive measurements, were used effectively to change a project's course multiple times. A summary checklist of key concepts learned in this chapter is shown in Section 5.7. And an Apply Now exercise in Section 5.8 allows continued practice in applying the concepts from this chapter.

Chapter Roadmap

Chapter 5 focuses on the type and use of Complex Systems Methodology measurements available to increase the chance of obtaining successful project outcomes. It specifically:

- explains the meaning of success against which to measure outcomes
- describes the underpinning measurements that are essential for achieving successful outcomes
- describes the types of measurements available
- provides process steps in implementing each of the measurements
- explains how to report measurement outcomes
- provides a case study to help put the concepts from this chapter into real-life examples
- provides a summary checklist of key chapter concepts for easy reference
- provides Apply Now exercises to assist in the application of the measurement concepts described in this chapter

5.1 Defining Success

The focus in this section is on the use of measurements to drive performance to successful outcomes.

- Fundamentally, a project that is not managed through the use of meaningful measurements has a high probability of failure.
- Success is a subjective measure based on many different things. Individuals, when discussing the same project, can assess success differently. And it can be defined in different ways across disciplines.
- Even if all of the processes discussed in Chapters 3 and 4 are implemented, a project that is not managed with measurements throughout the life cycle will most likely fail to achieve its stated objectives.

Both the stakeholder-focused and solution-focused processes discussed in Chapters 3 and 4 provide the basis for thinking about measurements that will help direct the project to a successful outcome. In this section, the definition of success and what it means to achieve successful outcomes is explored.

5.1.1 Successful Outcomes

The definition of successful outcomes can be different for each stakeholder but can also have different meanings, depending on where the project is within the life cycle. During research and development, for example, successful outcomes can come in the form of the publication of a paper, a scientific finding, a useable prototype, a technological breakthrough, or a new technique. In the construction or production phase, each milestone achieved, specification met, or product completed can represent a successful outcome. And during the operations phase, achievement of greater process efficiency leading to positive financial impacts, achievement of economic growth, or unencumbered performance can reflect a successful outcome. Figure 5.1[1] provides a visualization of how a project may be perceived by the stakeholders.

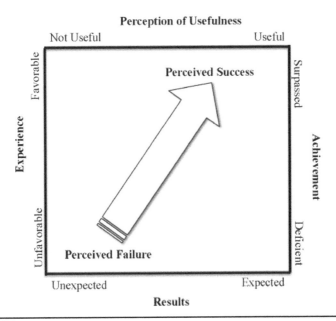

Figure 5.1   Perception Model (Reproduced with permission from Wingate, L. M. [2014]. *Project Management for Research and Development: Guiding Innovation for Positive R&D Outcomes.* Boca Raton, FL: CRC Press/Taylor & Francis Group.)

Each axis shows a spectrum of subjective assessments associated with the experience, the perception of usefulness, the achievement of the scope as defined in their stakeholder documents—such as the operations plan and use cases—and the results overall. If the results were as expected, the project met or exceeded what was planned, the stakeholders had a favorable experience during the project, and the outcomes of the project are perceived to be useful, then the stakeholders will deem the project a success. Conversely, if the results were unexpected and deficient, if the project outcomes are unusable, and if the stakeholders had an overall unfavorable experience, they will determine the project was a failure.

Success for the key stakeholders is measured by their satisfaction with the results and their willingness to accept the deliverables. Their need for value from the project will have been achieved. The clear and agreed-upon targets that the project manager and key stakeholders converge on—at the beginning of the project and as managed throughout the project via change control methods—are the most important way to achieve stakeholder satisfaction and outcomes success. Knowing that the achievement of successful outcomes for the stakeholders depends on these targets and alignment measurements that support that mission is essential.

Within that larger vision of achieving project success in the eyes of the stakeholders, both the project manager and the systems engineer aim to achieve successes within their areas of expertise. These discipline-specific successes are reflected from carefully chosen measurements taken throughout the course of the project life cycle and throughout operations of the system. Because each project and operational environment is unique, the identification of what constitutes success may differ widely.

5.1.2 Project Management Successes

Project success requires more than the successful execution of the scope, schedule, and budget baselines. A project, as a whole, can only be deemed a success if the key stakeholders' needs have been met and that they perceive that they have received the value they anticipated. In addition, the stakeholders, including the organization sponsoring the project, is expecting to receive a benefit from the project's successful performance in the form of gains, assets, or overall business value. The subjective interpretation of the project will be positive if the project outcomes reflect the desires of the key stakeholders. Because each project is unique, the project manager's responsibility is to carefully define the measurements that will reflect progress in a way that can be verified with data and information that supports the project's needs and provides the key stakeholders with the appropriate information. This will allow the stakeholders to validate the progress and outcomes throughout the project and ultimately to accept the system. The project manager must also ensure that all work that goes into a project is measureable and will lead to the outcomes that are expected.

Within the discipline of project management, successes can be measured across all process areas that are used to manage the project. Performance against planned baselines for costs, schedule, and scope are the outcomes that generally define success for a project. If the project achieves completion in line with the baselines, then the project manager can claim success. However, as mentioned previously, projects that meet all baselines may still be deemed unsuccessful if the stakeholders do not assess the project as successful or as providing benefits. It is often the case that when stakeholders are dissatisfied with the project outcomes, it is difficult to close the project. This dissatisfaction may manifest in a lack of sign-off on final documentation, payments may be in jeopardy, or additional requirements may be levied against the project before project closure can be achieved, driving up cost and impacting the schedule.

The most important thing that a project manager can do to ensure success is to obtain an agreement from the key stakeholders on the expected outcomes and on the definition of success for the project. This may be harder than anticipated, however. During project charter development and scope definition, the majority position of the key stakeholders is acceptable, and complete concurrence across all parties is not required. This could cause a situation in which agreement as to what success looks like, or even what outcomes are acceptable, is not a universal position with the stakeholders. The minority within key stakeholders may actually refuse to accept the project at close-out. Because of this risk, it is the project manager's responsibility to gain consensus as to what is expected from the outcomes of the project, even if there is not agreement as to the exact scope of the project.

Once this documented agreement defining the successful outcomes is in place, measurements can be identified and implemented that will validate the progress toward, and achievement of, the outcomes. Incremental or intermediate measurements play an important role in assurance and validation with the stakeholders that progress is being achieved and that it will ultimately meet their needs. These incremental measures are discussed further in Section 5.3.

5.1.3 Systems Engineering Successes

The identification of systems engineering successful outcomes may be more difficult than those within project management. The clarification of the key stakeholders' vision of what they need is often vague and intermingled with their wants and ideas of what can technically be accomplished within their budget and timeframe. As described in Chapter 4, a stakeholder needs analysis takes place during the mission-definition phase of the project life cycle. During this activity, the systems engineer works with the key stakeholders toward the resolution of the project vision through operational concept and scenario development. The subsequent understanding and agreement as to how the system is envisioned to behave during its operational life will drive the view of the project outcomes. Successful outcomes will be achieved when the project outcomes match this vision. The challenge for the systems engineer is to develop a solution that also meets the performance specifications, exhibits the appropriate emergent behavior, and meets the cost and schedule constraints.

As in the project management discipline, a documented agreement defining the successful outcomes identified in the systems engineering processes should become the baseline for that which is measured. Once this is in place, measurements can be identified and used to confirm the progress. As with project management, systems engineering incremental measurements play a critical role in verification of the capabilities and validation to the key stakeholders of the technical progress of the project.

5.2 The Use of Measurements

This section provides insight into how to obtain valid, timely, and useful measurements and use them wisely to drive toward positive outcomes.

- If used correctly, measurements can reduce defects and rework and can serve as an early warning to resolve problems before impacts are felt.
- There are overarching foundation support requirements that must be in place in order to perform robust measurements.
- The inaccurate collection or misuse of measurements will lead to poor insight and decision making.

Measurements serve a multitude of roles in a project. When applied appropriately and consistently, they provide timely insight into progress and trends and help detect issues while assisting in the understanding of root causes. They improve communications and understanding and provide valid and concrete evidence for decision making. In short, they measure activities critical for the project's success. Ultimately, measurements need to result in understanding, confirmation, decisions, and/or actions.

Seasoned project managers and systems engineers should be involved in measurement identification and establishment. They will incorporate input from individuals experienced in quantitative analysis, statistics, survey research, and operations research. Each of these disciplines is trained in identification of measurements that provide accurate information derived from data that comes from disparate sources. And they are knowledgeable about data integrity requirements, in addition to the misinformation traps that can be found by implementing and representing measurements incorrectly, such as the misuse of statistics to support a poor conclusion or the use of compositional fallacies based on extrapolating from outliers or single data point trends.

5.2.1 Supporting Foundations

Certain types of measurements are known to provide value in assuring projects reach successful conclusions. However, in order for a project to obtain that value, a set of supporting foundations must be in place. Measurements cannot be used effectively if the foundation is not in place for collecting, analyzing, and reacting to the information that emerges from the measurements. This may seem self-evident, but in many cases, organizations are not structured in a manner that provides all the supporting foundations for project success, but rather for organizational success. Therefore, identifying the gaps between what the organization foundation provides and what the project needs to facilitate project measurements is an important step. In particular, the organization should be aligned in the following ways:

- the organizational structure itself, as is related to the positioning of the project manager and the systems engineer
- the organizational communications protocols or norms

- organizational support for governance activities, including holding to established baselines with clear ramifications for nonperformance

Understanding the organizational environment as aligned to the project needs and addressing supporting foundation deficiencies will decrease risk to the project.

5.2.1.1 Organizational Structure

Organizations can be structured in many different ways. There are institutional processes, policies, and cultural behaviors that provide paths through which work flows. Who has authority and responsibility, the chain of command, information flows, and delegation norms—are all organization specific. As was described in Section 5.1, there must be a clear understanding of the roles and responsibilities that each participating organization, department, division, etc. brings to the project.

Each project is unique, and therefore an understanding and appreciation of the organizational environment, and all of the resultant constraints it finds itself in, is an imperative. Because many organizational constraints directly affect the ability of the project manager and systems engineer to perform their duties in an optimal way, an important activity is to develop a roles and responsibilities matrix that will identify all of the major roles to be performed and then align the individuals to those roles.

Although there are different schools of thought as to the most effective way to implement this type of roles and responsibilities matrix, as long as it includes the ability to identify a single role that is accountable (A) for the successful outcomes associated with each task, and the roles that are responsible (R) for performing the tasks that lead to the outcomes, then that should suffice. The (A) role, and the person that performs the role, can sit at different levels of the project hierarchy. They typically do not perform the work but approve the work product that comes from the (R) role(s).

A role (and the associated persons serving in those roles) identified within the matrix as responsible (R) will ensure that the tasks within the purview of the role are performed. They may or may not delegate or direct staff in the performance of the work. The reason more than one person can hold the responsibility for performing a task is that roles are interdisciplinary and may be spread across independent organizations, departments, and divisions. Each independent entity would identify an individual who is responsible for the performance of that part of the task. An example of an extremely simple roles and responsibilities matrix is shown is Figure 5.2.

In this simple version, tasks are placed along the left hand of the matrix. Major roles are placed along the top of the matrix. Within the intersections, an (A) and one or more (R) are inserted. Other designations can be added by the organization if desired, but for simplicity, only the two designations are used in the example. Keep in mind that these types of roles and relationship matrices can be as complex or as simple as needed to convey what is necessary, so that a clear boundary is established between the roles for all major tasks, particularly if there is ambiguity or disagreement as to ownership.

An important step once the initial matrix is completed is to obtain documented agreement as to the allocations of responsibility and accountability. Once this is complete, the roles and responsibilities matrix will be baselined and then fully socialized and communicated throughout the organizational

	Project Manager	Systems Engineer	Project Team Members
Project Status	A	R	R
Technical Status		A	R
Change Board	A	R	R
Technical Reviews	A	R	R

Figure 5.2 Example of Roles and Responsibilities Matrix

entities involved as well as throughout the project. This step is important, because even if the parties involved in the development of the matrix achieve concurrence, others in the larger organization or even the project may not be in agreement.

5.2.1.1.1 Project Manager Organizational Alignment

The project manager position may be aligned in many ways within the organization. However, some alignments typically produce better project outcomes. A project manager that is aligned within an organization in a manner that provides a clear escalation path up, and with clear authority to delegate project work across all disciplines as needed for project success, will be the most effective. This translates to a structure in which the project manager actually has both the authority and responsibility to implement the project as he or she sees fit. When a project manager is placed in the role, but not provided the authority and responsibility to execute the project, the risk rises on the project and the probability of achieving successful outcomes diminishes.

The level of controls that the project manager can exert on the budget, schedule, staffing, and scope development varies tremendously between projects and organizations, and there have been successful models that represent all varieties of levels of project manager control in these areas. There does not seem to be a single right way of aligning these responsibilities with the project manager; however, it is clear that the less direct control the project manager has in these areas, the more influence skills the project manager needs to compel resource owners to provide what is needed by the project. Influencing skills are often tied directly to leadership skills. An understanding of this will drive the type of project manager needed for the project.

The project manager's talents and skills can be a deciding factor in project performance. A talented project manager will obtain commitments, recruit appropriate talent, drive schedule, manage costs, enhance benefits, reduce risk, and communicate up, down, and across organizational boundaries. The better the alignment within the organization facilitates these activities, the higher the chance of project success.

5.2.1.1.2 Systems Engineer Organizational Alignment

The project systems engineer is responsible for processes that are complementary with project management, as well as unique systems engineering processes, as was discussed in Chapters 3 and 4. Optimally, the organizational alignment for systems engineering is in a direct reporting relationship to the project manager. In some organizations, the systems engineer is in a direct reporting relationship to the organizations' chief engineer, which is a much less effective alignment. The reason that the alignment with the project manager is preferred is that the project manager and systems engineer have complementary and heavily overlapping process-focused and overarching responsibilities.

The systems engineer provides an interdisciplinary or cross-discipline approach that aligns the projects activities into a systems context and perspective. The systems engineering discipline provides the structure to integrate activities and to ensure that the project technical solution performs as a whole. This mitigates the risks associated with having the systems engineer reporting up through the engineering discipline, where no single person has the comprehensive view of how each of the unique pieces fit together as specified, and ensures that each piece works as an integral part of the system as a whole. This is especially true for projects that span to other disciplines outside of engineering, such as software development, or for those systems that are expected to have emergent behavior.

In the role of project systems engineer, a management service is provided to the project manager to ensure that the project comes together as a whole. The systems engineer also provides the technical service to the project manager. Performing this technical service may require knowledge and expertise in the technical field of the design, or it may require process knowledge only. In the event technical expertise is required, the subject matter expert may emerge from the engineering field and be nominated to fill the systems engineering role. This is where the complexity emerges associated with lines of supervisory authority.

If there is debate about the proper alignment of the systems engineer, a careful review should be performed of the roles and responsibilities of the position, as described in Section 5.2.1.1. This eliminates ambiguity about which overarching process roles are the systems engineer's responsibilities and which cannot be performed when the systems engineering position is subsumed under a specific expertise discipline, such as engineering. It should also be noted that the systems engineer should be careful not to take ownership of the subject matter expert's technical responsibilities. Instead, assuring that adequate time in the schedule is allocated to the expert for that work activity is crucial.

5.2.1.2 Communications

Communications are an integral foundation activity in that, if not done well, they can cause significant risk to the project. Communications-related activities can account for up to 90 percent of a project manager's responsibilities.[2] Projects have stakeholders who must receive pertinent information in a timely and ongoing manner. Communications needs will typically span the breadth and depth of organizational boundaries, often crossing outside of the organization as well. Although detailed stakeholder communications processes were defined in Section 3.1.2 (page 69), communications are referenced here in supporting foundations, due to the fact that the majority of projects are not performed in isolation—that is, outside of an organizational construct. Therefore, there is an inherent relationship that needs to be addressed.

The organizational infrastructure used to facilitate communications should be reviewed and assessed for adequacy. Any unique project needs that are associated with communications across geographic, societal, language, culture, organizational, or departmental boundaries must be identified and compared against the current organizational capability. This assessment is the responsibility of the project manager in collaboration with the organization's information technology and communications departments or responsible staff members. The project manager will negotiate for additional capability as appropriate. The plan for utilizing the organizational communications capability will be folded into the project communications plan.

5.2.1.3 Governance

Project governance activities rely on the supporting structure of the organization. To the extent that change and configuration control and maintaining baselines against which change can be assessed is an integral part of the culture of the organization, then it will be straightforward and easy to initiate these processes within the project. If they are not part of the existing culture, or the project is cross-disciplinary and relies on performance within other departments and divisions, the project will experience higher risk that is due to the fact that it is unlikely the project will be able to drive the cultural change throughout all of the performing departments and divisions. Therefore, the project may not be able to expect conformance to the governance needed.

5.2.1.3.1 Baseline Management

As discussed in Chapters 3 and 4, the setting of project baselines for cost, schedule, and scope is a critical first step in every project. If the discipline of setting project baselines is culturally accepted and anticipated within the organization, the risk to the project will be low. If, however, an organization supports activities that erode, damage, or weaken the baselines—such as when there are soft-starts during which the project is allowed to evolve over time without a designated set of baselines, or when the project soft-ends, where the project is allowed to continue on past the official end date of the baselined project—the risk to the project intensifies. Whenever uncontrolled modifications to the baselines are made, or when there is a failure to operate within the agreed-upon baselines, the probability that the project will be unable to achieve its objectives is increased.

5.2.1.3.2 Change Management

As defined in Chapter 3, change management activities ensure that change impacts are all accounted for and understood. If the organization does not support the change management process, changes will be made without this rigor, and the project will quickly lose the ability to hold on to the baseline. Without change management rigor, the project manager will not be able to affectively analyze change impacts or to proactively address changes to the budget, schedule, or scope. It will also impact the project's contingency and the choice of which risks to mitigate. The effect of not managing change well in an organization and within a project is easy to spot. There will generally be confusion and lack of support or buy-in to changes once they are made known. Poor implementation resulting in higher costs, schedule impacts, and out-of-control scope will inevitably result.

5.2.1.3.3 Configuration Management

Configuration management is a core supporting foundation. Without this process, the full assurance that the current state of the system and all elements of the system design are known and demonstrable, through documentation and inspection of the articles, will be lost. A solid configuration management process ensures that no part of the system design is held only within the knowledge of individual team members or in unreachable systems. From auditing perspectives, configuration management allows access to clear information on previous versions of the design as well as reproducibility of the design.

Without good configuration control acceptance within the organization, the impacts of changes will be extremely difficult, if not impossible, to assess. There will be a direct impact to the quality of the project outcomes. Real impacts to the project baselines of cost, schedule, and scope include confusion throughout the team as to which versions of components, parts, subsystems, etc. are being used throughout the system. It will be impossible to associate a system version, with all known sub-element versions that make up the whole of the system.

An abundance of specification non-conformance waivers will emerge, and there will be a general lack of meaningful traceability within the system. Interfaces will be difficult to design because what is known about the interfacing designs will be suspect. It will also be impossible to test and verify changes before implementation, because there will not be a well-understood performance baseline to test against. Although the process for implementing configuration management is straightforward, it takes a firm organizational commitment to the processes, with impactful ramifications for departments or individuals in not following the processes.

5.2.2 Measurements Watch Items

Choosing measurements that will provide the most meaningful information for early warning, verification, and validation activities is both an art and a science. It is helpful and recommended to consult with individuals who have backgrounds in disciplines associated with quantitative analysis, operations research, and statistics. These experts can add real value, because it is all too easy to choose measurements that appear to be useful but in reality drive decision making in irrelevant or inapplicable directions. Choosing appropriate measurements requires heavy collaboration between a knowledgeable and experienced project manager and the measurement subject matter experts.

Measurements can be misleading. They require careful validation before and during use. For example, introducing a measurement will often influence the results. It is also important to choose measurements that will accurately reflect the current situation and that can be verified by multiple occurrences (not a single data point trend). There must be guards against manipulation of the data to reflect an alternative view of the current situation. Data quality and control is an imperative. These types of measurement mishaps (poor measurement choices, assumptions made on single data points, manipulation of data to fit a situation) are all-too-common occurrences that have been known to cause serious harm to many projects.

Measurements are used to keep a project on schedule, on budget, or within scope. They can serve as a gate at which work is approved before being allowed to proceed and to act as boundaries to assess the quality of work being performed. They need to accurately assess actual completed activities versus planned activities. Percentage completed on tasks is often qualitatively assessed and prone to error and manipulation. Technical debt can accrue if measurements do not expose unaccounted for tasks, tasks declared complete when they are not, and tasks completed but done incorrectly. Diligently monitoring and measuring activities that give the appearance of progress, rather than identifying actual risks, only wastes resources and makes it difficult or impossible to address real issues in a timely manner.

Measurements should be used to:

- minimize negative effects on a project
- ensure orderly and timely development as outlined in the project plan
- support effective decision making throughout all levels of the project
- verify and validate results
- provide confirmation that the project is ready to progress in the life cycle

Measurements should also highlight blocker situations and decisions that are not being made in a timely manner. Although each project will require an appropriately tailored set of measurements, a good rule of thumb is to identify those measurements that reflect information associated with the most important features of the project and that do not cost more to implement than the value of the information received about the status. Contractual requirements, the level of risk and complexity on the project, and the cost associated with the measurements chosen for implementation on the project must all be considered when establishing performance measurements.

Unless time, funding, and scope are of no consequence, even small informal projects require some level of measurements to ensure forward movement toward the intended outcomes. Once chosen, they must reflect accurate data that are implemented and interpreted correctly. Projects often fail even when metrics show no issues. This is due to faulty reporting or measuring the wrong things. The types of measurements most often used to drive positive project outcomes is discussed in Section 5.3.

Measurements should be chosen carefully based on the answers to the following questions:

- Who is the audience?
- What information will provide an accurate insight into the situation being reviewed?
- What is the assessment timing that will provide the most useful insight in a cost-effective manner?
- Who is responsible for providing, collecting, and managing the information?

5.3 Types of Measurements

In this section, the types of broad, strategic measurement methods that are available to project managers and systems engineers are discussed.

- Measuring is an important activity; however, it can easily consume all the resources and end up costing more than the value one would get out of the measurements.
- It is important to carefully choose measurements that are essential so that the information gathered is the most pertinent and provides true early warning for the project but uses only the resources necessary to collect.
- Each of the measurements ultimately selected for the project need to sit on a firm foundation, as was discussed in Section 5.2.1, and must also be based on essential measurements (Section 5.4) that act as a standard against which progress is assessed.

Measurements are made through measures, metrics, or a combination of both. Measures can be described as data that are in quantitative form, such as hours spent on a work package. Metrics are qualitative, compiled from data, and represent progress made against a stated objective such as quality or schedule. Figure 5.3 provides a model to assist in understanding the types of measurements and typical project activities that are measured.

Measurements can be quantitative or qualitative. Quantitative measurements refer to a quantity of a thing. It is easier to verify and replicate quantitative measurements, because they are mathematically based. Qualitative measurementss refer to the quality of a thing and can be difficult to verify and replicate because they are based on a sense of something that may be assessed on many factors that may differ from person to person. The bottom of the model shows a continuum from quantitative to qualitative. A combination of both quantitative and qualitative measurements may be used, for example, in quality assurance assessments.

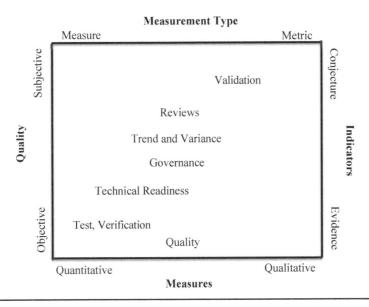

Figure 5.3 Types of Measurements

Measurements can be subjective or objective. Subjective measurements are based on opinion, judgments, and personal interpretations, whereas objective measurements are observable and repeatable regardless of who the assessor is. The left side of the figure shows the continuum between subjective and objective. It is usually (but not always) the case that a qualitative measure will also be subjective, and a quantitative measure will also be objective.

Measurement indicators can be based on evidence or conjecture. These are reflected along the right side of the figure. Evidence can be associated with verifiable results that can be repeated. Conjecture, on the other hand, is basically an opinion or speculation. Some measurements can fall within the middle range, such as when a model is used to quantify results; but then conjecture is used to extrapolate into a point of view associated with the meaning of those results. Often these indicators are referred to as leading, which means that current data can be used to predict future behavior. These types of indicators are likely to affect objectives, and information gleaned from them generally can be used in the short term to improve or influence the direction or trend. In contrast, lagging indicators refer to types of indicators that follow a trend and would require more significant and long-lasting or strategic changes to impact.

The types of measurements shown within the figure, and aligned to a location at which they are most likely to be found, include performance progress, communications, planning and control, and process and quality measurements. For example, performance progress measurements, such as reviews, are shown as combination measure and metric, quantitative and qualitative along the top and bottom axes. Along the right and left axis, reviews show as both more subjective and based more on conjecture. In contrast, test and verification activities, usually based on hard evidence, are shown in the quadrant of the figure that reflects quantitative, objective, evidence-based measures.

5.3.1 Trend and Variance Analysis

The type of Complex Systems Methodology measurements most often performed is trend and variance analysis. Trend analysis is based on historical data, whereas variance analysis is based on the difference between what was planned and what actually happened.

Both qualitative and quantitative information can be used to perform trend and variance analysis. Some examples of qualitative variance analysis include process auditing, management or team performance to objectives, and reconciliation of root causes. Quantitative variance analysis includes activities that assess the differences between planned and actual costs, schedule, and/or technical parameters.

Quantitative and qualitative trend analysis, on the other hand, is the activity of making an educated guess as to how things will progress, from the baseline plan to what the future holds, using past performance as an indicator of future performance. If the project manager and systems engineer have developed measurements that provide a range of performance parameters from which performance can deviate before impacting the project, it will be necessary to react only when performance falls outside of that range, not at every fluctuation.

Period comparisons of the chosen measurements are made by both the project manager and the systems engineer over the course of the project, evolving with the project as it moves through the life cycle as new situations emerge driving new measurement requirements. The periodicity is based on the needs of the project management and stakeholders.

5.3.2 Statistical Sampling Measurements

Many systems engineering processes, particularly associated with quality, are verified through statistical sampling. Sampling is a process where a random representative of a product, service, or process is

collected and analyzed. It is anticipated through statistical sampling that the sample is a true representation and falls within the actual normal quality range. A statistician should be consulted when using
statistical sampling, so that misuse of the many different statistical methods is not a risk and the data
generated by the sampling are a valid representation. But in essence, a sample or samples are chosen and
then inspected for compliance to the expected outcomes.

In addition, the use of a variety of methods can provide multiple lines of evidence, confirming and
verifying the message coming from the data, and substantiate expected performance. This can be especially useful early in the life cycle of the project. However, as stated in the beginning of this section, it
is important to ultimately choose only those measurements that are essential so that the information
gathered is the most pertinent and provides a true early warning for the project, using only the resources
necessary.

5.3.3 Key Performance Indicators

Project managers may decide that a project is complicated enough, or has enough risk, that special measurements, called *key performance indicators* (KPIs), should be identified and tracked. Key performance
indicators, also referred to as key success indicators, help define the most important performance on
a project. They are useful in focusing information to mitigate overload and confusion that can occur
when a significant amount of information is analyzed regularly. This is not to say that all of the other
measurements are not useful; however, when information is provided frequently and in significant volume, it can actually result in measurement paralysis, in which decisions cannot be made because the
true issues cannot be discerned.

Key performance indicators serve a function similar to the project charter. Each project will have
key performance indicators that are organizational and industry unique. As many or as few key performance indicators as needed will be chosen as appropriate to assist in analysis and valid decision
making. Throughout the project, these key performance indicators are assessed to ensure they continue
to provide meaningful information and are retired or replaced as necessary through the formal project
change control mechanisms.

Collaboration between the project manager, systems engineer, and key stakeholders must take place
in order to apply meaningful key performance indicators. Expected outcomes from the project are
reviewed, the risks associated with each project's products are assessed, and choices made between
which measures of performance should be selected. The periods for reviewing performance associated
with the key performance indicators are also determined. The key performance indicators are agreed
upon by the key stakeholders and baselined. They describe the attributes of the project that assume
the most risk and that are imperative for project success. These key indicators can be quantitative or
qualitative, objective or subjective; they can be a measure or a metric; and they can be evidence based
or conjecture based.

5.3.4 Technical Readiness

The use of measurements of technical readiness can assist a project manager and a systems engineer
in understanding the progress the project is making associated with its scope. Technical readiness is
assessed through different ways and throughout the life cycle of a project. Using these incremental
measurements, progress can be assessed before allowing advancement to the next level and stopped if
sufficient progress is not being made. Keep in mind that this measure is different from the measures of
effectiveness, which reflects the stakeholders' satisfaction with the project outcomes upon the completion of the project.

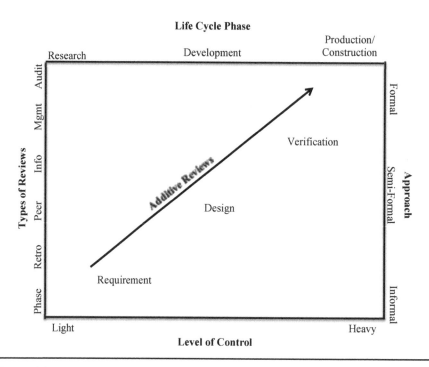

Figure 5.4   Additive Metrics (Reproduced with permission from Wingate, L. M. [2014]. *Project Management for Research and Development: Guiding Innovation for Positive R&D Outcomes.* Boca Raton, FL: CRC Press/Taylor & Francis Group.)

Technical readiness measurements require regular collection, assessment, and verification with the resultant timely decision making. The types of measurements evolve over the life cycle of the project. For example, during the research and development phase of the project, technical readiness to progress is reflected in successful test results. Later, development and production/construction phases are measured through quality metrics assessed on production/construction output. The more formal a project is, the more rigor will be applied to assess its readiness for progression, and therefore additional processes will be added. Figure 5.4[3] demonstrates the additive nature of metrics over project life cycle and formality parameters. As projects move through the life cycle, more metrics will be applied.

The use of a formal set of defined gates can simplify the process of assessing technical readiness and make it repeatable. The use of technical readiness levels (TRL)[4] provides a robust construct for this process to evaluate technological evolution and to ensure forward movement when it is ready to do so, and only then. "The concept of TRLs was originally developed by the National Aeronautics and Space Administration (NASA) in the early 1980s, and is now used internationally in a variety of research and development and innovation-related disciplines to systematically progress these activities along the life cycle."[5]

Depending on the industry, there can be any number of levels of readiness. A continuum of readiness moves from basic research through all phases and into operations. Prior to progressing from one level to the next, a readiness review is held to assess if the project has met the criteria that was established as part of the technical readiness levels process. Each organization can define its own readiness criteria. This methodology is one of the most useful for minimizing the risk associated with technical debt, as described in Chapter 2. Technical readiness levels generally follow this type of stratification:

- **Level 1** – pure research; outcomes = papers, early prototypes
- **Level 2** – research and development; outcomes = concept definition, prototypes, test results
- **Level 3** – development and early production; outcomes = complete design with all interfaces identified
- **Level 4** – production/construction; outcomes = first articles, final products, services
- **Level 5** – operations; outcomes = operationally ready

5.4 Essential Measurements

The Complex Systems Methodology measurements previously discussed in Chapters 3 and 4 are emphasized in this section. These measurements are essential for control of the project.

- These essential measurements help to identify areas needing improvement and to assess quality in two specific areas: key performance and life cycle phase progression readiness.
- Implementation of essential measurements requires a planned approach, inclusive of the definition of a measurement concept, a precise explanation of what will be measured and in what unit of measurement, and how often measurements are performed.
- Obtaining value from the application of essential measurements requires consistent application and follow through.

The application of measurements in a way that provides the most value to a project can be a time-consuming and resource-heavy endeavor. Many project managers will have a dedicated manager in place to oversee the activities related to the performance of all the tasks associated with the essential measurements, as well as the associated controls of baseline, resource, risk, integration, quality, and governance. The position description is referred to as a project controls specialist. All complex projects should have this specialist in place on the project, reporting directly to the project manager.

5.4.1 Baseline Control

A project that does not have a valid cost, schedule, and scope baseline, carefully controlled for change, will not reach its targeted objectives. That is not to say that the funding allocated to the project will not be spent, or the labor associated with the project will not be charged, or some level of the scope attained. What this means is that the activities that are known to be critical in the performance of the project as envisioned by the stakeholders have a low probability of being achieved if the project is not actively managed.

There are many reasons that project control is never attained or is lost. Most of those reasons have to do with the project's cultural environment. The project may not be in a supportive environment in which controls are accepted. The project manager and/or the systems engineer may also be misaligned organizationally, limiting their ability to provide the oversight necessary to perform careful controls. Or the organization involved may favor an uncontrolled environment, with an ill-defined target outcome that evolves without guidance during the course of the project. There are situations in which the only role of the project manager is to report on the progress of an uncontrolled project. As long as that project manager and systems engineer are not accountable or responsible for achieving any milestones or outcomes, and have the ability to solicit reasonable progress reports, such an organization may be satisfied. However, that is not generally a satisfactory position for a project manager or systems engineer, because the information needed to reasonably and methodically progress toward success is facilitated by controls. Finding themselves in this type of organization will generally be frustrating for both disciplines.

Because standard project management and systems engineering processes assume that controls can be applied, the following essential measurements are based on those standards and assumptions as well. The establishment of baselines—a standard in both disciplines—is an imperative. The intermediate measurements that are used by the project manager and systems engineer will depend on the desires of the key stakeholders, the complexity of the project, the size, the associated budget, and the scope life cycle phases that are involved. Both disciplines, during the tailoring of the project (discussed in Chapter 6), will settle on the exact measurements to be used and the tempo and timing of measurement collection and communications (Section 5.5).

In Sections 5.4.1.1–5.4.1.3, measurements associated with the baselines of schedule, cost, and scope will be reviewed. Some of these have been discussed in Chapters 3 and 4 and will be referenced accordingly. All are shown here for completeness.

5.4.1.1 Schedule Control

The schedule is a time-based plan for achieving all of the project objectives, as defined in the project documentation. The schedule is inclusive of any products provided through contracts, subcontracts, partners, collaborators, or other project participants. All project activities are captured in one overarching integrated schedule. A project schedule is dynamic and must be constantly monitored and adjusted.

Schedule control requires two knowns. One is a baseline schedule with an anticipated or planned value for the timeframe identified, broken into suitable segments (such as monthly, quarterly, semi-annually, or annually). The other is a comparison of actual data across the same categories of work and during the same timeframe. In more loosely controlled projects, key milestone percentage of completion versus planned is used instead of the entire schedule baseline. What the project manager wants to achieve with schedule control is to be able to accurately predict future behavior based on past performance, resolve blockers to schedule progress, and address schedule contingency needs in a methodical way in order to protect the overall schedule baseline.

As was discussed in Chapters 3 and 4, project managers use the critical path method and schedule contingency to clearly document the estimated time it will take to complete each activity captured in the schedule. In the simplest description, the critical path is an end-to-end look at the planned tasks estimated to take the longest to complete, although overlapping tasks must also be factored in. If all of the concurrent and sequential tasks have been identified, and the critical path is known, then the amount of float (slack) in the schedule can be calculated. This is the amount of time that can be applied to non-critical–path sequential tasks before impacting the critical path—and the overarching schedule. In other words, additional time can be allocated to those tasks without impacting the overall project schedule. An important measurement that must always be applied is to understand the remaining float available to the project and the status of performance to plan. The critical path method can provide that pertinent information.

As described in Chapter 4, scheduling measurements will come from standard reporting (spreadsheets, Gantt charts, etc.) or through earned value management reports. Standard reporting usually focuses on critical path methodology (CPM). As the use of critical path methodology for forecasting generally only provides a straight-line trajectory, the schedule risk assessment (SRA) method is a good supplement for forecasting clarity. Schedule risk assessment uses Monte Carlo simulation techniques to provide useful forecasts through sensitivity analysis. An operations research analyst, statistician, or risk manager can provide the necessary subject matter expertise to complete this type of analysis.

Both types will provide comparisons between planned and actual/earned labor, material, and services across predetermined intervals. The benefit to using an earned value method for measurements is that there are pre-defined formulas available that, when applied, offer indices that are meant to provide robust decision-making capability. These indices provide summary snapshots about project performance, both negative and positive. Useful decision making with indices relies on accurate data, interpreted correctly.

Some of the more common calculations and indices used to manage baselines include the baseline execution index (BEI) and earned value management. The baseline execution index measures the percentage of tasks completed relative to baseline task percentage anticipated to be complete at a moment in time. When using earned value management, schedule variance can be used to measure progress. The value of the work expected to have been performed during the time period—that is, its planned value—is compared to the amount of work that has actually been accomplished, leading to an earned value calculation: $SV = EV - PV$, where a negative schedule variance reflects a behind-schedule position. The schedule performance index (SPI) can then be calculated using the formula $SPI = EV / PV$, where a schedule performance index value of <1.0 indicates the behind-schedule position, and 1.0 reflects an on-schedule position.

A caution should be noted for a positive schedule variance or schedule performance index, in that these can give a false sense of security. Positive schedule variance and schedule performance index can indicate that project tasks are ahead of schedule; however, a careful look at the critical path progress should be done to ensure that those tasks are performing to plan, because non-critical–path tasks can drive a positive schedule variance while the critical path tasks fall behind. This is a common problem when critical path items are difficult to achieve (such as with research and development tasks that are on the critical path).

To address this deficiency, earned schedule (ES)[6] can provide a more useful measurement. In this method, the cumulative value of the work is measured in time completed versus planned. Schedule variance (time) = ES / Actual Time. An index for SPI (time) = ES / Actual Time can then be calculated at any time during the project, irrespective of formal time increments. If the number generated is positive, the schedule is ahead of plan. If the number is negative, the project is behind schedule. Because the application of earned schedule can be performed on the critical path (in isolation from the other project elements), the most impactful schedule variances—the ones that can impact the schedule baseline—can be monitored and addressed.

As a reminder from Chapter 4, earned value, if used on a flexible project or on a project with flexible work packages embedded within them, should take care to ensure that the earned value reporting collected from those work breakdown structures are collecting and providing useful and valid metrics to the project manager. These include measurements that are time bounded and measure completed useable features. Earned value, planned value, and schedule variance have the same meaning. The schedule performance index can also be calculated in the same way as the traditional method of project management, except including the time-bounded completed features for the time period of interest for the overall project.

5.4.1.2 Cost Control

Every project has an allocated total budget. The determination of which cost management methodology to use is the responsibility of the project manager, as described in Chapter 4. The project manager sets the proper budget contingency level for the project and performs cost estimation and analysis, often with the help of subject matter experts responsible for project controls. The project manager is responsible for all project-related costs associated with external entities in binding agreements to the project. The overall project budget will reflect all budgeted activities for a complete picture of the project from a budgeting perspective.

The total budget is decomposed through the work breakdown structure work packages so that the costs can be easily captured and rolled up into summary cost-to-budget reports. Within the work breakdown structure work packages, the distribution between the budget categories is captured, including labor and materials and services; associated procurements and contracts, etc. Total budgets are esti-mated using various methods, such as itemized costing or historic costing from projects that are exactly like, or similar to, the project being estimated. Using an itemized method is sometimes referred to as bottom-up estimating. It is the most commonly used method of budget development.

Each work breakdown structure work package manager will estimate the costs for that work pack-age based on historical records, vendor bids, and procurement contracts. An integration step will be for each manager to ensure that the labor costs, which are being included for each work breakdown structure budget, take into consideration that the costs associated with the number of labor hours for each individual have been scheduled, that the accurate hourly rate is associated with each individual, and that all of the indirect costs and fees are appropriate. The work breakdown structure budgets are referred to as control accounts—the level above the work package. The overall budget that will be baselined will include a roll-up of all work breakdown structure budgets, indirect costs, and any other organizational costs that are appropriate. Careful attention must be paid during this process to account for budget contingency that is included in each work breakdown structure budget, so that there is not an overabundance of contingency assessed on the project. Budget contingency is the amount of funding that will be used to mitigate risks throughout the project. Risk control is discussed in Section 5.4.3.

Another important factor for effective cost control is associated with the rigor of maintaining and collecting cost information. Organizational norms, processes, procedures, and culture affect the qual-ity, accuracy, and level of detail of the collected cost information. Organizational data quality norms for cost collection and reporting must match the needs of the project, at the right levels, and in align-ment with the timing requirements. If there is a mismatch, the project manager will not be effective at cost control. Assuming that there is parity, or, alternatively, if the project manager can compile the data outside of the provided systems, then cost control can be performed between the baselined budget and the costs as they are accrued.

Project cost performance trend and variance analysis can be performed in alignment with standard organizational tempos (e.g., monthly, quarterly) or more often if needed by the project. Depending on the organizational resources that support the collection of cost data, some analysis may be provided by organizational reports from labor and cost accounting systems, and some may require collection and analysis by the project manager outside of these systems. The data that will be used to assess budget per-formance must be of sufficient quality that the information gleaned from it will be error free, because important decisions will be made based on the analysis.

Many projects use earned value management to control costs when a more formal approach is desired. If earned value management is used to control schedule, it will also be used to control costs on a project, and vice versa. As was described for earned value management schedule management, the budget can also be managed using planned value and earned value. Planned value is the baseline budget, and earned value refers to the value of the work that has been performed. The actual cost of a project reflects the actual spending and can be used in both standard and earned value management budget analysis. A cost variance (CV) can be used to explain the difference between the actual cost and the earned value, using $CV = EV - AC$. An example of earned value reporting, including planned value, earned value, and actual cost, is shown in Figure 4.8 (page 119). Keep in mind that not all earned value management activities are described here—only the most essential for driving toward positive project outcomes.

Similar to the earned value management related to schedule indices, a cost performance index (CPI) can also be calculated. By executing the formula $CPI = EV / AC$, the cost performance indicator can be cal-culated and analyzed. A CPI > 1.0 indicates efficient cost performance. Two indices that are used—the to-complete performance index (TCPI) and the estimate at complete (EAC)—can be used to assess if the overall project is at risk in meeting its cost and schedule outcomes. The formula $TCPI = (BAC - EV) / (BAC - AC)$ can be used to calculate the expected actual cost to complete the planned work, where budget at completion (BAC) is expected to be met. TCPI = (BAC − EV) / (EAC − AC) should be used when the budget is expected to be exceeded.[7]

As described in Section 5.4.1.1, if earned value is being used on a flexible project or on a project with flexible work packages embedded within it, the earned value reporting collected from those work breakdown structures must include measurements that are time bounded and that measure completed useable features. Earned value, planned value, and actual cost have the same meaning as they do when using traditional project management methods. The cost performance indicator can also be calculated in the same way as the traditional method of project management, except including the time-bounded completed features value to date divided by the actual costs to date for the time period of interest for the overall project.

When using indices, if the project manager assesses within <25 percent of project completion that the budget is anticipated to be exceeded, and the schedule float will be insufficient to ensure the project reaches its goals in the timeframe scheduled, the estimate at complete should be calculated. In fact, by the 15–20 percent project completion point, an experienced project manager can generally assess whether or not the project will complete on schedule and within the budget. If the data suggests that success is not likely, the earned value data can provide the evidence needed to determine whether or not to proceed with the project. In some cases a rebaseline of the project would be implemented.

To rebaseline a project implies that the baseline has been lost, and the project manager and systems engineer must reestablish a new baseline. This typically occurs when any of the three baselines—cost, schedule, or scope—has become so seriously eroded that the project must be reimaged. This situation often results in a reduced scope, because schedules and budgets are generally less capable of being significantly modified once a project has begun. If a rebaseline of the project is required, a bottom-up review can be performed to provide insight into changes that might be implemented so that the project can be realigned to perform to a revised set of outcomes. Of course, because any rebaseline has direct and serious implications and impacts on the original scope, schedule, and budget of the project, this must only be undertaken with full participation of the key stakeholders.

5.4.1.3 Scope Control

Each project has a defined scope that includes a solution definition that meets stakeholder requirements and forms a baseline that is approved by the key stakeholders. This baseline—which contains all technical scope definitions included in the solution—is expected to result in successful project outcomes. It includes the scope statement, the work breakdown structure, and the work breakdown structure dictionary. Measurements associated with controlling scope are therefore critically important. It is sometimes the case that the project manager will focus on overall scope performance inclusive of all activities across disciplines (scope requirements inclusive of product, processes, training, etc.), while the systems engineer focuses on the cross-discipline technical performance. This is an acceptable division of duties between the two disciplines and plays to their strengths. The determination of the appropriate measurements associated with scope control is the joint responsibility of the project manager and the systems engineer.

The project manager identifies the measurements that help control all scope elements, including the product activities internal to the project, and that are performed through external entities (binding agreements, etc.). This includes ensuring that all process and supportive materials development (e.g., transition plans to next phase of the life cycle, technical training plan, maintenance manuals, etc.) are being developed as per the schedule. The systems engineer is responsible for selecting the measurements that are the most meaningful for reflecting the true state of the technical development activities throughout the project life cycle. As they work with subject matter experts to capture the solution set specifications and parameters, they will also work with those experts to identify the optimal way to reflect and control progress so that, for example, the project does not progress into a new life cycle phase without passing through an official gate.

One of the most prevalent drivers of cost and schedule overruns is uncontrolled scope. Having firmly established intermediate milestones and gates is key, so that technical scope assessment can occur and decisions can be made about the readiness of the technology to progress and to verify that it is meeting the intended specification range and reducing risk before moving to the next phase. Ultimately, before the project moves into the production/construction phase, all risk should be retired, and all technical specifications must be met. The same is true as it moves from production/construction into the operations phase. Section 5.4.3 addresses risk control. As explained in Chapter 2 and Section 5.2.2, technical debt can accrue if this does not occur, which will impact the success in the next phase of the life cycle, because this debt will need time and funding to resolve outside of the production/construction plan for a clean transition from the design and development phase.

A complicating factor in measurements associated with scope performance is the challenge of collecting verified and useful intermediate measurements. It is common that highly technical teams and subject matter experts are involved in many, if not all, of the work breakdown structure work package development. In addition, any external organization that is providing technical products or services for the project have subject matter experts involved as well. As the cross-disciplinary governor, the systems engineer is not expected to have technical depth across these disciplines. Conversely, subject matter experts do not always have the integrated system perspective in mind and have a focus only on the development of their primary technology. Ensuring that the technical teams provide the necessary coherent input (converted to management summaries of non-technical-speak) and are willing to support the necessary processes to ensure the project-wide scope is controlled is an imperative. Therefore, measurements need to be identified that will help the technology experts convey their successes and assist them in achieving continued forward momentum. Those measurements, once identified and agreed to, are provided to the project manager for integration into the overarching project processes for collecting and communicating the results of the measurements.

Because the project manager and systems engineer generally are not the technical experts across all areas of the scope development landscape, it is necessary to convene external-to-the-project subject matter expertise for verification of results throughout scope development. A natural point with which to convene these external resources is during gate reviews, although they can be brought together at any time. The only constraint to the number and type of reviews held is the funding and schedule availability.

Project scope performance measurement collection and assessment at the system level is performed in alignment with, and on the same tempo as, the overall project performance measurement collection and assessment. Scope measurement for components, parts, and subsystems, interfaces, and integration performance as assessed against the baseline can occur at any time and as often as deemed appropriate and needed by the systems engineer. The life cycle progression of all work breakdown structure work packages is assessed by the location within the life cycle of that work package. For example, one could be the purchase of an off-the-shelf component part, for which little is needed other than verification that it meets the specifications, whereas another could be a product that requires heavy research and development, and therefore requires close review at regular intervals. The systems engineer will put together the measurement plan that describes the measurements and timing that will be applied to each part of the work breakdown structure, and also describes the overarching measurements and timing for review of the system as a whole.

Typically, if a project is using earned value to track cost and schedule metrics, it will also want to track technical performance through the use of the measures of effectiveness, key performance parameters, measures of performance, and technical performance measures (reference Figure 2.2: Measuring Technical Performance, page 40). These measurements are a useful benchmark for comparing progress against the planned performance. A requirements traceability matrix for the project defines specifications and acceptable ranges (reference Figure 4.9: Example of Requirements Verification Matrix, page 124). Through the implementation of the technical performance measures, the critical technical parameters can be identified within the project scope and tracked (reference Figure 3.10: Example Technical of Performance Measurements, page 92). Technical performance measures provide the ability to perform incremental comparisons of the predicted value, tolerance levels, parameters, and thresholds, baselined in the scope against verified performance.

5.4.2 Resources Control

Project resources include not only the project team, but also all the resources that are responsible for performing tasks on behalf of the project. Maintaining control of the resources associated with a project is challenging. Resources include all the labor, time, materials and services, and all efforts provided by formal agreements (internal organizational direct and matrix reporting relationships, external organizational contracts and subcontracts, memoranda of agreement, etc.) and informal organizational agreements, such as staff that are loaned to the project for certain tasks. Labor resource availability will affect the schedule, labor salary and benefits costs, and any associated material and services costs, such as computing equipment and travel costs, that will also have a direct effect on the budget.

The riskiest project labor-related situation is when individuals work on various projects at the same time. This adds complexity to the project, particularly if critical and/or rare skills are allocated for only a portion of time on a project. Planning must be done carefully to consider the time allocation of these staff members, any conflicting work schedules, timing of work that the project needs, and time off for vacations or work trips associated with a non-project area of responsibility. This situation is more difficult for the individual as well, in that it is challenging to split time between one or more projects. Essential measurements include the monitoring of the actual hours worked versus the plan and the actual material and services planned against actuals. These two pieces of information will confirm to the project manager that the project resource spend is on a positive path.

The support of labor and materials and services tracking from organizational systems can provide readily available data if the organization's culture supports the collection of the data in the form and tempo that is needed by the project. Even if a system is in place, sometimes the tracking of employee labor charges is not completed at the necessary work package level. A project manager may not always have access to reports generated about labor charges associated with individuals who are not in their direct supervisory line, even if they are on the project. Whenever data are not available electronically, the project manager must compile the information by hand for each individual, by name and hours charged, and keep track of any additional time spent on or off the project that was not originally anticipated and planned in the schedule, which can be challenging for larger and more complex projects. Ultimately, the project manager must collect enough timely and detailed data to analyze the comparison between hours allocated by a named person for that time period and in aggregate so that any underspend or overspend trends can be resolved.

Another unknown associated with labor changes that must be controlled has to do with increases in salaries, such as for raises, which will directly affect the bottom line of the project. The mechanisms must be in place so that any planned changes to salaries for project staff who have direct line reporting outside of the project are communicated proactively to the project manager so that the cost impact to the project can be assessed. In some cases, a yearly increase across all staff can have such a dramatic impact on the project that it puts the baseline in jeopardy. Project managers must consider this inflation factor when compiling their original budgets, although it is a common occurrence to miss this fundamental requirement. In addition, the cost of recruiting, hiring, relocation, and other associated human resources activities that are directly associated with the project and not included in the general overhead factors must be planned for and monitored. In comparison, materials and services are relatively easy to control. Any unintended items can be managed within the budget either using unused budgeted funds—such as when a labor resource was hired later than planned, when a material costs less than anticipated, or through contingency.

5.4.3 Risk Control

Risk is a naturally occurring part of projects. Risk is prevalent regardless of whether it is managed or not. To control risk on a project is to actively manage situations that emerge, both positive and negative, throughout the life cycle and to take specific actions to reduce known technical risks in the design of the project scope. Chapters 3 and 4 discussed the processes associated with risk management, but there are some specific measurements that should be used to ensure that risk management and reduction are managed as a normal part of project activities.

The project manager will assemble the key stakeholders to review project risk on a regular basis. During this review of project risks, the highest risk items and the opportunities that have been presented are assessed and decisions made as to what, if any, actions should be taken. If risks are mitigated, then funding from the contingency may be used, drawing down the total amount available. New items may be brought forward, while some existing items may be reprioritized. This is a highly interactive process, and the decisions that are made are entered into the risk register by the project manager. If no decisions are being made, or if risks keep appearing that are not identified in the risk register, then the process is not functioning correctly, and an independent subject matter expert in risk management should be brought in to review the situation and correct any deficiencies. This type of active management will control the potential impacts of risk on the project.

The activities associated with technical risk reduction must also be carefully controlled to ensure that progress is achieved in time to meet the technical milestones and so that the project is not negatively affected. As described in Chapter 4, these technical risk reduction strategies include trade studies, prototyping, and modeling and simulation.

The controls and critical measurements associated with using prototyping as a risk reduction strategy can be described as ensuring that the form, fit, or function meets the specifications. Early versions of a product representation, in a physical form or in software code that provides an electronic representation of a product, are prototypes that can be measured against the known specifications. It reduces risk by providing a way to envision a system or elements of the system before substantial investment is made in the design. It ensures that the design will meet the specifications before investing in production by demonstrating functionality in a basic model to check form, fit, and function, and it mitigates the potential to overinvest if a capability is deemed unnecessary. Prototypes can also be used to test decomposed elements independent of the system. Changes that are driven through prototyping must go through change control and be communicated to the design teams.

Trade studies provide a risk reduction strategy associated with decision making between options. Using trade studies to evaluate alternative designs means that a clearly defined set of parameters is known and is used to evaluate solutions. The outcome of operating effects on the design can be assessed so that an optimal solution is chosen. Optimal means that the solution meets the technical needs, within the cost and schedule constraints of the project. Trade studies are typically concluded early on in a project, before the design is finalized. Measurements associated with progress on completing trade studies and documenting and analyzing the results are most effective. Decision making associated with trade studies must be communicated back to the development teams through careful change control.

The risk reduction that comes from modeling and simulation activities depends on how the technique is used. In its simplest form, it can demonstrate the capability of a particular part to confirm performance. In its most complex form, a model can be built to represent an entire system and to simulate its performance in a wide range of operating environments so that operations impacts can be addressed in the design and the system as a whole are optimized. Any version between these two examples can be deployed, depending on the project's needs. The most critical measurements for using this method are based on data integrity. The data that are used must be robust, defendable, current, and repeatable. Measurements using modeling and simulation techniques for risk reduction will include comparative results and change control submissions affecting the design.

The measurements that will be used on risk reduction depend on the project; however, they will be associated with quantitative and qualitative measures associated with the defined product definition and specifications. The most important things to remember about risk reduction strategies is that they are meant to test and verify performance to specifications, to learn about potential behaviors so that the design can be optimized, and to provide information for decision making associated with optimization of the design. Measurements must be made often throughout the life cycle and compared against previous measurements so that trends and performance parameters can be established. Performance parameters provide a range of acceptable outcomes, because results may not be identical.

5.4.4 Quality Control

As described in the previous chapters, quality is assessed both quantitatively and qualitatively to verify and validate the results of project activities. Quality control is the action taken to ensure that quality is measured and achieved throughout the life cycle of the system. During a project, quality controls will measure conformance of the design to the specifications, as well as progress toward meeting the measures of effectiveness, key performance parameters, measures of performance, and technical performance measures that have been defined. Measures of effectiveness reflect how the customer will perceive quality, and therefore, quality controls associated with measures of effectiveness must be applied as intermediate validation steps throughout the project, as well as when the project has come to its natural closure.

During production/construction activities, additional quality controls are added to ensure that the produced or constructed elements meet specifications. They may also measure process efficiencies such as manufacturing cycle time, yields, inventory turns, and number of defects or non-compliance events, to name a few. During operations, other quality control measurements are applied, such as time operating versus non-operating and the number and type of maintenance preventative and corrective actions performed.

All quality measurements are documented in the quality management plan. These quality measurements inform the controls that will be applied. Although all projects are unique, there are standards for quality that can help define the appropriate controls for each phase of the life cycle. The American Society for Quality provides guidance on the use of International Organization for Standardization (ISO)[8], Total Quality Management (TQM)[9], and Six Sigma and Lean. Depending on the type of project, software, hardware, process, or service, the quality measures that are chosen will inform the quality controls that will be applied.

In general, quality controls will take the form of test, verification, and validation. These three activities have different objectives but work together to ensure that progress toward outcomes is controlled. Test controls generally are in place to ensure that both real and simulated components, parts, subsystems, and the system being tested meet the expected performance specifications and tolerance levels and contain the appropriate attributes within controlled conditions. Verification controls ensure through modeling and simulation, inspection, audit, or physical examination that there is reliable evidence supporting the performance that is specified and that the performance is repeatable and stable. Validation controls ensure, through inspection, that the stakeholders concur that the performance demonstrated in the verification processes are suitable, are satisfied with the final outcomes, and that the end product meets their needs.

Whenever a quality control activity identifies shortcomings, these become risk items. The project manager and systems engineer must work to resolve the issue before progressing. Resolution is through standard risk management for the project and can result in different mitigation strategies. If changes are required to address the impact through the revision of scope, schedule, or budget, the change control process is engaged. In this way, these processes are tightly integrated. Throughout the system life cycle, quality is controlled across all process areas in many ways. The two methods that are identified as essential are the use of audits and gate reviews, because they have been shown to be highly effective in assuring quality.

5.4.4.1 Quality Audits

Audits are an independent activity used to assess conformance and compliance to processes and procedures, to verify the quality of products, or to assess the effectiveness of services being performed. Project managers and systems engineers perform audit activity throughout the project life cycle to guide the project toward positive outcomes. If independent subject matter experts are needed to perform an audit, such as when the subject matter is highly technical or outside the expertise of both the project manager and systems engineer, individual subject matter experts from outside the project are asked to perform the audit and report back to the project manager.

An audit can be used to control quality through the investigation and reporting of a lack of compliance to the established processes, procedures, or tolerances that have been established for the project. A report of this type is generally a written draft of findings and recommendations. The project manager will provide a formal response—which includes mitigations to resolve the issues—and is responsible for implementing the recommended changes within the project.

5.4.4.2 Quality Gate Reviews

Gate reviews use the evidence provided from audits, modeling and simulation, tests, and verification to obtain objective concurrence that the project is ready to progress within its life cycle. Reviews perform a critical role in the active management of a project. They provide a venue to thoroughly evaluate, discuss, and assess the true state of the project. Having this information allows well-informed decision making. Depending on the phase of the project, reviews can be informal conversations all the way to formal events. There is a significant body of knowledge associated with reviews in both the project management and systems engineering disciplines. Standard entrance and exit criteria provide a solid and accepted basis for which decision making can occur without debate. Deviating from these standards allows doubt to form as to the quality of the gate review outcomes and should be avoided at all costs. The recommended review steps outlined in the following sections are based on these standards.

Both the project manager and the systems engineer are involved in these reviews. The project manager leads the project management reviews, and the systems engineer leads the technical reviews. Because of the interaction of the scope on the budget, schedule, and risk, the project manager holds ultimate responsibility over the actions taken that are associated with the technical reviews. Any changes are formally reviewed through change control.

5.4.4.2.1 Project Management Reviews

The project manager holds quality and audit reviews associated with the performance of the project. These include performance to budget, schedule, and scope, along with benefit realization progress. In general, these reviews provide evidence that the work is being performance as planned, within the scheduled time frame, and within the budget parameters. Risk is assessed as well during these reviews. The objective of these reviews is to understand the current progress, to obtain enough information to make prudent and timely decisions, to correct issues that are expected to drive the project off course or impact the ability of the project to achieve its outcomes, and to remove blockers to progress.

The number and formality of these reviews is project dependent; generally, there is a pre-published agenda and attendees list, the minutes of the meeting are taken, all action items and decisions resulting from the review are documented, and assignments to the appropriate teams are distributed. Before the next review, these action items are completed, and the status is reported at the next review.

5.4.4.2.2 Systems Engineering Reviews

There are a significant number of reviews available to the systems engineer. These reviews are typically initiated to evaluate the progress and quality of the technical work packages. They demonstrate technical progress, technical risk reduction, and conformance to specifications. A project is typically not allowed to move from one phase to another until it has demonstrated that it has met its objectives for that phase of the project and is at the appropriate state of readiness to proceed. Quality control is met when criteria to enter and exit the review are set following standard systems engineering processes. Failure to utilize standard processes increases the risk of approving a project to progress before it is technically ready. As described in Chapter 3, this situation will increase the technical debt in the following phase.

These reviews are generally hosted by the systems engineer and attended by the project manager, all key stakeholders, as well as any independent-from-the-project subject matter experts if needed. The systems engineer will use these review outcomes to signal a readiness to progress from one phase to the next or to hold the project from progressing until concerns are addressed sufficiently.

The essential reviews that are discussed here are those that will assist in the assessment of quality and the controlled progression of scope development over the course of the life cycle. For those purposes, the critical systems engineering reviews include:

- System definition review (SDR)
- System requirements review (SRR)
- Conceptual design review (CoDR)
- Preliminary design review (PDR)
- Critical design review (CDR)
- Production/construction readiness review (PRR/CRR)
- Operations readiness review (ORR)

Figure 5.5 demonstrates how these reviews would be applied over the course of a project life cycle. These reviews are further described in the following sections.

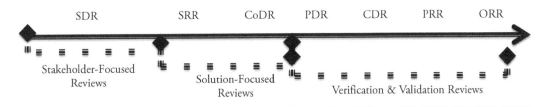

Figure 5.5   Continuums of Reviews

Systems Definition Review

The systems definition review is the most important stakeholder-focused review that a project will facilitate. The objective of the review is to confirm with all the key stakeholders what the project team understands are the needs that the project will fulfill and the benefits to be realized. Ideas about how the stakeholders will operate the systems, their descriptions of use cases and scenarios, their measures of effectiveness, and the resulting derived stakeholder requirements and validation criteria are confirmed at this review.

The full design team must be in attendance, because they need to be fully exposed to the expectations of the stakeholders so that the optimal design can be crafted. At this review, key performance parameters are also confirmed. The completion of this review and the resulting aligned design strongly increase the probability that the project will be able to successfully achieve its outcomes to the satisfaction of the stakeholders. If this review is not completed, it will increase the risk to both the stakeholders for getting a viable system out of the project that contributes to business value, and the project team for being able to develop the solution within the budget and schedule allocated for the project. This will lead to what will become a high-risk change environment associated with scope evolution, because there will not be a way to control the expectations of the stakeholders.

Systems Requirements Review

The systems requirement review will be used to assess the solution-focused project scope descriptions against the configuration-controlled system definition as described during the systems definition review. The review is held once the entrance criteria are complete. The entrance criteria include the design criteria as they have been converted to technical requirements. Performance specifications, specification tree, measures of performance, a system functional architecture (a top-level box diagram describing the functions), and an as-allocated graphical representation of the product tree are critical entrance criteria. The review will confirm with the stakeholders that their needs were heard correctly and give the project team confidence that what they technically envision is a solution that will lead to a valid outcome.

Conceptual Design Review

As in the previous reviews, the conceptual design review plays an important role in quality control. Its purpose is to confirm that the design as envisioned is reasonable, realistic, and attainable. Flow down from the system definition to the system requirements and to the conceptual design should be robust and verified. Prior to the conceptual design review, the physical architecture design, including the positioning within the architecture of all known interfaces, will be drafted.

During the conceptual design review, supporting documentation associated with the designs are made available to the review participants and will include conceptual design drawings, technical assessments, trade studies results, modeling and simulation results, and analysis of interfaces, along with the documentation compiled during the system definition and system requirements reviews. The requirements for test, verification, and validation associated with the measures of effectiveness, measures of performance, and key performance parameters are provided as well. It should be clear from the documentation where each of the project products aligns into different life cycle phases. Risks associated with the conceptual design should be clearly articulated in the risk register, along with their prioritization, ownership, and mitigations.

Exiting from a successful conceptual design review establishes the project functional scope baseline. This will allow the project team to start their requirements flow down, continue to reduce risk for those early life cycle system elements, and invest appropriately in the resources necessary to reach a viable preliminary design.

Preliminary Design Review

The quality gate of the preliminary design review ensures that the design as allocated—with appropriate requirements flow down—has been accomplished and that the preliminary design of the system meets specifications. The risk portfolio will have changed from conceptual design review to preliminary design review, and a revised understanding of the risks going forward is imperative. The entry criteria for this review include all documentation from the previous reviews plus all supporting materials that have emerged since the conceptual design review. These will include additional trade study and modeling and simulation results, as accepted and approved through formal change control.

During the preliminary design review, an assessment will be made as to the completeness of the material specifications for all levels of the system as designed to date. This assessment can be based on supporting analysis, results of studies, prototype successes or failures, or any other evidence that supports the as-allocated design. Technical performance measures will be presented, approved, and then used for development from preliminary design review through to the critical design review.

Exiting from a successful preliminary design review establishes the project technical as-allocated scope baseline, which includes specific detailed design work packages for each configuration item in the system. As discussed in Chapter 2, a configuration item is a product of the project that is tracked through the assignment of a unique identifier. These are typically items that evolve with design, are carefully configuration controlled, and are critical to the success of the project.

The preliminary design review is conducted once all major design issues have been resolved. The review is meant to ensure that all requirements have been allocated and flowed down in a manner that will be sufficient to test, verify, and validate that performance. The preliminary design must be demonstrated to be feasible within the risk profile and within the cost and schedule constraints. As described earlier, if the preliminary design review does not uncover gaps in the technical design and the preliminary design review is passed, this technical debt will increase the risk associated with the budget and schedule investment that must be made in the following phase. Detailed design of the system commences from this point and should progress with the least amount of technical debt possible, due to the fact that between preliminary design review and the critical design review, the maturity of the design must be such that full-scale production or builds can progress without technical debt.

Critical Design Review

The critical design review is the final design gate before the project products enter production/construction. At the critical design review, the as-designed baseline across all supporting disciplines is set. The maturity of the design includes detailed specifications that meet all the performance requirements that have been tested, verified, and validated throughout the previous phases. The design at this point will be built within the established parameters of budget and schedule. Risk should be minimal, with all impactful risks to the design retired.

Upon successful completion of the critical design review, the designs are evolved to the point at which they can confidently be used to inform manufacturing, production, and/or implementation teams and provide artifacts required to fabricate, build, or code the project products. Any final risks must be mitigated at this point. These detailed drawings and specifications associated with the configuration items form the initial product baseline for the build.

Production/Construction Readiness Review

As discussed in Chapter 2, the production/construction team, rather than the systems engineering team, usually hosts the production/construction readiness review. In some cases, in projects with limited production/construction requirements—such as with a software or process product—integrated teams may not hold a production/construction readiness review but will progress directly to an operational readiness review. However, for projects that have a heavy reliance on production/construction, a production readiness review is a necessity.

The production/construction readiness review establishes that the build-to design that is provided is ready for the production/construction phase. These build-to design documents should be adequate for subcontracting to vendors or other parties associated with the project. The planning on the manufacturing or build side that is necessary to produce the design must also be in place. For example, certain build and support equipment, such as molds, platforms, tools, test facilities, etc., may need to be acquired and installed. All fabrication, assembly, integration, and test capability must be in place. Support equipment that will be used in maintenance activities must be identified and procured or built. The determination of the need for first production articles, or a phased approach starting with a low-rate initial production, to confirm the design before the full production/or construction can begin must be made. And the processes associated with integrating engineering changes and addressing discrepancies and latent defects during the production/construction must be put in place. And finally, a measurements plan must be in place and an auditing process implemented to verify the quality of the production/construction outputs. The production/construction readiness review will determine if the project is ready to move into this phase.

Operational Readiness Review

The operational readiness review is the most critical review for stakeholders. The system, as-built and ready to move into operations, is presented. All supporting operations documentation, including maintenance, training, operations manuals, specifications for support equipment, etc., must be included.

At the operational readiness review, the project team confirms that the system is physically ready to operate, and stakeholders have the chance to establish their satisfaction with the capability. The measures of effectiveness will be reviewed against the delivered system (the as-built), and the performance specifications that have been verified and validated throughout the design and build processes will be confirmed. Established operational measurements are confirmed. The key stakeholders will formally sign an acceptance certification, and the system turn-over will be complete. The systems as-maintained configuration is then managed through operations.

5.4.5 Governance Measurements

Governance activities, by nature, are controlling functions. To call them out separately in this chapter is just to reiterate that the measurements associated with governance activities are essential to the successful performance of the project, and for risk reduction that will greatly increase the probability that successful project outcomes will be achieved. As discussed in Chapter 4, governance is applied to the integrative activities across all disciplines that are associated with the project. Governance measurements typically come in the form of change and configuration records that track all evolution against the project baselines that have been established.

5.4.5.1 Change Control Measurements

During the early phase of a project, a change management plan is developed by the project manager to define how change control will be performed on a project. It describes the processes for proposing, reviewing, approving, implementing, and communicating change throughout the multiple disciplines involved and across the full life cycle of the project.

A basic change control process will start with a change that is nominated and presented by the person requesting the change. If the change is minimal, the project manager may approve it without the need to consult with others. Otherwise, the project manager convenes a change control meeting and will distribute an agenda and supporting documentation to the change control participants ahead of a decision meeting. Because changes often have unintended consequences to far-reaching disciplines, a cross-discipline team of reviewers, including the proposer and any subject matter experts, is invited to review the potential change.

As described in Chapter 3, individuals that oversee the baselines and the process owners of any effected area must staff an interdisciplinary change board. During the change control meeting, a configuration controlled change log is updated to identify the request, record the actions and binding decisions, document the approvals, assign the change implementation ownership, and communicate the change to the project team. If necessary, the risk register will be updated as well. A decision on a change request may be deferred to perform additional research and analysis; however, these situations should be rare. In those instances, a firm time limit should be placed on the additional activity and a return date scheduled during the change meeting.

Measurement of change control comes in the form of the long-standing record of proposed and approved change requests, and the audit results from reviewing the implementation of the approved changes. This change process requires careful attention and process discipline from the whole team to be effective. If this process is effective, there will be few instances of surprises, and the design process should proceed in a stable manner. If this process is not effective, however, there will be ongoing and significant impacts from surprises, out-of-control scope, unintended impacts to budgets and schedule, and lost baselines.

With the use of rigorous change control processes and effective use of measurements, at the end of the project, the baselines plus the accepted changes will equal the completed and accepted project specifications. In addition, the difficult decisions that generally result from poor performance in projects have been made, such as revising the specifications for a technology development once it is clear a specification cannot be met. The use of contingency and the impacts on risks will have been carefully controlled.

5.4.5.2 Configuration Control Measurements

The systems engineer is responsible for the configuration management plan, measurements, and implementation. This is a cross-discipline activity that requires participation from every discipline in reviewing and understanding the potential version change impacts, just as was described in change management. The impact of losing configuration control on anything (e.g., documents, prototypes, experiments, design of a configuration item) can have far-reaching consequences on a project. At a minimum, when configuration status is lost on a configuration item, there may be an impact to the schedule and the budget in trying to resolve the loss of configuration before the project can proceed.

A configuration status accountant maintains the configuration baselines, changes, and implementation of those approved changes across the project team. He or she is responsible for recordkeeping, reporting, calling reviews, maintaining the record of the technical specifications, and providing documents of the correct design versions for review.

Configuration audits are performed to evaluate adherence to configuration management processes and procedures. A measurement of successful configuration control comes from the knowledge and assurance that every project participant has the most current version of a product, drawing, document, etc. available. Approved configuration changes, as well as audit results from a review of the configuration items, are maintained by the systems engineer or configuration status accountant. This configuration process requires commitment across the project and, if it is effective, little time will be lost in resolving configuration issues. If the process is not effective, ongoing impacts will drive risk, affect quality, cause confusions, and ultimately end in specification non-conformances, causing a lack of traceability throughout the project.

5.5 Communicating Measurements

The display of measurements so that they convey pertinent information in a timely manner is a project imperative. This section describes the methods that are useful in communicating progress associated with the essential measurements.

- Misunderstandings of communications are common, and team members, as well as stakeholders, can unintentionally misconstrue the messages about project progress.
- A display, plot, diagram metrics, measures, or other visualization of each measurement is helpful in the dissemination of information that may be difficult to explain otherwise. Visualizations provide simplifications of text descriptions and help put the situation into an understandable context.
- Conveying a situation through visualizations comes with a requirement to ensure that the representation is accurate. This is particularly critical if the display of measurements drive decision making.

The type of communication visualizations usually associated with the project management and systems engineering essential measurements are generally the summary type of charts, such as histograms, and control charts showing progress against the plan (earned value, schedule and budget, and technical performance measures). In many cases, data matrices, such as those created in a spreadsheet, are used to display information that is multidimensional and complex. Spreadsheets however, are often difficult to view, difficult to put in context, and almost impossible for anyone other than the person putting in the data and building the formulas within the spreadsheet to recreate, no matter how simple it appears on the surface.

Although there are any number of diagrams and charts that can be used to convey project status and health, there are two types of communications charts that can provide a snapshot of essential measures. These are the foursquare and stoplight charts.[10] These high-level summary communications charts can contain whatever information the project manager and systems engineer feel is the most useful in conveying the status information that will drive appropriate decision making. Typically, they contain some combination of information about budget, schedule, and scope performance, risk, benefits, and earned value graphics and indices when appropriate. An example of a foursquare chart is shown in Figure 5.6.

A summary status of all work breakdown structure work packages can be tracked using a stoplight chart. This type of chart does not provide any detail associated with the activities identified, but it provides a quick look at performance so that attention can be focused on those areas that are posing the greatest risk. This method helps focus management's attention on the activities that are posing the highest risk to the project. Follow-on discussions and investigation into what situations are driving the performance issues are then held. The stoplight chart identifies those areas to be tracked. In the simple example shown in Figure 5.7, the stoplight chart shows performance to cost, schedule, and scope parameters.

Both the foursquare and the stoplight charts are compiled monthly or quarterly as appropriate for the project. Whichever methods are used to communicate measurements, the intent is to drive the appropriate behavior that will allow accurate and timely decision making to occur so that the project is able to achieve the outcomes intended.

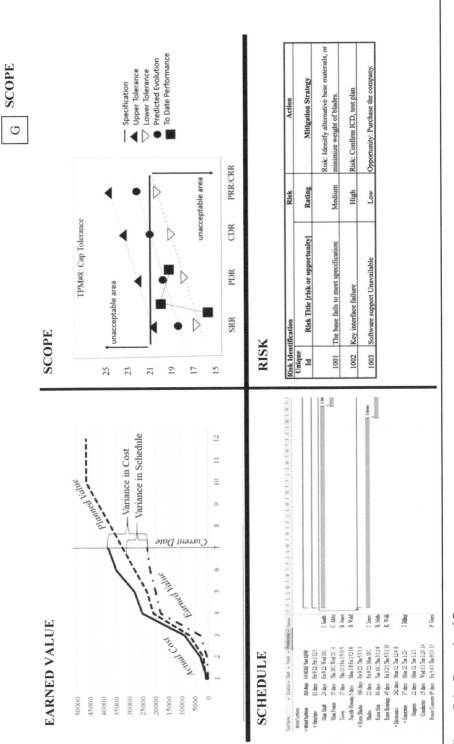

Figure 5.6 Example of Foursquare

Project	WBS		Task	Cost Performance	Schedule Performance	Scope Performance	Suspected Issue
1.0 Wind Farm Project				Y	G	G	Interdependency with task 3 impacing cost
	1.1	Structure	Structure Fatigue Testing	G	G	G	
			Bearing Axial Stress Test	G	Y	G	Staff vacation impacting schedule
			Drivetrain Rotor Loads Stress Test	R	G	G	Early purchase of materials and equipment
	1.2	Rotor Blades	Blade Stress Tolerances Fatigue Test	G	G	G	
	1.3	Electronics	Braking Tolerances Performance Test	G	G	G	
			Voltage Profiles Performance Test	G	G	G	

Legend	>$100k = Red	>40 days = Red	>5% of Spec = Red
	>$50k = Yellow	>20 days = Yellow	>3% of Spec = Yellow
	up to >10k +/- = Green	up to >3 days +/- = Green	up to >1% of Spec +/- = Green

Figure 5.7 Example of Stoplight Chart

5.6 Case Study: Alva-Green Coaching Group

All case study quotes are from interviews held with the individual presenting the background. In this next case, the interview was held with Pat Alva-Green, and all quotes are attributed to her.

In 2004, Pat Alva-Green took a different career path than the one she had been following for over 30 years in corporate America. She decided to turn her passions into a business to help women professionals and executives balance their professional and personal lives. She chose this path when her life changed, and she found herself in a place of options and opportunities. Alva-Green explained, "Rebuilding a career after a major corporate downsizing and after surviving breast cancer taught me that with intention, clarity, focus, and action anything is possible."

Although she had been coaching as a side job and had not up to that point had enough confidence to go into the field full time, she decided that she would not allow herself to be so dependent on working for an organization, but would "hang my shingle and go into this full time." She wanted to be independent and follow her passion, while helping other women achieve their goals. "I am passionate about helping women discover and put into action the tools, plans, and resources that create extraordinary results and provide greater personal fulfillment," Alva-Green said.

Her expertise as a coach was developed through coursework, experiences, and training in the corporate world. She explains, "I developed my expertise through my years with companies such as IBM and Lockheed Martin, where I coached teams, professionals, managers, and executives, and my own personal talent."[11] She found herself as the go-to person on the teams she worked on for anything interpersonal or lifestyle related. Her easy manner, "whole person approach," and deep listening skills allow her to assist her clients with their most challenging issues. Providing this support in the workforce was such a successful model that she started offering her coaching abilities outside that environment as well. Developing her own company to offer lifestyle coaching was a dream she had held for a long time that was now going to be a reality.

5.6.1 Project Charter and Plan

Alva-Green's plan was to develop a coaching company to assist her clients in creating the lifestyle they desired through coaching and retreats. She would also promote the business and enroll clients through speaking engagements and teaching professional development courses. She planned to incorporate coaching and project management into the courses that she already offered in mediation and in Polarity and Reiki energy therapies, for which she held certifications. Because she is bilingual in English and Spanish, she would market to both communities of speakers. Alva-Green explains, "I believe that deep down my clients have the answers they seek, and that through deep listening and asking powerful questions, I provide the space for my clients to discover their own answers. My clients become aware and know who they really are and are able to achieve these changes in their lives."

In discussing her desire to focus on women in the workplace, Alva-Green explains, "My experience has shown that 75 percent of clients are women—they are the ones who are willing to pay for the coaching, and they acknowledge that in order to transform themselves they require help." She further clarifies that they often do not feel like their talents are being utilized effectively or that they have skills that are being suppressed. "Clients desire clarity on why things aren't working, first and foremost," Alva-Green says.

5.6.2 Stakeholder and Communications Management

Determining who the stakeholders were for her business, and the types of communications that would be required took some research. As Alva-Green explained, "I assessed who my ideal client was and what services and solutions they needed, along with which social media they used. Identifying the channels that are needed to get my message out is critical to success." She explains, "Although social media is a great vehicle to get your message out, community involvement is still key. Entrepreneurs need to spend at least 60 percent of their time marketing. Your message must be clear and in their language." She decided to use face-to-face meetings and the internet to reach her stakeholders. Alva-Green established an internet presence and publicized the opportunities to meet with her face to face. Communications were distributed through multiple social media sites[12,13] and a dedicated website.[14]

5.6.3 Scope Management

To incorporate the planned scope, Alva-Green needed to have a path forward to implement each of the core focus areas of her business. The coaching, retreats, and marketing efforts all had different requirements and paths to implementation.

5.6.3.1 Architectural Design

Alva-Green had a vision as to what she wanted to provide to her stakeholders. This vision was a design that included four separate pathways associated with helping women "make a difference in the world and be fulfilled in their lives," Alva-Green explained. The coaching activities would include, for example, topics such as clarifying career goals, preparing to make bold moves into new work positions, and finding practical ways to address interpersonal issues in the workplace.

Speaking engagements and opportunities to teach professional development courses depended on the requirements specified by the event planners or the professional organizations that would contract with her. However, they would all showcase her talents and solicit interest in her counseling retreats. As Alva-Green explains, "I would receive requests to give a presentation or run a course based on topics of interest to my clients and for which I have expertise." Although these activities were important, her passion was in the area of providing retreats that would help women see a way forward in their lives and their careers.

5.6.3.2 Requirements Management

To initiate her business, Alva-Green explains, "I needed to have a place to hold retreats, as well as to obtain retreat materials, including material and content to share, business cards, presentations, and other lecture materials. It was important that I had certifications in the fields of study and that I have a solid reputation." She would also require an internet presence that would reach her intended stakeholders.

5.6.4 Acquisition Strategy and Integration

As part the process she went through to set up this project, she needed to decide on the type of business she wanted, what she would call it, how to register it, and what type of support she would need. For example, the help of a certified public accountant, an illustrator to develop a logo, technical support to set up a website, printing for business cards, a business coach, and other start-up needs all required planning, funding, and then implementation.

5.6.5 Risks and Opportunity Management

One of the most pressing risks was that when retreats were planned, all costs must be recovered. Alva-Green initiated a cancellation policy that would ensure at least four participants were booked, eight desired, to host a retreat. As she learned the business and experienced some hard lessons, as a risk mitigation step she started separating out the costs for lodging, food, supplies, time, etc. and having the participants pay for lodging and food directly. Alva-Green explained, "Minimizing my cost risk meant transferring some of the costs, mostly the highest dollar value costs such as lodging, directly to the client." Then the only costs that would be at risk were the retreat supplies and her time. She now reserves the block of rooms, but attendees pay for their own lodging and meals directly.

Coaching risks are usually minimal; Alva-Green explains that she would see "consistency in the problems that women are experiencing in the workplace" and build modules that address those needs. For example, she often sees that women do not set good boundaries. So, one of the modules she offered is learning how to set boundaries and to deflect challenges to those boundaries. Alva-Green helps the clients with an approach to present how they want to be treated. She also provides coaching modules on creating and maintaining a positive mindset, time management, de-cluttering, and developing rituals that will help balance their lives. "[You] can't let success come in if you have a very cluttered life," Alva-Green explains.

5.6.6 Quality Management

Quality management is mostly assessed through client interaction and bookings. "When you get the feedback that the client wants to schedule a coaching session or a retreat, or they stay to the end of the webinar, that is when you know you have accomplished your goal," Alva-Green says. She specifically monitors which videos get the most responses and tracks the questions that she raises in her talks or through responses to the newsletters that are posted to her website. She is looking for things that resonate with her audience, particularly those that elicit a response that results in booking an event or coaching session. Alva-Green says, "Quality is directly correlated to how these women are relating to what I am saying or writing." Journaling to keep track of what works and what does not is a good quality management technique.

In addition to attracting the clients that support her business, Alva-Green also measures her success in that she "must have fun at what I am doing! If I'm not enjoying the experience, I am not successful." She also explains, "I have a passion to move people forward. I love what I am doing. Being able to provide the space for women to come in and accomplish what they are wanting is success."

5.6.7 Test, Verification, Validation

To ensure the highest success rate for enticing new and recurring client participation, Alva-Green follows a methodical approach and "changes only one thing at a time so that it is clear what worked. If you change too many things at once, you will not know which factor was the one that drove the outcomes," Alva-Green said. She explains that it is important to test how things are said, how opportunities are positioned, and indicates that it is necessary to "test and tweak a process to make sure it is optimized." Most importantly, she exclaims, "I refuse to be afraid! Even if I put something out there and it fails, I'm looking for the one thing that I can change that will provide a different result."

5.6.8 Governance

Although Alva-Green initially started her business focusing on women in the corporate world, she would like to embrace the entrepreneurial community. Her intuition led her to take advantage of the larger market. She is investigating what it will take to evolve to a more general business focus, rather than one that is strictly women focused. This is a change that will affect her offerings to women and requires a new strategy to attract the broader client base into the events and coaching that she is interested in hosting. The way the business is currently configured, change of this magnitude will require re-branding and refocusing of the marketing to reach the broader audience.

5.6.9 Outcomes

Alva-Green feels that the outcomes she has experienced as an entrepreneur, focused on helping women make changes in their lives that move them forward, has been largely successful. However, as she has discovered over time and with experience that her passions are evolving. Her initial focus on helping women achieve their objectives is developing into wanting to help people who are interested in starting a business find their way forward.

5.6.10 Lessons Learned

- One of the major lessons learned from implementing and executing this business is that it is not possible to do it all. As Alva-Green explains, "You must determine what you are clearly good at, and then hire out the rest so that you can focus on bringing your best forward."
- Journaling is a critical skill and activity. As she describes, "You need to log those lessons so you don't keep doing them over and over again."
- "Questions are powerful!" Alva-Green emphasizes. "No one has all the answers. It is important to trust yourself, but to ask questions that allow the answer to surface within you." She further suggests that phrasing the question is helpful, such as, "'What does the life I desire to live require of me today?' This question opens you up to possibilities."

5.6.11 Case Analysis

This case describes the implementation of a small business with evolving and growing requirements. It reflects an application of systems engineering processes in ways that help define, elaborate, and implement processes that were directly aligned to the outcomes that were experienced and could be traced in a manner that evolution could occur. Risk management is an important component of this project, and the process was directly applied to minimize impacts to business costs. As the scope of the business continues to evolve, these processes will help Alva-Green focus on the best value for her efforts.

5.7 Key Point Summary

Chapter 5 explores what success means and why it is important when choosing measurements. It also describes the various Complex Systems Methodology measurements available and explains their use in driving decision making and in communications across the project and with the stakeholders, both incrementally and throughout the life cycle of the project. This chapter focuses on essential measurements that are most effective in supporting active management of the project through the repetitive assessments performed against project baselines.

These concepts apply the learning from Chapters 3 and 4, which described the complementary and unique stakeholder-focused and solution-focused activities that are required when using project management and systems engineering processes in a project. Key concepts from this chapter are described below. Key terms are provided in the Glossary.

5.7.1 Key Concepts

- Project processes, as defined in Chapters 3 and 4, must be actively managed using meaningful measurements to decrease the risk of failure.
- Success is a subjective measure and is dependent on perception. Because of this, the project manager and systems engineer must carefully define what success means and then choose the appropriate measurements that inform that view.
- Measurements can serve as an early warning system associated with triggers that drive risk within the project.
- To successfully use measurements to inform decisions, some foundation support requirements must be met associated with organizational structure, communications, and governance.
- Measurements must be based on accurate information and then summed up in such a way as to eliminate the distracting noise level associated with information overload. Measuring too much, measuring the wrong things, or using inaccurate data to draw conclusions are important things to consider.
- Choosing measurements includes the consideration of the impact on resources in order to ensure that the value obtained by the measurements is worth the cost.
- Essential measurements assess quality in the areas of key performance and life cycle progression readiness. Similar to all measurements, they require active management throughout the project and throughout the life cycle to realize the value.
- Effective communications require that measurements be displayed in coherent, accurate, timely, and understandable visual formats.

5.8 Apply Now

Defined in this chapter are the necessary steps to apply measurements to the project management and systems engineering processes that have been presented in Chapters 3 and 4. The Apply Now section provides a summary table (see Table 5.1) to help visualize the chapter material, along with questions that prompt the reader to understand the use of measurements and to choose appropriate measurements for a project that they are experienced with.

1. *Project success can be defined in many different ways; however, the most important perception of success comes from the stakeholders.*
   Think of a project that you have completed in the past in which the stakeholders were not satisfied with the outcomes of the project and yet the project team was satisfied. Describe which measurements could have been applied to mitigate that risk on the project. Mark in the appropriate boxes in the table.
2. *Using measurements successfully requires some supporting foundations be in place—for instance, the organizational structure, communications, and governance activities supported in the organization must be in alignment with the project needs.*

Table 5.1 Apply Now Checklist

COMPLEX SYSTEMS METHODOLOGY™ SM (CSM™ SM)

APPLICATION FOR ALL PROJECTS

Questions to Ask	Responses										
	Stakeholder	Communications/ collaborations	Needs analysis	Scope baseline	Quality						
Are complementary stakeholder-focused documents/processes in place?	Register/ charter	Communications/ collaboration plan/ issues register	Stakeholder requirements/ use cases, concept of operations, scenarios	Validation	MOEs, KPPs						
	Cost baseline	Schedule baseline	Functional architecture	Scope baseline	Quality	Resources	Physical architecture	Risk	Governance		
Are complementary solution-focused documents/processes in place?	Baseline budget/ budget methodology	WBS/baseline schedule/ schedule methodology	Functional architecture diagram/ interface definition	Technical requirements/ baselined scope/ interface definition	MOP/ TPM/ reviews/ RVTM/ KPI	Acquisition/ procurement/ labor, materials, services/ workforce management plan	Performance specifications, integration requirements, ICDs/design solution	Risk register/ management/ prototypes, modeling & simulation, trade studies			
									Change control	Configuration control	Integration/ test, verification, validation
Which measurements will be used?	Cost and schedule baseline	Scope baseline	Resource controls	Risk control	Change control	Configuration control	Reviews	MOE, KPP, KPI, MOP, TPM	Change control	Configuration control	Integration/ test, verification, validation
	Baseline, 4-square, stoplight	Baseline, 4-square, stoplight	Contracts, agreements	Risk register	Change meetings	CI	SRR, CoDR, PDR, CDR, CRR, ORR	MOE, KPP, KPI, MOP, TPM			

COMPLEX SYSTEMS METHODOLOGY™ SM (CSM™ SM)

APPLICATION FOR ALL PROJECTS

Questions to Ask	Responses								
Are complementary stakeholder-focused documents/processes in place?	Stakeholder	Communications/collaborations	Needs analysis	Scope baseline	Quality				
Are complementary solution-focused documents/processes in place?	Cost baseline	Schedule baseline	Functional architecture	Scope baseline	Quality	Resources	Physical architecture	Risk	Governance
Which measurements will be used?	Cost and schedule baseline	Scope baseline	Resource controls	Risk control	Change control	Configuration control	Reviews	MOE, KPP, KPI, MOP, TPM, TRL	

Describe a situation in which a misaligned organizational structure, lack of communications infrastructure, or lack of support for governance activities led to poor outcomes on a project. Explain how this risk could have been mitigated based on your learning from the chapter. Mark in the appropriate boxes in the table.

3. *Essential measurements include baseline control, resource control, risk control, and quality control.*
   Thinking of a project that you have been a part of, which of these essential measurements were employed? What was the outcome of the project? Which essential measurements could have led to better project outcomes had they been implemented? Mark in the appropriate boxes in the table.

4. *Change and configuration measures ensure that evolution is carefully controlled.*
   For a project that you have been involved in, explain how change and configuration controls were implemented and measured. What was the project outcome? Could different implementation of change and configuration controls have led to a different outcome? Mark in the appropriate boxes in the table.

5. *Communication of measurements is important to project success.*
   Describe the communication mechanisms that have been used on a project you are familiar with to share measurement results. What worked well, and what worked poorly? Would the foursquare and stoplight charts provide any additional visibility into the measurements? Mark in the appropriate boxes in the table.

References

1. Wingate, L. M. (2014). *Project Management for Research and Development: Guiding Innovation for Positive R&D Outcomes,* Boca Raton, FL: CRC Press/Taylor & Francis Group.
2. Rad, R. F., and G. Levin. (2006). *Metrics for Project Management: Formalized Approaches, Management Concepts.* Vienna, VA: Management Concepts, Inc.
3. Wingate, L. M. (2014). *Project Management for Research and Development: Guiding Innovation for Positive R&D Outcomes.* Boca Raton, FL: CRC Press/Taylor & Francis Group.
4. Defense Acquisition University. (2001). *Systems Engineering Fundamentals.* Fort Belvoir, VA: Defense Acquisition University Press.
5. Wingate, L. M. (2014). *Project Management for Research and Development: Guiding Innovation for Positive R&D Outcomes.* Boca Raton, FL: CRC Press/Taylor & Francis Group.
6. Lipke, W. (2003). "Schedule Is Different." *The Measurable News.* College of Performance Management of the Project Management Institute.
7. Fleming, Q. W., and Koppelman, J. M. (1996). "Forecasting the Final Cost and Schedule Results." *PM Network* (v. 10) 1:13–18. Newtown Square, PA: Project Management Institute.
8. International Organization for Standardization. (2013). *ISO Standards.* Geneva, Switzerland: International Organization for Standardization. Accessed December 5, 2013, from http://www.iso.org/iso/home/standards.htm
9. Deming, W. E. (1986). *Out of the Crisis.* Cambridge, MA: MIT Press.
10. Wingate, L. M. (2014). *Project Management for Research and Development: Guiding Innovation for Positive R&D Outcomes.* Boca Raton, FL: CRC Press/Taylor & Francis Group.
11. Alva-Green Coaching. (n.d.). "About Alva-Green." Accessed July 9, 2017, from http://www.alva-green-coaching.com/About.htm
12. Alva-Green Coaching. (n.d.). Accessed January 1, 2018, at https://www.facebook.com/PatAlvaGreen/
13. Alva-Green, P. (n.d.) "Patricia Alva-Green." Accessed January 12, 2018, from https://www.linkedin.com/in/patricia-alva-green-437b471
14. Alva-GreenCoaching.com. (n.d.) "Pathway to Profit Through Productivity." Accessed January 12, 2018, from www.alva-greencoaching.com

Chapter 6

Tailoring

Tailoring is important for all of the tasks described in Chapters 3 and 4. Understanding the processes associated with project management and systems engineering is the first step in applying the processes appropriately to optimize project outcomes. This chapter describes the Complex Systems Methodology for determining the appropriate amount of rigor when applying each of the processes. This rigor is referred to as *tailoring*. Tailoring decisions depend on the:

- life cycle stage of the project
- life cycle stages of the products within the project
- impacts of risk drivers

Section 6.1 focuses on the definition of tailoring. The activity of tailoring is often misunderstood, even though it is an essential activity in both project management and systems engineering. This section provides a common language and understanding for applying tailoring so that it is defensible, repeatable, and understandable to the stakeholders.

Which factors need to be considered when making tailoring choices and how the life cycle phase of the project drives tailoring requirements is explained in Section 6.2. The many factors to be considered make it essential that the person responsible for tailoring has the knowledge and experience to make decisions that do not have detrimental long-term effects on the project. Tailoring requires an accurate understanding of the project needs and the life cycle phases of the project and all the products being developed as part of the project.

Implementing project management process tailoring is described in Section 6.3. Special consideration for essential processes is noted, along with information on the impacts that improper tailoring might have in those process areas. Suggestions for tailoring up and tailoring down are provided. Section 6.4 follows the same construct, but for the tailoring of systems engineering processes.

In Section 6.5, an approach to decision making and communicating tailoring decisions is outlined. Communications of the tailoring to the project team is an imperative, so that there is a common understanding as to what processes need to be applied and how they should be applied. Following this is the development and implementation of a tailoring plan and auditing of the actions throughout the project life cycle.

A case study using a non-traditional project in which tailoring was applied is described in Section 6.6. As shown in the case study, project management and systems engineering processes were used effectively to deal with the complexity of an ever-changing environment.

Key concepts learned in this chapter are provided in Section 6.7, and an Apply Now exercise is offered in Section 6.8 that allows the reader to put the materials learned in this chapter into practice.

Chapter Roadmap

The focus on the art of tailoring is described in Chapter 6. Appropriately applied tailoring of project management and systems engineering processes can provide a solid foundation for project success and can save the project cost and schedule impacts from unnecessary process applications. This chapter specifically:

- defines what tailoring is and how it is used
- describes the factors that influence tailoring choices
- introduces the concept of tailoring by life cycle phases
- provides direction on how to apply tailoring
- explains the risks associated with tailoring
- provides a case study to assist the reader in applying the chapter lessons
- includes a summary checklist of key chapter concepts
- provides Apply Now exercises to assist in the understanding of the tailoring concepts described in this chapter

6.1 Definition of Tailoring

This section provides a description of Complex Systems Methodology tailoring that can be used on all projects.

- Tailoring is a standard process within both the project management and systems engineering disciplines.
- The need for appropriately scaled rigor of the project management and systems engineering processes is referred to as tailoring.
- Project managers and systems engineers use tailoring to ensure that the right amount of process is applied to the project processes.

Chapters 3 and 4 describe the stakeholder-focused and solution-focused project management and systems engineering activities that will be tailored at the beginning of each project and when transitioning from one life cycle phase to another (e.g., development to production).

Tailoring is an important concept that is often misunderstood by casual practitioners of the project management and systems engineering disciplines, as there is not a single, comprehensive description between the disciplines, nor within them, that provides enough information to apply tailoring appropriately. In Complex Systems Methodology, tailoring does not mean eliminating or ignoring a process. It refers to how much effort is put into the definition and implementation of a process.

Each project is different, and tailoring is an art best applied by knowledgeable and experienced experts in each of the disciplines, as they must use their judgment to determine what amount of detail is needed in order to ensure that they provide sufficient measurement information so that the project can be controlled and the outcomes achieved. This is because improper tailoring can dramatically increase the risk to a project. By applying too little process or too much process, either the cost to the project for the process implementation will be higher than necessary (for process management), or significant risk to the project will be allowed through lack of appropriate control mechanisms. The tailoring decisions, and the resulting extensiveness of the plans and management rigor, will depend on the scope, complexity, stakeholder needs, and other factors that will be assessed by the subject matter experts.

The benefits of applying proper tailoring will be realized in effective project management without excessive process burden and in achieving the project goals. In general, performing tailoring involves comparing project management and systems engineering processes to the project needs and determining the amount of effort that should be placed in implementing and managing the processes, so that cost and labor impacts are appropriate and the processes are rigorous enough to drive the project to successful outcomes. Another way to say this is that the processes applied must optimize project performance at the least cost associated with process efforts.

Tailoring can be implemented in several ways. In general, these include things such as adjusting the formality of reviews, or the amount of detail that needs to be tracked and documented. In some cases, a process is completely unnecessary, so therefore it would be wasteful to spend any time considering the process. For example, if a simple project does not have any procurement requirements, this will negate the need for a procurement management plan. Tailoring can mean something as simple as including a paragraph into the project plan as to how a process will be implemented and then lightly managing that process throughout the life cycle. Or conversely, it could mean requiring a stand-alone document that outlines a series of steps that will be performed, then aggressively managing that process throughout the life cycle.

The project manager and systems engineer work with their teams to assess the project requirements and apply the process discipline in the appropriate form. The benefits of a properly tailored project are realized in achieving the project goals, reducing risks, and meeting stakeholder needs within the cost, schedule, and scope parameters of the project. Specific process tailoring recommendations within each of the disciplines of project management and systems engineering are described in Sections 6.3 and 6.4.

6.2 Applying Tailoring

This section describes the factors, circumstances, and influences that must be carefully evaluated when making tailoring choices, as they have a direct contributing impact on the project risk profile. Based on having an accurate perception of the needs of the project and stakeholders, tailoring requires:

- an understanding of the maturity of the project requirements
- an appreciation of the risk tolerance of the stakeholders
- the consideration of the life cycle phase of the project
- an understanding of how to tailor within different management methods

All of the stakeholder-focused and solution-focused processes discussed in Chapters 3 and 4 are considered when tailoring for a project. They provide the basis on which to assess the process rigor that will be needed to drive the project to successful outcomes. This section explores those important considerations in process tailoring for projects, including tailoring associated with requirements maturity, project risk, the life cycle phase, and the project management methods being used.

Tailoring provides the opportunity to minimize time and effort spent in implementing processes where there is not a need within the project. Knowledgeable and experienced project managers and systems engineers must implement tailoring. These are individuals who understand all of the downstream needs that are fulfilled by the implementation of each process. Not performing some of these processes without understanding how they are used is a critical error. On the other hand, it makes sense to only use the resources (labor, budget) that are necessary to develop the processes, using the appropriate rigor so as not to be wasteful. Tailoring affects the depth of the content development. The justification for all tailoring decisions are documented for reference and are communicated throughout the entire project team. The processes, as tailored, are audited during the project implementation to ensure that they are being applied appropriately. At each phase gate review, the tailoring decisions for the next phase are made, documented, and implemented.

6.2.1 Tailoring by Requirements

In order to tailor effectively, a clear and accurate picture of the project and stakeholder needs is required. Questions about what the project entails, why it is being done, where the project will be performed, who is working on the project, and how the project will be completed must all be answered. These inquiries are generally satisfied in the early phase of a project. Confirmation from the stakeholders that the needs have been captured correctly and are valid, as well as understanding how stable the requirements are or if they are still evolving, is part of both the project management and systems engineering processes.

For example, the level of known requirements will be markedly different for a highly creative research project focusing on a new concept or idea, versus one on a full production run of a mature, build-to-print part. The research project may only have a small number of known requirements and therefore will apply only the most impactful processes that will help knowledge advancement, such as providing a basic scope statement against which internal reviews of progress can be performed. In comparison, a full production may have a significant amount of process implemented in order to ensure that a large number of requirements are successfully met.

Formality requirements can also come from the customer, the organization, and other key stakeholders. The more formal the approach on a project, the more process rigor is generally required. Figure 6.1 demonstrates the increase in process rigor within a variety of factors. For example, a project that is fulfilling a need for a tightly controlled set of specifications may require a much more rigorous

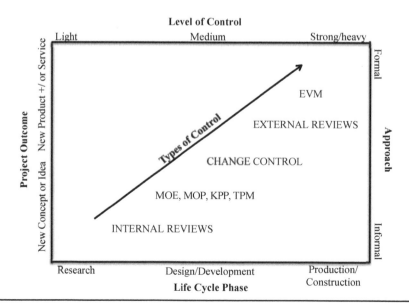

Figure 6.1   Additive Measurements

implementation of the processes than a project that can tolerate an evolution in specifications through-out the life cycle. This model shows that, as the project evolves along the life cycle (bottom), and the maturity of the design develops (left), the level of control moves from light to strong or heavy (top), and the formality of the approach increases (right).

As the control requirements increase, the amount of formal project management and systems engineering processes applied to the project will also increase. This is because it is the full implementation of the processes—untailored—that must be applied to minimize the risk and provide the strongest controls. Every project is different, and there are many situations that will affect what tailoring should be applied. The project manager and systems engineer should be thorough in their investigation of the factors so that the proper tailoring decisions can be made.

6.2.2 Tailoring Risks

Tailoring is a widely misunderstood process; however, it is one that is critical to the appropriate application of project management and systems engineering. It is often the case that tailoring is seen as a way to eliminate unnecessary processes and to focus only on the processes that appear to give the most value to the project. How much process tailoring is too much will be subjected to debate, opinion, and disparate views. In reality, all processes are important and should be reviewed for inclusion. Everyone may have an opinion, but for a project to be successful, the final determination must be made by a knowledgeable subject matter expert in the project management and systems engineering process disciplines.

Indeed, the most important tailoring factor is the qualifications of the project manager and systems engineer responsible for tailoring. The knowledge and practical experience from managing past projects provides a solid basis for effective tailoring. This is because a thorough understanding of how each process impacts the project and is used throughout the life cycle is an imperative for effective tailoring decision making. To the inexperienced observer, processes may appear on the surface not to have much impact later in the phases, but in reality they may be critical. An example of this would be, by not having a fully implemented test, verification, and validation plan, the ability to prove that specifications have been met may be adversely affected or may impact the ability of a project to progress into another life cycle phase. The risk of tailoring inappropriately, therefore, increases the general risk of the project.

Tailoring inappropriately may also affect the ability to manage overall project risk. As discussed in Chapter 1, risk refers to something that might occur and, if it did occur, would cause an impact on the project. If, for example, the project manager makes a decision to ignore the process of risk management, then the project is likely to experience constant unexpected events that must be resolved with unplanned budget and schedule. The level of risk that the organization is willing to accept, and the level of risk that the project manager is willing to accept, will both be deciding factors in tailoring decisions. The more risk tolerant those stakeholders are, the fewer controls will be required; the less risk tolerant, the more controls will be required. As a project progresses to the production environment, the risk generally decreases and controls are increased to ensure that the most costly phases of the project (late development and production) are managed in accordance with the budget and schedule.

As mentioned in Chapter 5, unless time, funding, and scope are of no consequence, even small projects require some level of project process rigor so that it is known what was actually achieved in scope, that the budget was not exceeded, the schedule was adhered to, and the risks were effectively managed. These are very basic, but necessary, pieces of information. Indeed, tailoring activities should never exclude project risk management. A risk-averse organization or project manager will generally react strongly to perceived failures associated with lack of controls. In this environment, it is wise to minimize tailoring and to implement the full suite of processes.

A good rule to follow for tailoring is to review the most important activities of the project that are needed to achieve the project outcomes (identified in the work breakdown structure) and to determine the level of risk associated with meeting the requirements. Then the project manager can apply the processes at the appropriate level to drive the project to successful outcomes. Whether or not the level was appropriate may not be known until the final outcomes of the project are known, but in general, one will know the tailoring is effective by the evidence that controls are adequate to ensure forward motion in the project at the expected pace and cost, without wasted effort, and that there is enough useful, pertinent, and timely information so that effective decision making can proceed.

A simple litmus test is to review the documentation coming from each process. If a decision could be made from the evidence shown in the documentation (a change could be evaluated, or a trend could be recognized), then the process is useful. If documentation is being produced from a process that is not driving decision making, it is unwisely using project resources that could otherwise be engaged elsewhere. Project teams that expend energy compiling countless reports and documents often find themselves with too much information, distorting their ability to assess what is truly happening within the project and hampering their ability to make decisions. It is usually easy to spot these, as the project team will be complaining about spending too much time writing and not enough time doing.

Once tailoring decisions have been made, two final actions are absolutely required. The first is the documentation of the tailoring decisions. This is so it will be clear what decisions were made and why. And second, this document must go through an approval and sign off on the tailoring from the appropriate stakeholders. These steps are critical in providing context for the decision making, both during the project and after the project closes and the outcomes are known.

6.2.3 Tailoring by Life Cycle Phase

Wherever a project is within the life cycle, it is necessary to have some level of project management and systems engineering process discipline applied in order to provide visibility, guide and control activities, and maintain boundaries of scope, schedule, and budget. The life cycle phase will identify the maturity of the design, development, or production, leading to different tailoring choices. Having a clear understanding of the project work breakdown structure products that will be designed and built and where each of these products fall within the life cycle is the first step in effective tailoring.

Process tailoring is completed at the beginning of a project and at each phase gate of the project. Tailoring a project requires a view of the overall project life cycle framework, along with an understanding of the different life cycle phases that the different components, parts, and subsystems may be in. As discussed in Section 6.2.1, a significant driver of process scaling is related to controls. Project products that are earlier in the life cycle phase may require processes that ensure progression rather than performance. This is because, during the early phase, exact specifications of technical parameters may not be known, and work will be focused on understanding the nature of the product that can be built. It is important to apply processes that will highlight the learning, provide the necessary communications, and track change. As products move through the life cycle from design to development and then to production, the design becomes more stable, and heavier controls will be employed to ensure performance to the defined specifications and to cost and schedule parameters.

Depending on the location of the element within the life cycle, project management and systems engineering processes may be applied from a tailoring-down approach or a tailoring-up approach. A tailoring-down approach, generally used for late development and production/construction, starts with an assumption of a full implementation of process discipline across all processes on the project, and as each process is evaluated, a decision is made to reduce the rigor of the process application. For research and development projects, a tailoring-up approach starts with an assumption of no implementation of process discipline across the processes, and as each one is evaluated, the decision on the level of process rigor to apply is made. Figure 6.2 provides an example of a life-cycle tailoring decision matrix.

Project	WBS	(Chargeable Control Account)	Task code	Life Cycle Maturity Phase	Requirements Stability	Risk Tolerance	Tailoring
1.0 Wind Farm Project			100000				
	1.1	Structure	111000				
		1.1.1 Main Shaft	110001	Late Design	Stable	Low	Tailor down
		1.1.2 Main Frame	110002	Late Design	Stable	Low	Tailor down
		1.1.3 Tower	110003	Late Design	Stable	Low	Tailor down
		1.1.4 Nacelle Housing	110004	Late Design	Stable	Low	Tailor down
	1.2.	Rotor Blades	120000				
		1.2.1 Blades	120001	Late Development	Stable	Medium	Tailor down
		1.2.2 Rotor Hub	120002	Early Development	Evolving	High	Tailor up
		1.2.3 Rotor Bearings	120003	Late Development	Stable	Medium	Tailor down
	1.3	Electronics	130000				
		1.3.1 Generator	130001	Mid Design	Evolving	Low	Tailor down
		1.3.1.1 Magnets	131001	Mid Design	Evolving	Low	Tailor down
		1.3.1.2 Conductor	131002	Mid Design	Evolving	Low	Tailor down
		1.3.2 Power Converter	132001	Research & Development	Evolving	High	Tailor up
		1.3.3 Transformer	133001	Research & Development	Evolving	High	Tailor up
		1.3.4 Brake System	134001	Research & Development	Evolving	Low	Tailor down

Figure 6.2   Example Life-Cycle Tailoring Decision Matrix

It is important to assess both the overall project life cycle and the work breakdown structure product life cycles, as products that make up a project are often in different phases of the life cycle.

When considering life cycles, research, design, and development activities—both software and hardware—can be particularly high risk, and the tailoring should address the implementation of processes that will provide structure that will allow and foster progression, while tracking change. When these types of products are being developed for insertion into a project that has stable or existing elements, such as when the project is building upon an existing capability, interfaces are going to be a key risk item.

6.2.4 Tailoring for Different Management Methods

Tailoring of processes is most often associated more with traditional methods of project management and systems engineering than with flexible methods. The traditional method of project management requires a series of activities be performed that document, implement, execute, and close a project in a mostly sequential series of steps.

In comparison, the processes associated with flexible management methods, for example Agile and Spiral project management methods, use a series of sequential, concurrent, and iterative processes to drive toward outcomes. Systems engineering also requires that a series of steps be performed in sequence, starting with the lowest level part, evolving into a product that is integrated into a system over a course of tested, verified, and validated steps. However, within the process itself is a cyclical series of steps that allow for elucidation and elaboration of requirements, synthesis, and evolution throughout the full process life cycle.

The cornerstone of how each of the disciplines work, regardless of the project management method used, is in establishing and maintaining a baseline description of what will be done, how much budget is allocated to the activity, and over what time period the work is to be completed. Always knowing where one is in the process is an imperative. For the systems engineer, understanding the full traceability between the highest level requirements through the lowest is also critical and must be maintained at all times (explained in Section 6.4).

As described in Chapter 2, a baseline is a defined level, timeframe, or set of attributes that serves as a control and comparison point. This is the agreed-to version of something that is under formal control at all times and only changes when the pertinent stakeholders (the ones impacted) agree to a proposed change and it is implemented into the project. Progress and performance are measured against baselines. In flexible types of project management, a methodical, iterative approach is applied that drives the design and development to the ultimate outcomes desired by the stakeholders, who may not be able to envision the final product that will ultimately meet their needs. Software development as well as products in the research, design, and development life cycle phases often utilize this iterative, looping/recursive type of approach. Management of these activities aims for a progression of successful outcomes that often build upon the previous successes.

In project management, there are many flexible methods to choose from, with new ones emerging regularly. When used in combination, project management and systems engineering processes provide the most effective set of processes for projects that are managed using flexible methods, as these types of projects often are technically challenging, thereby benefitting from the combined processes of both disciplines. The tailoring of the processes is still a necessary step. The project manager and systems engineer will review each process in relation to the needs of the project. For example, if the project is in its early life cycle phase and is expected to evolve, a flexible project management approach may be preferred. This will include a shortened development tempo, excellent requirements capture processes (such as described in Chapter 4, stakeholder-focused processes), robust change and configuration management methods, risk management with a focus on prioritization, and strong test, verification, and validation execution. This is in addition to the standard overarching project budget, schedule, and scope management activities that are valid for all projects.

During planning activities, the type of project will drive the tailoring requirements. Highly technical projects that are controlled using a flexible approach (such as in Spiral development) will most likely include systems engineering processes, such as modeling and simulation, trade studies, and prototyping, where experimentation and testing are essential. The tailoring of these processes is necessary, and again, the project manager and systems engineer will review each process in relation to the needs of

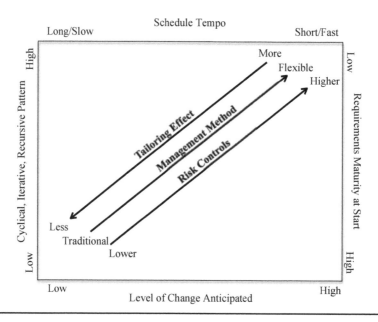

Figure 6.3   Management Methods

the project. Careful attention will be paid to ensure that all necessary processes that provide structure to the assessment of product maturity are included. Formal entrance and exit criteria may be defined to assist in the product maturity assessments.

While tailoring decisions should include an understanding of the different management methods that may be involved in the project, it is imperative to understand that they are just one of several factors for consideration. When the management methods, the life cycle phases, the risks, and the requirements are all taken as a whole, from a systems viewpoint, the project manager and systems engineer can determine the right tailoring approach. This can be seen in Figure 6.3.

Projects that are highly cyclical, iterative, or recursive in their development methods (left axis), have a short or fast scheduling tempo (top axis), have low requirements maturity (right axis), and a high level of anticipated change (bottom axis) will most often use a flexible management method and will likely require higher risks controls. This will affect the tailoring decisions so as to more precisely address the needs of the project and ensure that the processes necessary for successful outcomes are not tailored out.

6.3 Tailoring Project Management Processes

Project management processes serve the purpose of guiding project activities throughout the life cycle of the project. When used appropriately, these processes increase the probability that a project will achieve its desired outcomes and that the stakeholders will be rewarded with results that meet their needs. Tailoring of project management processes requires a careful approach. In this section:

- Essential project management processes are identified.
- The effect of tailoring on each standard project management process is described.
- A suggested approach to tailoring up and tailoring down is provided.

Tailoring is an art. The project manager uses knowledge of the project management standard processes and previous project experiences to compare and contrast the needs of the current project, then makes decisions regarding the level of rigor that is required to manage and control the project activities. In the following sections, the tailoring landscape for each project management process area that was identified in Chapters 3 and 4 is reviewed.

6.3.1 Stakeholder Engagement

Project managers are acutely aware that they must meet the key stakeholders' needs in order to be seen as having provided a successful project. They aim to achieve this success through understanding the full context of what the key stakeholders need at the beginning of the project and maintaining a careful progressive evolution of those needs throughout the project. To ensure that this happens, an essential stakeholder engagement process is used to identify all the stakeholders associated with the project and to document their expressed needs and anticipated level of engagement in a stakeholder register. In practice, all stakeholders that could impact or be impacted by the project should be identified regardless of the size and complexity of the project. This process is simple and the list may include any number of stakeholders, from one to 100 or more. This information will be used in the communications planning, so if not complete it will impact the project manager's ability to effectively convey project progress, to capture change initiatives, and to manage expectations of the stakeholders.

Tailoring affects the documentation requirements of the stakeholder engagement plan and associated communications management plan (Section 6.3.2). If the project is simple, a separate plan (other than the register) may not be necessary, or a short paragraph explaining the approach for engaging stakeholders could be written in the project management plan. This is a tailoring-up approach. If the project is complex and large, the tailoring-down approach would start with a requirement for a stand-alone document that outlines all of the steps used to identify the stakeholders, key stakeholders, a description of their needs, communication requirements, tempo of interactions, formal sign-off requirements for phases of the project, and any other pertinent information associated with these relationships. As the project decreases in complexity, the requirements for a separate document, or a full description of the steps that will be taken, subsides, and the necessary information can be inserted into the project management plan as appropriate.

6.3.2 Collaboration and Communications

Understanding and effectively addressing the collaboration and communications needs of a project is a key success factor and simultaneously presents a significant risk. As discussed in Chapter 3, collaboration refers to project interactions that leverage the strengths of individuals, organizational entities, and organizations. These binding agreements are usually documented in a manner that outlines the specific commitments as well as the communications methods that will be employed during all relevant phases of the project. Any project with significant collaboration should tailor down, starting with a full set of documented binding collaboration agreements that define the scope of work, schedule, budget allocation, resource commitments, accountability, decision-making and escalation paths, risk mitigation process, and any legal information needed to amend or terminate the agreement.

These documented agreements are essential. If these agreements are processed with less rigor—for example, accepting a handshake agreement rather than a documented and signed agreement—the project manager's ability to control the scope, schedule, and budget of the project will be in jeopardy. Tailoring down can occur in situations where internal departments or divisions are providing services, since there is a failsafe where the organizational management will be able to intervene and exert influence in the event of nonperformance. This is not the case generally for external organizations, where performance cannot be controlled from the project.

In the process area of communications, the primary importance is in managing the expectations of the stakeholders. Stakeholders will often assume that if nothing is being communicated, nothing is being done. Without consistent communications, stakeholders' natural inclination for their needs to evolve and change will not be addressed and dealt with within the project in a controlled manner. Tailoring activities should never exclude an initial assessment of stakeholders' communications needs, which can be documented within the stakeholder register. Understanding each key stakeholder's preferred communication method and frequency is essential.

It will be important to tailor down in complex projects with many key stakeholders and collaborators, starting with a full implementation of a project communications plan that includes how pertinent information will be identified, collected, processed and screened, distributed, and archived. This process should also include information on how to address questions, concerns, and issues that arise from the communications process. For simple projects, the basic information needed to understand the communication that will be used can be addressed within the project plan.

6.3.3 Total Project Scope

Understanding the project scope generally takes the form of definition within the project charter and the scope statement. Any project that cannot extract clear statements of purpose will experience serious difficulty in achieving successful project outcomes. The project team will find itself chasing a moving target of needs, which will ultimately be impossible to hit. Therefore, an essential process that needs to be implemented is to identify and gain concurrence on the charter and scope of the project.

Tailoring up can occur on a simple project where the project scope is straightforward or in an early evolutionary stage. This charter and scope statement, in its limited form, can be documented in the project plan without any additional elaboration in separate documentation. In more complex projects, tailoring down requires starting with a separate charter and scope statement documented outside of, and prior to, the development of the project plan.

6.3.4 Total Project Schedule

Development of a total project schedule is an essential task, although this is an area where there can be great variability in implementation. For small, simple projects, tailoring up can be accomplished by simply identifying the total project timeline start and end dates. Additions to this would be to account for the timelines associated with completing specific tasks, and then to include the resources required for each of those tasks.

For complex projects, a tailoring-down process would start with a schedule management plan that describes all processes to develop and maintain a fully integrated master schedule, inclusive of all activities, resources, duration estimates, and critical path and associated float (slack) identified. A workforce or resources plan would also be initiated to ensure the project's ability to recruit, train, retain, and maintain the appropriately skilled resources to perform the project scope.

6.3.5 Financial Management

The processes associated with addressing project-related financial information, such as costing, budgeting, forecasting, and acquiring and procuring materials and services are essential. These activities are generally auditable, and lack of adequate controls in this area can lead to legal and punitive impacts, as well as affect the project's ability to perform. Funding organizations want to know that the project manager is conscientiously managing the project expenditures and will not overspend. Mismanagement of funds has a direct result on the project's ability to achieve the planned outcomes. Therefore, the processes associated with any type of financial controls are considered essential.

Simple projects may use existing organizational processes to estimate project costs, build budgets, and obtain performance reports. In a tailoring-up approach, budget tracking for these types of projects may use a basic spreadsheet in combination with these organizational inputs as a starting point. As the complexity of the project increases, more detailed financial information will be required to be maintained and controlled. These processes can be documented in the project management plan until the point where it makes sense to move the information into stand-alone documentation.

Projects with a significant budget should tailor down, starting with a cost management plan that describes the processes used to estimate costs, develop and track project budgets, and implement and monitor contracts, acquisitions, and procurements. Large projects with heavily evolving requirements, such as with a significant research and development element, are inherently risky and therefore should tailor down appropriately to address the risks that are driven by the unknowns that could directly impact the budget (rather than tailor up, as might be appropriate for early phase research and development projects).

6.3.6 Risk Management

Risk management is a process that is essential. A project, throughout its life cycle, will always experience some of the risks that are identified during the initial and sequential phases of the project. These risks are identified and documented in the risk register, along with their probability of occurrence, impact of occurrence, and mitigation plans, as described in Chapter 3. The project will also experience the realization of risks that were not previously identified. These risks can occur in any of the process areas, but generally they impact one or a combination of budget, schedule, and scope. For this reason, project managers typically retain a portion of schedule and budget to address these risks in their contingency or reserve.

Tailoring of risk management generally refers to the level and rigor of documentation that is provided and the resulting active management that must occur. For small projects, a risk register may be maintained and used to inform decision making without any additional documentation requirements and will use a light management approach. For large or complex projects, a risk management plan and a risk register are implemented and actively managed. This plan describes the processes that are used to identify, track, assess, and mitigate risks depending on their priority.

6.3.7 Quality Management

As described in Chapter 3, project quality management is mainly associated with the ability to meet the stakeholders' expectations. It can also focus on the process integrity within the project itself. In Sections 6.3.1 and 6.3.2, the essential processes of identifying the stakeholders and understanding how to communicate progress throughout the project were described. These are foundational processes that if done correctly will lead to quality outcomes. If done poorly, quality will not be achieved.

Tailoring associated with quality management addresses the rigor applied to the quality activities, such as the number, type, and tempo of measurements that will be defined and implemented on the project to control quality. Tailoring down requires the starting point to be a stand-alone quality management plan, which outlines the inclusion of regular programmatic reviews, formal change and configuration management, and measurements that demonstrate trends used for decision making. Small projects that are not complex may have a summary statement in the project management plan that describes the measurements and reviews that will be utilized.

6.3.8 Integrative Management

Plans that describe cross-project life cycle processes, such as the project management plan and change control plans, are essential. As described in Chapter 5, change is constant, and the need to always know where one is within the project is an imperative. Tailoring down starts with separate stand-alone plans, each describing in detail the processes that are employed throughout the project life cycle. These plans are used to actively manage the project to ensure that the project outcomes are achieved. It would be expected that inclusion of all project management processes across the project life cycle, and all associated documents and management activities, will be implemented and managed.

Tailoring up for small, simple projects could require a paragraph to be inserted into the project management plan describing how the project will be managed and change controlled. Then the simplest of management activities employed, perhaps a one-page project plan with a paragraph documenting the change control process or an ad hoc meeting to discuss and agree upon a change, would suffice.

6.4 Tailoring Systems Engineering Processes

Systems engineering processes serve the purpose of guiding technical activities throughout the life cycle of the project. The systems that are being engineered are typically multidisciplinary and interconnected. The complex nature of integration can affect both the individual products associated with the system as well as the integrated system itself in unintended or unanticipated ways. In addition to the risk in the development of systems, tailoring activities carry a great deal of risk in themselves, due to the interconnectedness of the processes. Tailoring of systems engineering processes requires an approach that takes into careful consideration the long-term impacts (throughout the entire life cycle, including operations) of each decision. In this section,

- Essential systems engineering processes are identified.
- The effect of tailoring on each systems engineering process is described.
- A suggested approach to tailoring up and tailoring down is provided.

As is true in project management, the application of tailoring in systems engineering is an art. The systems engineer uses his or her knowledge and experience with systems engineering standard process implementation and execution on past projects to decide how to apply appropriate rigor so that the project technical outcomes are optimized. The tailoring landscape for each systems engineering process area that was identified in Chapters 3 and 4 is reviewed in the following sections.

6.4.1 System Stakeholder Engagement

The systems engineer understands the importance of building relationships with the key stakeholders. These stakeholders will be involved in the processes of technical development throughout the life cycle. The system stakeholders provide the vision of how the scope of the project transforms into an operational concept that meets their needs. An essential process includes the development of this joint understanding of what the stakeholders need and how they will measure the effectiveness of the system that is provided to them upon delivery. Without this fundamental meeting of the minds as to what will be developed and how it will be assessed, achieving a good outcome will be problematic.

This is not to say that the stakeholders must know exactly what they want and how they intend to use it at the beginning of a project, as this is unrealistic. Often one does not know what is wanted until it is revealed. This does not pose a problem for the systems engineer, who is used to cyclical and iterative design processes. It is in the development of the long-term relationship with the stakeholders, and in the identification of the vision of how things will work and what constitutes a good system, and then maintaining that relationship throughout the evolution of the system design, that is the value added by this process.

As the design matures, stakeholder requirements may change and the specific performance of the system will solidify. As long as the change is controlled and the stakeholder's expectations are managed along the way, and it leads to a final result that is anticipated and consistent with the stakeholder's vision as it has evolved, the stakeholders will be satisfied and the project will be deemed successful.

Because of the need to build this long-lasting relationship with the stakeholders, tailoring is not recommended. However, if the project is simple, additional information needed by the systems engineer in the design and development of the system, such as the system stakeholder engagement processes, or on the stakeholders themselves, can be added to the project stakeholder register and project plan rather than maintaining a separate register.

6.4.2 Communications and Decision Support

Just as the project manager needs to consider carefully how to communicate with the stakeholders, systems engineers also need to have a plan as to how to communicate with the stakeholders on the technical design and development. Most importantly, an approach to decision making needs to be considered.

This is particularly important when stakeholders drive design evolution. This essential process used for decision support needs to interact with the project change management process. If this process is not tailored properly, the most likely result is stakeholder dissatisfaction and wasted effort on the part of the design and development team.

Tailoring up the communications and decision support processes for small, simple projects starts with providing input into the project management plan, the communications plan, and the stakeholders register. Additional processes can be added to manage decision support. Tailoring down for complex projects includes a technical communications plan that describes how technical information will be conveyed, where the documentation is stored, and how decisions on technical change will be made. The process should include methods to collect and address technical questions that arise from these communications. Information as to if, or when, external independent reviewers or evaluators will be brought in to assist with decision making should be included.

6.4.3 Technical Scope Management

Of all the systems engineering processes, technical scope management is one of the most interconnected. For example, actions taken in stakeholder engagement processes inform the design solution that the project team brings forward and evolves. Tailoring out of any of these processes increases the probability of impacting activities downstream and should be carefully assessed before making a final decision. Inappropriately tailored processes during the mission definition phase of the project could lead to missing key enabling system requirements, such as not arranging for the use of enterprise requirements management software, or the appropriate set up of the work breakdown structure in the organizational financial system.

If the life cycle determination is not made in the early phases of the project, a more (or less) robust process may be applied to certain efforts that will affect the activities far down the life cycle. As the life cycle progresses, improperly applied tailoring may manifest itself in a lack of specifications and in issues associated with the architecture, compatibility, interoperability, interfacing, design synthesis, operations, and maintenance activities. The failure to achieve a producible or constructible design, or failure to experience the anticipated desired behaviors of the system, are also potential issues if tailoring down has been too aggressive or tailoring up has been insufficient.

In technical scope management for complex or large projects, stand-alone documentation, such as a systems engineering management plan and specification documentation that can maintain traceability of the design from the highest levels to the lowest, is critical. These documents define measures of performance and contain clearly defined integration requirements. They will inform the design of the architecture and physical solution and will provide a solid design baseline from which to build, operate, and maintain the system. In a small, simple project, the tailor-up approach can start with input into the total project scope documentation and result in a light scope implementation.

6.4.4 Interface Management

One of the most important, and often insufficiently addressed, processes is interface management. Interfaces are often left to a future phase, but if not planned carefully into the design, will likely cause downstream rework, with additional cost and schedule implications. For all projects, this is one process that should only be tailored up if the product is expected to be completely stand-alone (no connection to power sources, no wireless connections to other devices, etc.). Any other project should tailor down. A suggested approach to tailoring down is to develop an interface charter during the mission definition phase that identifies all known interfaces of any kind. With an interface management plan clearly outlining the approach to verifying compliance to interface requirements, the actual interface requirements specifications are then documented in interface control documents, which require approval signatures from the originating and the terminating functions. During test and verification, these interface control documents will be used to confirm that specifications have been met.

6.4.5 Technical Schedule Management

In many small, simple projects, a separate technical schedule is not required. Technical milestones and technical effort assessments are often provided to the project manager for inclusion in the project schedule. The provision of this information is an essential process that cannot be bypassed, as this detail about the proposed schedule is associated with the scope and will drive the overarching project schedule and contingency.

Depending on the life cycle of the activities within even the smallest project, technical schedule and the associated effort can be the highest risk. For example, in cases where early development is required, technical milestones may be understood, but the schedule required to meet the milestone may be unknown. Or sharing subject matter experts between projects may be required, and their time allocation may be uncertain. Understanding the anticipated schedule and effort may lead to additional schedule contingency being allocated, and may even lead to an addition to the risk register.

For large or complex projects, technical schedule management should be tailored down according to the needs of the project. Starting with a separate systems engineering management plan, which contains the information on the processes used to control technical activities, the handoff of the technical milestones, anticipated schedule, resource requirements, etc. between systems engineering to project management must be documented and agreed to by both parties. The technical schedule becomes part of the project integrated master schedule that is actively managed throughout the project.

6.4.6 Acquisition and Procurement Management

The activities associated with acquisition and procurement management offer the most opportunity for tailoring. This is due to the nature of these activities, which may be subjected to significant organizational constructs. In small projects, there may be no such activities. All project technical design, development, production, or construction may be handled entirely within the organization with materials and services readily available internally.

In most cases, since not all organizations are completely self-contained, some level of purchasing or acquisitions is required, and a tailoring-up approach can be used. In these cases, the essential process is to identify the needed materials and services in a timely manner and to provide that information to the project manager so that the overall project needs can be provided to the organization responsible. Not considering this process, or tailoring it out completely, will result in schedule and cost impacts and could result in impacts to the technical development if the resources are hard to find, difficult to obtain, or otherwise not available when they are needed.

In large or complex projects, a tailoring-down approach should be used—inclusive of both an acquisition or procurement plan—and an acquisition or procurement integration plan should be developed in collaboration with the project manager and the acquisition or procurement manager. The organizational department or division (or the project manager if appropriate) will work closely with the systems engineer in developing the statements of work and contracts and agreements. If these processes are not followed, risk increases significantly for the project, and it can be expected that the scope, schedule, and budget may be impacted, as expediting and obtaining scarce resources could be a direct consequence.

6.4.7 Risk Reduction

Technical risk reduction is another area that can be tailored significantly depending on the project. This is because some projects may be producing or constructing a simple or established build-to-print system. The projects that require a more rigorous type of risk reduction typically contain research and development, or early design and development. The system as a whole could contain one type of risk, or certain elements of the project may contain risk while other elements are stable and mature.

A tailoring-up risk reduction process is generally acceptable in most circumstances. A tailoring-down approach would be used for projects that contain critical technologies or are under development. In these cases, a technical assessment and risk reduction strategy and plan should be put in place and tracked throughout the project. Models, simulations, trade studies, prototypes, and other activities that provide fidelity about the performance or specifications of the product will drive risk down and mature the design. In production or construction, a phased approach may be used to compare the test articles to an initial implementation so that any discrepancies or defects can be detected before significant funds are expended. If tailoring down is not applied appropriately, it could result in the project failing to produce a valid design, it could cause a design failure, or it could lead to latent defects that will ultimately impact the project budget and schedule.

6.4.8 Quality and Measurements

Quality and measurements are the most popular processes to tailor out of a project, because the project team often sees them more as added work than added value. However, from a strategic perspective, these processes are essential if the project is to progress in a methodical way, such as in a phased gate approach. Holding the project at predetermined times in order to assess the progress to plan and to assess the technical performance measures and quality that has been achieved is an imperative. If the project has a technical scope, then tailoring down should be used.

Each performance measure and review should be carefully considered and only the most appropriate ones scheduled. The rigor associated with each of these activities—for example, the number of technical performance measures employed, the number and type of reviews (both internal and external)—will depend on the size and complexity of the project. The only time a tailoring-up method should be used is in simple technical projects, or those that are in the research and development phase of the project, as tests, verification, and reviews will be project dependent. In any and all cases, a knowledgeable and experienced systems engineer should assess the appropriate application of the processes. Similar to risk reduction, implications of inappropriate tailoring will be experienced later in the life cycle, when the project fails to produce a valid design or when significant latent defects that must be corrected in the production or construction phase are discovered.

6.4.9 Test, Verification, Validation

The processes associated with test verification and validations are essential. The level of effort and rigor that are put into each of these activities can vary widely from project to project; however, they should never be ignored. Testing provides the opportunity to verify the performance of a particular activity against the expected performance. Verification is the step that corroborates the accuracy of the performance measured, while validation provides affirmation that the performance is as expected and desired (typically performed by the key stakeholders).

If a project tailors out these processes, the lack of confirmation and evidence provided by the steps will become an issue when the project is moving from phase to phase. In addition, the project can vector off course in its application of testing or in the pursuit of capability that is not required or needed by the project. For example, a design capability may fall within the specifications needed by the project; however, the design engineers have a desire to increase the capability to a higher level. If the project does not have a test and verification plan that describes the "good enough" capability, labor and budget may be expended on activities that are outside the scope of the project. If successful, and the stakeholders are made aware of the increased capability, this may become a revised stakeholder specification that must be reviewed through change control processes. In a traditionally managed project, the cost and schedule impact of both the additional design, the test and verification activities that have already been done, plus the impacts to the baselined budget, schedule, and scope, must all be revised. This negatively affects the project, although it might positively affect the overall system. The way to mitigate the probability of this occurring is to follow the processes in developing and managing to a test, verification, and validation plan.

On projects that are being managed by a flexible method, and those that are in the early life cycle phases, including research and development, the test and verification activities may be focused on exploring the realm of the unknown and on learning. In these cases, it may not be known what falls within the realm of possible, and it is to the advantage of the project (and the stakeholders) to provide broad parameters for tests and verifications. This way, the most likely designs can be captured and do not negatively impact the overall project scope as defined at the beginning of the project (or as formally changed).

A tailoring-down approach should be applied in all projects. An initial requirements verification and validation matrix that is developed in the mission definition phase of a project provides the target for which tests, verification, and validation activities are recorded and approved. The documentation required to capture results from testing and verification can vary widely from project to project, depending on the complexity and life cycle phase of the project. Research and development projects may require that experiments, tests, and verification of test results only be documented in a laboratory notebook. Large complex projects will require formal capture and documentation. Ultimately, as in all projects, whether experimentation, design, development, or production/construction, it must be clear that the results are repeatable and stable to be usable in the next project life-cycle phase.

6.4.10 Governance

As discussed in Chapter 5, governance is an essential measurement and is used to control all of the integrative activities within the project. Tailoring is an essential activity, and therefore the tailoring out of any governance activities is not recommended. Chapter 4 explained in depth that governance generally refers to those activities that track evolution and movement against defined processes. This includes change control used to assess and integrate change methodically and configuration controls used to identify items using unique identifiers and sequential versioning (of any document, product, design, etc.).

As inappropriate tailoring of the governance processes could have dire effects on the project (loss of configuration control, out of control scope, ad hoc schedule slips, budget overruns), it is suggested that the project manager and systems engineer, and even subject matter experts themselves, verify with an independent subject matter expert that their tailoring approaches are appropriate. In some organizations, tailoring requests must be approved by a higher authority. This provides a level of assurance that the tailoring will not negatively affect the project.

Inserting appropriate controls into a new project can be relatively easy if the organization(s) involved are supportive. However, in projects that are applying project management or systems engineering processes into an existing non-supportive culture, or into the middle of an existing system's life cycle (e.g., as in an upgrade or evolution of a capability into an operational system), it becomes more of an imperative to implement the appropriate level of process without tailoring erroneously. This is due to the cultural establishment that often exists around operating the system and the challenges that that culture can exert onto a project.

As the project manager and systems engineer develop their plans and implement process tailoring, special consideration should be made to put the appropriate (and possibly higher) level of control for governance in place. The implications of adding, removing, revising, or otherwise integrating into an existing system can be controversial and more technically challenging, depending on the state of the information available on the original design. For example, many operational systems are not carefully change controlled, so the versions of all components, parts, and subsystems are known only within the field of experienced people who work on the system. Implementing change on those systems may inadvertently break them. This must be considered through more rigorous change control processes so that these concerns can be mitigated through reverse engineering, testing, and verification prior to implementation.

Another significant consideration in tailoring governance for both new systems and modified systems is the systems engineering requirement for full traceability from the highest level of requirements to the lowest level. The decomposition activity, which forces requirements definition and capture, operational and maintenance views, and stakeholders' perspectives on how they will use the system, may seem pointless to engineers that are used to bottom-up engineering (a process of designing, prototyping, and testing), which aims to enhance a system from the lowest levels up to the system level. This can happen in projects in which the stakeholders do not have a clear vision of what they need. Rather than helping the stakeholders define their need and then designing a system to meet those needs, they may move forward with the design in a way that the stakeholders are not aware of, or they may desire to evolve a system in a manner that does not match the stakeholders' view. When cost and schedule are not unlimited, this problematic situation can be resolved through governance.

Based on these considerations, the preferred method for tailoring governance activities is to tailor down using a minimalistic approach. Enough process must always be in place to know how, when, where, and to whom reporting will occur on changes affecting processes and content. Large, complex projects will require separate stand-alone documents to describe the change and configuration management processes and to maintain the definition of the product tree, as allocated, as designed, as built, and as operated and maintained configurations. The level of effort that goes into these documentation efforts and the resulting active management levels of engagement depends on the size and complexity of the project.

6.5 Communicating About Project Tailoring

Making decisions about project tailoring and then communicating those decisions so that they are understood and actionable is an important part of the tailoring process. This section explains the methods that can be developed and used not only to ensure effective communications associated with making tailoring decisions but also while using tailored processes throughout the project life cycle. Communication of tailoring includes ensuring that tailoring decisions are:

- understood, and take into consideration the views of the team, stakeholder, and other subject matter experts as appropriate
- documented, and made available to the project team and stakeholders
- are conveyed at the beginning of the project and through the implementation of the tailored processes
- are accepted (and auditable) by the project team and stakeholders

The project manager and the systems engineer are responsible for assessing the level of process tailoring that should be applied to the project; however, they typically do not make those decisions alone or without input from stakeholders (including interested organizational entities) and project team members. Independent subject matter experts may also be consulted. However, the ultimate decisions as to the tailoring for the project is the responsibility of the project manager.

The knowledge of project management and systems engineering processes across the life cycle is imperative to make good tailoring decisions. As described in other sections and chapters, tailoring is best done by a knowledgeable and experienced project manager in collaboration with a knowledgeable and experienced systems engineer. The lack of understanding of the processes, why they are needed, or how they work across the project life cycle to effect outcomes will cause issues when trying to incorporate processes and can lead to poor tailoring decisions. Tailoring is often equated to not doing. This is dangerous, and it is not accurate.

Tailoring decisions are based on risk avoidance through properly applied process. A tailoring model can be used at the starting phase of a project, and then adjusted at every phase gate. The project manager can use the model to point out the potential impacts of tailoring decisions. The model does not provide one right answer but provides decision-making context and a target range of rigor that should be applied based on risk. A model can help display the information that is difficult to explain. They provide simplifications of information, particularly important when decision making depends on the interpretation of the message, and can be used to guide discussions about risk exposure and the mitigating process controls. A model can document the association between the risks to the project and the processes that will be utilized to minimize those risks and ultimately provides the statement of record for tailoring decisions.

The use of a tailoring model will help the project manager make decisions about tailoring that are defensible, realistic, relatable (to real risks and activities), intuitive (make sense), repeatable, and explainable. A well-built model will help communicate the decisions that are made to all interested stakeholders.[1] It can demonstrate and clarify the reasons tailoring is done and explain the risk-related inputs that drive those decisions. A spreadsheet can be used to create a basic model as shown in Figure 6.4.

Category	SubCategory	Risks	Risk Rating Score Low=1, Medium=2, High=3	Sum Assessed Risk
Programmatic Risks	Collaboration Level	Significant Number of External Collaborators	3	HIGH
		Collaborators-Project Management Expertise	2	MEDIUM
		Collaborators-Systems Engineering Expertise	1	LOW
	Organizational Risks	Coherence of the Team	3	HIGH
		Clear Project Ownership	1	LOW
		Acceptance of Project Management Processes	3	HIGH
		Acceptance of Systems Engineering Processes	2	MEDIUM
	Project Risk	Life Cycle Phase of Project	1	LOW
		Project Definition Risk	1	LOW
		Stakeholder Engagement	2	MEDIUM
Cost Risks	Budget Adequacy	Labor, Materials & Services Coverage	1	LOW
	Contingency	Percentage Adequacy Based on Risk Register	1	LOW
	Financial Change Control	Ability to Control the Cost Baseline	1	LOW
Schedule Risks	Schedule Adequacy	Issues with Critical Path	2	MEDIUM
	Contingency	Adequacy of Float	2	MEDIUM
	Schedule Change Control	Ability to Control the Schedule Baseline	2	MEDIUM
Scope Risks	Requirements	Clarity and Completeness of Requirements	1	LOW
	Traceability	Ability to Maintain Traceability	2	MEDIUM
	Scope Change Control	Ability to Control the Scope Baseline	2	MEDIUM

Figure 6.4   Example Tailoring Risks

The model can be built to reflect any risks that have been identified for the project. There are several steps that can be employed to build and use the model. These include:

- **Step 1:** Identify the programmatic risk, cost, schedule, and scope risk. This can include any risks identified by the project. It might include risks in any of the project process areas. In the example shown in Figure 6.4, the project risks are divided into programmatic risks, cost risks, schedule risks, and scope risks, then further subcategorized to describe the risks associated with collaborations, organizational risk, project risk, budget and schedule adequacy, contingencies, change control, and requirements and traceability. These risks may or may not be different for every project, but these provide a reasonable assessment starting point.
- **Step 2:** Score the risks as low, medium, high. Project managers may decide to add more fidelity to this model. This simple example demonstrates that each risk area as defined in Figure 6.4 can be a 1 for low risk, a 2 for medium risk, or a 3 for high risk. The risk by area is calculated by averaging the scores associated with that category.
- **Step 3:** Then the overall risk score is computed and an average determined. The tailoring model, if built on a spreadsheet, can sum and average automatically. This makes it easy to use on any project needing tailoring. As shown in Figure 6.5, an overall project assessment of medium means that risk associated with the project averages to medium in intensity and that a tailoring-up or tailoring-down approach is appropriate, depending on the project's needs. An overall project assessed at low risk would allow tailoring, but it also might not tailor at all. In fact, the project manager and systems engineer should be careful to choose only the process rigor that is essential for the success of the project. A project assessed overall as high would not likely tailor or would do so carefully and would be expected to apply all the processes with the appropriate level of rigor.
- **Step 4:** For each subcategory of programmatic, cost, schedule, and scope, the risks scores are computed and an average identified. These averages will provide insight into where risk reduction processes are the most needed. Mitigating risks in these areas will also mitigate the overarching project risk.
- **Step 5:** Next, review the tailoring recommendations. These recommendations are based on standard processes. Each risk to be mitigated with a process should have a project management and systems engineering process or processes assigned that are known to reduce risk. If the project manager does not have experience in one or more areas, a subject matter expert can be consulted to verify that the right processes are being selected. Figure 6.6 shows an example of tailoring alignments that can be used.
- **Step 6:** The project team must perform an overall review of the project management and systems engineering processes that are chosen for the project and any concerns or questions addressed. It is not required that everyone be in agreement, but it is required that the project team has had a chance to review the recommended tailoring and understands the reasoning behind it.

AVERAGE SCORE TOTAL PROJECT	2	MEDIUM

Category	RISK BY AREA	Assessed Risk
Programmatic Risks	2	MEDIUM
Cost Risks	1	LOW
Schedule Risks	2	MEDIUM
Scope Risks	2	MEDIUM

TAILORING RECOMMENDATIONS:
Risk is Low, Tailoring is Allowed
Risk is Medium, Tailor Up or Down As Appropriate for the Project
Risk is High, Tailoring Disallowed or Discouraged

Figure 6.5 Example Tailoring Treatments

Category	SubCategory	Risks	PM Process Risk Mitigation	SE Process Risk Mitigation
Programmatic Risks	Collaboration Level	Significant Number of External Collaborators	Stakeholder Engagement, Agreement Commitments	Technical Stakeholder Engagement
		Collaborators-Project Management Expertise	Governance, Escalation Path	
		Collaborators-Systems Engineering Expertise		Governance
	Organizational Risks	Coherence of the Team	Communications Management	
		Clear Project Ownership	Charter	WBS
		Acceptance of Project Management Processes	Governance	
		Acceptance of Systems Engineering Processes		Governance
	Project Risk	Life Cycle Phase of Project	Requirements Verification & Validation Matrix	Requirements Verification & Validation Matrix
		Project Definition Risk	Risk Management	Technical Risk Reduction
		Stakeholder Engagement	Stakeholder Register	
Cost Risks	Budget Adequacy	Labor, Materials & Services Coverage	Budget Baseline, Basis of Estimates	Statement of Work
	Contingency	Percentage Adequacy Based on Risk Register	Modeling and Simulation	
	Financial Change Control	Ability to Control the Cost Baseline	Change Control	Configuration Control
Schedule Risks	Schedule Adequacy	Issues with Critical Path	Schedule Baseline, Critical Path	Technical Schedule Milestones
	Contingency	Adequacy of Float	Schedule Analysis	Schedule Analysis
	Schedule Change Control	Ability to Control the Schedule Baseline	Change Control	Configuration Control
Scope Risks	Requirements	Clarity and Completeness of Requirements	Requirements Verification & Validation Matrix	MOE, MOP, KPP, TPM
	Traceability	Ability to Maintain Traceability	WBS	SEMP, Review Gates
	Scope Change Control	Ability to Control the Scope Baseline	Change Control	Configuration Control

Figure 6.6 Example Tailoring Alignments

- **Step 7:** The final step taken by the project manager is to develop a tailoring plan that meets the needs of the project based on the assessment. For small projects, this might consist of a paragraph added to the project management plan. For large or complex projects, a stand-alone document might be required.

Once these steps are complete, the project manager and systems engineer will implement the tailoring plan and execute the project. Through the communications processes, the stakeholders and project team should receive notice of the tailoring decisions and should be given an opportunity to submit questions on the process and outcomes so that any issues or concerns can be addressed. Throughout the project execution, informal and/or formal audits of the process should be used to verify compliance to the processes and to gauge quality.

6.6 Case Study: Greenland to Scotland Challenge— In the Wake of the Finnmen[2]

All case study quotes are from interviews held with the individual presenting the background. In this next case, the interview was held with Olly Hicks, and all quotes are attributed to him.

Olly Hicks, an experienced adventurer,[3] made the decision to trace a route from Greenland to Scotland by sea kayak. The impetus for the expedition was to attempt to verify the feasibility of crossing a 1,200-mile expanse of ocean by kayak—a theory postulated in Scottish tales in the late 17th century of men making their way via small boats to the Scottish shores.[4] These people were known in Scotland as Finnmen.

One story in particular sparked Hicks's imagination. It was of a solitary man who, in the 1700s, appeared in a small skin boat, landed on the shores of Scotland, and after three days, expired without divulging any clues as to his origins, having spoken in a language that could not be interpreted. There were many hypotheses about these visitors. Legend has it these travelers could be Finnish or Inuit tribe members.[5] Wherever they came from, they would have had to cross a cold and dangerous spread of water. Hicks wanted to discover for himself if it could be done. In planning for the adventure, Hicks explains, "He would attempt to follow the most logical route between Greenland to Scotland, a route that the Finnmen might have taken."

6.6.1 Project Charter and Plan

The Greenland to Scotland expedition was carefully planned over the course of five years, and in the sixth year, Hicks quit his job in order to finalize plans, procure the modified kayak, train, raise the necessary funds from various sponsors, and choose a partner. Timing was important, so he planned to leave in July, when he knew they would be able to paddle by the light of the midnight sun and would have the best opportunity for fair weather. The path to be taken included East Greenland to Iceland, Iceland to the Faroe Islands, then on to the northwest shores of Scotland. This would amount to approximately 1,200 miles. He anticipated completing the expedition within six weeks, if he paddled for around 12–17 hours per day.

6.6.2 Stakeholder Engagement/Communications Management

In this expedition, there were many stakeholders with different communications needs. There were 28 sponsors; three were financial sponsors and the rest were product sponsors. Key sponsors were Virgin®, Red Bull®, and Newscape Capital. The number of stakeholders involved meant that there were different agendas and needs. "Trying to keep a lot of people happy required a lot of coordination across many moving parts," Hicks said. He needed to be able to provide information for uploads to social media feeds, which formed the marketing strategy for the sponsors. Red Bull also was providing a film crew to take photos and video of the expedition. The plan included transmission of information about the progress of the team through a satellite phone to Hicks's wife. She had taken the role of developing and posting to social media using previously captured photographs to supplement for effect, as Hicks did not have the capability to send photos during the adventure.

6.6.3 Scope Management

6.6.3.1 Architectural Design

The design of the project was very straightforward. The basic elements of the functional architecture were (1) the sea kayak, slightly modified to meet some unique requirements, (2) the proper clothing and gear, and (3) food and water. Other important considerations directly aligned to the ability to succeed in this endeavor included the physical and emotional stamina to complete the journey and favorable weather. The plan included paddling between 12 and 17 hours per day on a predetermined defined route. Each hour would conclude with a five-minute rest. At the culmination of five cycles, he planned to take a longer break to consume a meal and rehydrate. The project was also phased into distances traveled:

- **Phase 1:** Arctic Circle, Greenland, across the Denmark Strait to Hornvik, Iceland; est. 40 hours, approximately 138 sea miles.

- **Phase 2:** Iceland, approximately 25 miles per day around the Icelandic Fjords heading east along the north and eastern coast to Neskaupstadur. This location minimizes the distance to the Faroe Islands.
- **Phase 3:** The Faroes Crossing, approximately 4.5 days, the longest crossing of almost 300 miles.
- **Phase 4:** Faroe Islands to Scotland, paddle down the middle of the islands to Torshavn, then 180 miles to Scotland.
- **Phase 5:** Landing at Balnakeil Bay, Scotland.

6.6.3.2 Requirements Management

The architecture sub-element requirements that could be controlled and impacted by design decisions included the specifications for the sea kayak, clothing and gear, and food and water. There were also requirements for the physical stamina needed to complete the journey and for access to environmental information. It would be impossible, of course, to influence the weather. However, having situational awareness about the weather was a key requirement.

6.6.3.2.1 Sea Kayak

Requirements for the kayak included the ability to contain two sitting individuals, with an option for both to lie down for sleeping. There would need to be adequate watertight spaces to store sufficient expedition supplies and sealable cockpit canopies that would cover the kayakers to allow rest and sleep. The kayak required two collapsible sails that could be used in the wind to propel the craft forward. The kayak had to have an option for stabilization while sleeping, to minimize the risk of capsizing while they were in a horizontal position. Hicks mentions that he "wanted to sleep in a horizontal position for physical comfort, boat stability, and safety reasons." Experience had shown him that if one were to fall asleep sitting up, as happened to a previous partner, the kayak could be capsized, which could have catastrophic consequences. Safety was a key design parameter. This was one of the reasons that an electric pump was required in each cockpit to ensure the swift extraction of water. The kayak, mostly sitting below the water line, also had to assist in the retention of body heat in arctic-cold water (3 degrees Celsius/37.4 degrees Fahrenheit).

6.6.3.2.2 Clothing and Gear

Due to the storage area and weight considerations, a limited amount of clothing and gear could be accommodated. Each kayaker would require a kayaking paddle and cooking materials including utensils, pot, fuel and stove, and a lighter. Experience dictated the use of a dry suit, which could be vented to minimize condensation within the suit. Each paddler would require synthetic fleece clothing to wear under the dry suit, and in addition, a supply of large chemical "Blizzard" heat pads that could be put inside the dry suit to retain heat for up to eight hours. A personal floatation device was also required for each paddler, although these were carried, not worn.

6.6.3.2.3 Food and Water

The longest anticipated unsupported time during the expedition was expected to be approximately 6–7 consecutive days. Hicks anticipated carrying all required food, but resupplying water during the stops along the way. Requirements included water bladders and three liters of water per day per person. Food needed to be stored in dry sacks. The plan included cooking one meal per day before sleeping in an effort to optimize comfort and warmth and facilitate a sound sleep. Therefore, this drove a requirement for dehydrated food, which would be cooked each day. No other consumables required preparation.

6.6.3.2.4 Situational Awareness

To drive Hicks's decision making associated with weather patterns, there was a requirement to have access (electronic) to real-time information throughout the expedition. This requirement translated to a need for satellite phone and battery. He also required a global positioning system (GPS) for coordinates and a compass for direction finding.

6.6.3.2.5 Integration and Interface

The key integration and interface elements in the architectural design were associated with the kayak and the tent canopy, the kayak and the sails, the humans and the cockpit. These interfaces were extensively tested prior to the launch of the expedition. For example, one of the first versions of the tent system was a prototype version that went just over the paddler's head. It was rejected because the venting and insulating properties were very poor. Three variants of the kayak were developed and tested before the fourth one was accepted. Using information gleaned from each test expedition, revisions to the design were implemented in each new variant. The final design met the requirements for lying down, although in practice, there was no room to move on either side or above once the paddler was reclined, which led to some feelings of claustrophobia.

6.6.3.3 Research, Development, and Design Synthesis

This project included a full technical life cycle inclusive of research, design, development, acquisition and supply, integration, and identification and development of appropriate interfaces, as well as quality checks, test, verification, and validation. This was a five-year journey for Hicks, an "apprenticeship of learning skills, understanding what works and what does not, learning about the equipment, and the limitations of living in a boat."

It was important for the success of the expedition that everything be designed and tested to carefully chosen specifications. If the kayak did not perform as anticipated, or the other equipment failed to perform as needed, it could create a life-threatening situation for the adventurers. This was made difficult due to the lack of information as to what were the "correct" specifications, since very few expeditions such as this had previously been attempted. The standard sea kayak, which by design is not intended to support extended living on board, underwent four prototype iterations prior to the final version being selected. Not only did the hull have to be modified to support the requirement of two prone individuals, but a tenting system had to be designed that would allow for the transfer of fresh air while still maintaining a waterproof environment. After several design iterations, a tenting system that met the requirements was implemented which would attach to the outside of the cockpit rim. Also included were deployable air bags on each side of the kayak for each paddler and a sea anchor.

The sails were another essential element of the system in that they could be a force multiplier to the paddling. "The wind would inflate the sail and, although it wouldn't make the kayak move faster, paddling could be sustained for longer periods," Hicks explained. It was not a new concept, but had been used since the earliest explorers crossed the oceans. However, depending on the design, it added risk. In the event the kayak capsized, the kayak could not be rolled back up into position if the sail were up. For the tests, different versions were used. A collapsible version was the safest in the event of capsizing, however it was not very robust and might not hold up under stronger winds. A more rigid version might actually tip the kayak over in a strong wind, although there was a line that could be manipulated for tension. In practice, they were never tipped over by strong winds during testing. Hicks ultimately chose a collapsible sail as a standard, off-the-shelf, product. This met his need for convenience, performance, and timing.

The dry suits had been modified to include a rubber pipe around the neck for easy access to water for drinking on demand without stopping, and zips to vent in order to keep the suit from accumulating moisture during the long sessions of exertion.

6.6.4 Schedule Management

Overall, the expedition was considered to be a success. Hicks and George completed the crossing from Greenland to Scotland and proved that, as difficult as it was, it could be done (notwithstanding they were using modern equipment). Hicks spent five years planning the expedition, which was anticipated to take four phases across six weeks, paddling for a minimum of 12 hours per day. In reality it took seven phases across nine weeks.

In the first phase, they left Greenland and reached Hornvik, Iceland, as planned. Phase 2 included making their way from the western to the eastern coast of Iceland to Neskaupstadur. They spent four weeks moving through the fiords in relatively short hops of up to 50 miles, and then standing by in Neskaupstadur while waiting out severe weather and gale force winds.

The third phase had the team attempting the ~300 miles' ocean crossing to the Faroe Islands during a weather window. They expected this phase to take five to six days, based on Hicks' previous experience during a 200-mile trip from Shetland to Norway. However, they made a determination to turn back on guidance from the crew on a fishing boat concerned for their safety, as well as their inability to get concurrence from their weather advisors on the safety of the section they were about to embark on. It took an additional week before they would have another opportunity to try for the Faroe Islands.

The fourth phase (planned to be the final) ended up being a longer stop waiting out weather and working on the fishing vessel out of the Icelandic village of Stöðvarfjördur. Ironically, this was the same ship that had advised them to abandon the phase 3 crossing. The fifth phase then became the long open ocean push to the Faroe Islands, planned to take five to six days, but actually taking only 4.5 days. They landed at the northern-most island of Streymoy in a bay with a small settlement called Tjornuvik at the top of the Sundini Channel, which runs through the middle of the Faroe Islands. The previous delays in schedule were problematic, in that it was getting to be late in the season, more hours of the day were spent in darkness, and the weather would continue to get more unpredictable (worse). And indeed, prevailing winds and weather did keep them from progressing on for another 20 days, although they were able to paddle the kayak south through the islands and base it at the southernmost island of Suderoy, ready to depart for Scotland once the weather cleared.

The sixth phase, replanned to be the final leg, was a 180-mile section from the Faroe Island to Scotland. After one false start when the weather teams told them to abandon the departure from Suderoy to Scotland due to a last-minute change to the forecast, at the first weather window they had, at best 48 hours of reasonable weather, they attempted the crossing. Upon consultation with their weather experts, however, it became clear that they would not make it before the unfavorable weather was upon them. After discussions regarding the best approach, they decided to head to the nearby island of North Rona, which was an uninhabited island 50 miles north of mainland Scotland. They reached this island just in time before a force-10 Atlantic gale hit. This is where they remained for an additional six days until the weather once again subsided enough to proceed.

The final phase of their expedition was a 50-mile section, which would land them at Balnakeil Bay in Cape Wrath, Scotland. They took advantage of a 24-hour window and completed their journey 66 days after they started. Overall, the expedition took from July 1 through September 4, with a total of twelve nights at sea.

6.6.5 Cost Management

Purchases included the two-person modified kayak, sails, anchor, paddles, dry suits, gloves, other personal fleece clothing, personal floatation devices, cooking stove, pot and cutlery, fuel, large chemical heat pads, communications and navigation equipment and services, food, and water storage bags. Acquisition costs for the expedition included personal expenses that are described in Section 6.6.5.1. In addition, sponsors provided resources without disclosing the amount of associated funding.

6.6.5.1 Acquisition Strategy and Integration

The final kayak design was a modified 26-foot Inuk Duo 7.4-m sea kayak made of carbon fiber with Kevlar in the bottom of the hull for protection against ocean ice. Two extendable/retractable skegs on the boat, one in front and one in back, were included for steering efficiency and tracking. A pedal-operated rudder, which was operated from foot pedals in the rear cockpit, allowed for accurate steering. The extended boat provided enough room for the paddlers to lie flat on their backs within their cockpits. Sealable cockpit canopies (tents) were designed that operated somewhat like a convertible automobile. Deployable 40-liter air bags for each paddler, which would be used daily to stabilize the craft and thereby facilitate a stable sleeping platform, electric pumps, Karitek® sails, and a sea anchor were also procured. Additional clothing and gear included Kokatat Gore-Tex™ Expedition Dry Suits, Kokatat Paddling Jackets, and Kokatat SEA02 personal flotation devices.

6.6.6 Risks and Opportunity Management

Since very few people (less than 10) have completed an extensive ocean crossing (over 150 sea miles) in a kayak, there was not a lot of historical information to draw on. A kayak is by nature not a vessel meant for multi-day offshore expeditions. The risks associated with this adventure were significant and life threatening. The one crossing that they were trying to replicate did not end well for the participant. Hicks needed to ensure that his expedition led to a safe crossing, so he carefully considered each risk and his strategy for risk reduction. An important part of that was including multiple shorter expeditions before this one to test equipment and processes. Each of the risk reduction expeditions resulted in changes that were incorporated to make the next experience safer and/or more comfortable.

Weather would always be a serious threat, and an unpredictable one. Being caught in an "intimidating bit of water" during a serious storm could be catastrophic, so understanding the weather was critical. This was a lesson well learned in the first phase of the expedition. As Hicks related, "We were elated at first when we arrived in Iceland after a relatively benign crossing, which I thought would be much harder, and did not check the weather after we made our way around Iceland. But then we were hit with a big gale and that quickly rectified the complacency that had crept in, and emphasized the importance of staying informed of impending weather changes."

The main perceived risk before the trip was the cold and how to stay warm during paddling, during sleeping, and in preparation for any emergency situation in which the kayak overturned or might take on water. Experience had demonstrated that if the kayak did tip and acquire water, one could experience hypothermia in minutes. Therefore it was necessary to keep the dry suits on. However, as paddling progressed during the day, the suits would become damp underneath from perspiration. To mitigate the hypothermia risks (not eliminate them), dry suits would be kept on and the design would be modified to include zips to allow airflow exchange, thus limiting moisture accumulation. To keep warm, large chemical heat pads that would last eight hours were placed within the dry suit, and the paddlers would drink and eat hot food before trying to sleep every day. The kayak itself was lined with closed cell foam to assist in the retention of the heat of the occupants.

The application of the risk mitigation for hypothermia actually led to an additional risk. Hicks decided that he would boil water for one meal and a hot drink a day using a JetBoil® system. What makes this particularly risky is that using the cooking system would require the use of fire in a very small space in a vessel that was flammable, while wearing a flammable dry suit, and while being rocked by waves and wind. Risk mitigation was minimal, other than the application of extreme care and common sense, as there was not much that could be done. Other options for heating meals were considered, but they only partially fulfilled these requirements and were more costly, less effective, or too difficult to employ; they could use a ready-to-eat meal that was chemically heated, however there was no way to heat water for drinking; therefore, that was an unappealing option.

Other significant risks involved the kayak. The stability of the kayak associated with a sleeping configuration and during the sail use needed to be addressed. Hicks wanted to sleep lying down in a reclining position for safety and stability, as well as general comfort. His sleeping ritual included processes to "inflate the airbags and put out the sea anchor to steady the kayak, put the tent up, and lie down," Hicks explained. But having enough space to lie down in the kayak came with a price. The risk of designing the boat with enough room for two people to recline was that there would be a significant amount of open space in the bulkhead that, if capsized, would take in a much higher volume of water. The risk mitigation for this serious survival feature was the installation of a sealed electric bilge pump system, which could empty the kayak in 80 seconds. Hicks felt that he could survive a serious storm lying prone in the kayak in this configuration for six to 10 hours, but that "invariably he and his partner will suffer from stress, exposure and fatigue." They would be able to continue paddling for short durations but would retire to their sleeping configuration for longer durations and "float like a bottle on the sea" until the storm abated.

An additional kayak risk involved the steering system. The configuration required steering from the back of the kayak: two skeg fins on the boat, extended and retracted with a quick-release stainless steel cable. During launch and landing, these needed to be folded up and out of the way, as they extended lower in the water than the hull of the boat. There were no replacements for these in the event of a break. That being said, during the expedition it was often the case that the paddlers forgot or neglected to pull the skegs up out of the water and they dragged on the bottom. Fortunately, the breakage risk was not realized in spite of the extended "rigorous testing." From a safety perspective, this was an issue that was fortunately not materialized.

Another significant risk was illness or injury. For example, if the rear paddler was injured or had some other problem, that would make it extremely difficult for the other to continue steering. Since the rudder controls were in the back, there was no way the front paddler could steer the boat except through sheer force of paddling. In reality, the issue of injury was a real consideration. Both physical and emotional stamina were required to complete the journey. In the event of a crisis situation, Hicks would need to use his satellite phone or emergency position-indicating radio beacon to initiate a rescue operation.

Other risks included navigation. As Ollie and his partner were paddling on straight-line trajectories in general, the navigation was fairly straightforward. But as Hicks stated, "The first time you paddle out past the sight of land is quite disconcerting. There is a psychological effect once one loses sight of land." Once out in the open ocean, sea life could have potentially caused a risk. And as they neared land, polar bears and walruses, known to be in the area, might have been interested enough to come closer. However, the risk did not materialize during the expedition. They saw whales, dolphin, and sea birds, but no other wildlife.

There were minor risks associated with the accessibility of food and water, especially during the long-distance segments of the fiords and ocean. The option to use a desalinator was considered, and it was part of his equipment on board. As this was a laborious task, manually cranking for one hour to obtain one cubic liter of water, and he needed three liters per person per day, he had decided to carry all that would be needed and resupply at the landing spots along the way. He was never at risk of running out of water during the expedition.

Hicks planned to always carry enough food for the expedition phase. The risk came in the event that the circumstances of the weather held him at a location longer than the food stores were able to sustain him. During the crossing from the Faroe Islands to Scotland, the adventurers encountered serious weather that rerouted them to a small, unpopulated island where this was tested. Upon exploration of the island, they found a refuge hut stocked with food. With this addition to the food they had with them and by scavenging local natural resources, including sea birds and a sheep, they were able to sustain themselves for the additional unplanned six days.

6.6.7 Quality, Test, Verification, Validation

Key quality, test, verification, and validation steps included the kayak design, the cooking system, and clothing.

For the kayak, the sleeping system was a modification to the basic structure of the boat that required extensive test, verification, and validation. Testing for safety (including roll recovery), comfort, and maneuverability were completed, and the craft was deemed seaworthy for the expedition.

For the cooking system, conditions were expected to be difficult; standing water affecting any potential cooking surface, and wave motion plus proximity issues of any open flame to flammable materials, such as the dry suits and the boat itself, added to the hazard. The requirement for a warm meal was met by the use of a compact, portable open flame stove and pot system. Other options were considered and tested, such as chemical heat packs from military MREs, but the stove was found to be most efficient and served all their needs.

Hicks was not originally planning on using dry suits during the expedition. During an early 180-sea-mile test crossing from the Shetland Isles to Norway, a paddler fell asleep and tumbled into the water, causing a serious safety issue for himself and the other paddlers. By the time he was rescued, a matter of minutes, he was experiencing hypothermia and severe loss of dexterity. Dry suits became a necessity from a safety perspective. In addition, other clothing was tested, and layers of Reed Chillcheater® "Transpire Fleece" outperformed other materials for warmth, especially when the paddlers were static. Sitting still while paddling, with the majority of one's body under the water line, and then during sleep, entirely under the water line, results in a significant risk of hypothermia.

6.6.8 Governance

Any expedition of this caliber, especially when taken with another individual, comes with complicated decision making and a need to agree upon changes in a less than ideal environment. The necessity to make a decision or change often proceeds from an event that is stressful and risky. Therefore, it is important to be able to communicate well and make decisions effectively between the expedition members. Hicks understood this well, and before he agreed to take on his partner, George, they decided to undergo evaluations of their decision-making and management styles. They worked with Sandy Loder's performance consultancy, Peak Dynamics, to understand their individual styles of decision making and found that they were similar in that they are both highly self-critical but did not like criticism from other persons. They shared their results with each other and were equipped to avoid the obvious trigger points.

Both Hicks and George were experienced adventurers, so it was inevitable that there would be situations in which they felt that their opinion had a greater weight in decision making. As Hicks explained, "I was the project manager and leader so had that responsibility. I used a didactic style of teaching and coaching, providing the required theoretical knowledge to back up my decision making, since George was experienced in endurance sports, but not in long kayak voyages across open sea. Whenever the occasional important decision would come up, I would lay out the scenario and options to George, get his input, and as the project manager, would make the decision. Invariably we would agree. We got on very well and were task focused—although not blindly so—i.e., it was never a case of success at any price. I well know that where the ocean and Mother Nature are concerned—discretion is ALWAYS the better part of valor."

6.6.8.1 Measurements

The success measures used during the planning phases of the project were focused on standard project management activities, including scope, schedule, and cost management. Since Hicks was managing the project, it was his responsibility to ensure that the kayak was built on time and to the specifications. He was also responsible for securing sponsor funding and for all the logistics, such as transportation for the kayak and supplies and delivery to the start point.

The success measures used during the expedition were simple. Did they complete the miles and reach the destination they were aiming for? Did they have enough food and water? Were they able to sleep? Did they keep warm? The ultimate success was landing in Scotland at the finish.

6.6.9 Outcomes

The sponsors were satisfied with the expedition. Red Bull was able to get sufficient film footage to complete their one-hour documentary, *Voyage of the Finnmen*. Other sponsors were satisfied with the social media posts that pointed out the benefits of their products in the course of their use during the expedition. Hicks explained, "Depending on the amount of donation the sponsor put into the adventure, there was an expectation that their contribution would be recognized in social media. More donation equaled a requirement for more advertising exposure through social media."

6.6.10 Lessons Learned

During the risk reduction stages of expedition preparations, there were significant lessons learned, which folded back into the research and development cycles. Once the expedition was complete, additional overarching lessons learned also emerged. The most significant of these include the knowledge that a kayak is an extremely small, generally unstable, floating apparatus that, when crossing a large body of open water, includes a level of risk that cannot be mitigated. As Hicks explained, "There were significant limitations as to the weather we could paddle in. This was particularly true for the 300-mile segment from Iceland to the Faroe Islands, which was 1/3 longer distance than I had previously paddled. Through this journey, and on these long distance segments on the open ocean, I became more aware of our fragility and exposure." The general vulnerability was only exacerbated by the increased risk during strong, unfavorable weather conditions.

In addition to the lessons learned from the physical risks of the adventure, there was a growing appreciation of the effort it would take to manage the stakeholder requirements. Adventure stories need to be shared with the general public to generate enthusiasm and support. In turn, advertising is the reason why companies support these types of adventures. And since there were no photographs that could be taken and uploaded during the expedition, a series of pre-expedition photographs that could be shared during the expedition would make it more real for the stakeholders.

A key lesson was that, just as there were dedicated weather resources, there needed to be a dedicated social media storyteller for this expedition. Ideally, a media strategy should be planned. That strategy would include thematically produced photographs, such as pictures of the kayak in the water being paddled, use of the sails, cooking, etc. The media storyteller needed to be close enough to the expedition to make meaningful comments and posts and to understand and be able to address the requirements of key stakeholders, such as sponsoring organizations that needed specific messaging across social media.

Finally, an important lesson had to do with goodness coming from adversity. During the attempted crossing from Iceland to the Faroe Islands, when the adventurers were turned back from their open ocean crossing due to poor weather, they remained in the Icelandic village of Stöðvarfjördur, where they took on work on a fishing vessel. A fortuitous encounter with a fishing vessel and a crew that was insistent that they come back to shore turned into an experience that Hicks "cherished as the fulfillment of one of his lifelong dreams, working on a fishing vessel in Iceland." He continues, "Of course we were reluctant to give up the 60-odd miles, but we would have looked pretty stupid if things had gone wrong, and it was their local knowledge against someone working off a computer-generated forecast. Ultimately, it wasn't a difficult decision to make, and they took us and the boat on board. Less than an hour later, the fog was back and the storm set in, and they probably wouldn't have found us and certainly couldn't have gotten the boat on board. I couldn't help thinking this was more than serendipity. We then had a week working on the boat hauling in 20 tonnes of cod and living out a boyhood dream."

Finally, lessons learned regarding the film crew. "Yes, it really changed the expedition dynamic having film crew there, but it was a blessing on the one hand, because they didn't factor in ground support in the planning, but because they were there and had supplies they were willing to share. On the other hand, it changed the dynamic of outside influence, since they couldn't understand what we were going through or enduring. It could be quite intrusive. Of course they were there for one reason and we were there for another, so we had differing priorities and objectives, which could clash occasionally." Hicks said, "The film crew needed a local guide to get them in position, as they regularly missed opportunities to get great footage due to being in the wrong place at the wrong time."

6.6.11 Case Analysis

This expedition shows the benefits of careful planning, research and development, the identification early on of risks, and a planned risk reduction strategy that included test, verification, and then further research and development that would be implemented based on those results. It describes the appropriate use of a tailored systems engineering life cycle approach based on risk identification and reduction, and the application of lessons learned.

In this case, a project was well defined, the architecture was straightforward and well understood, and the requirements were well established. The risks that could not be mitigated, such as the weather, were high but well understood. Change that came about through changes in the weather were addressed during the expedition through efficient dialog between the participants, with the final decision resting on the expedition leader. The explorers addressed their stakeholder requirements and delivered as promised, although they recognized this area has room for improvement. Most importantly, they finished safely and achieved what they had set out to do, even though they exceeded their schedule.

6.7 Key Point Summary

The ideas explored in Chapter 6 focus on what tailoring means and how to apply it responsibly and in a manner that reduces project risk. It describes several factors that must be taken into consideration, such as the life cycle phase of the project and the project management environment. A thorough review of the project management and systems engineering processes that are tailored is provided. Within these reviews, the processes that are known to be essential and should always be done are identified. Tailoring-up and tailoring-down suggestions are provided as well. The Complex Systems Methodology tailoring process as described in this chapter is applicable to the project management and systems engineering processes described in Chapters 3 and 4 and reflected in Sections 6.3 and 6.4.

In addition to this important construct for the application of tailoring, a model is described that can assist the project manager and systems engineer in tailoring and decision making, and then in communicating that information to the project team and the stakeholders. In the next section, key concepts are provided.

6.7.1 Key Concepts

- The need for appropriately scaled project management and systems engineering processes is referred to as *tailoring*. Tailoring is a widely misunderstood process; however, it is one that is critical to the appropriate application of project management and systems engineering. The Complex Systems Methodology provides an integrated approach to tailoring that can be used for any project.
- Project managers and systems engineers use tailoring to ensure that the right amount of process (rigor) is applied so as not to be a burden on the project, yet be effective in reducing project risk.
- Factors must be carefully evaluated when making tailoring choices, as they have a direct contributing impact on the project risk profile.
- In order to tailor effectively, a clear and accurate picture of the project needs is required. This includes an understanding of exactly what needs to be done, why it is being done, where the project will be performed, who is working on the project, and when the project will be performed.
- Wherever a project is within its life cycle, some level of project management and systems engineering process discipline is required to provide visibility, guide and control activities, and maintain boundaries of scope, schedule, and budget. The life cycle phase will identify the maturity of the design, development, or production, leading to different tailoring choices.
- Making decisions about project tailoring and then communicating those decisions so that they are understood is an important part of the tailoring process. Communication of tailoring includes ensuring that tailoring decisions are implemented as planned and are followed throughout the life cycle.

6.8 Apply Now

The application of Complex Systems Methodology tailoring is important to the success of the project. If tailoring is applied without careful thought and a clear understanding of the impacts, the probability of the project achieving successful outcomes is at risk. This chapter describes the considerations and specific steps that can be used to tailor a project and to convey the decision-making process to the project team. This Apply Now section provides summary questions that allow the reader to apply tailoring methods to a past project experience.

1. *There is often a lack of concurrence as to the definition of tailoring.*
   Consider a previous project that experienced a poor application of project management and/or systems engineering processes. Describe how tailoring could have been described and demonstrated in a way that might have mitigated the risks that were experienced.
2. *Applying tailoring requires an understanding of requirements, risks, life cycle phases, and management methods.*

Describe a situation in which tailoring was misapplied within a project, based on your understanding of these factors. Explain how this risk could have been mitigated, based on your learning from the chapter.

3. *Tailoring of the project management discipline requires an understanding of the essential processes that should be considered in any event.*

   Considering a project that you have been a part of, which of these essential processes were tailored out or not performed? What was the outcome of the project? Which essential processes could have led to better project outcomes had they been implemented?

4. *Tailoring of the systems engineering discipline requires an understanding of the essential processes that should be considered in any event.*

   Considering the same project, which of these essential processes were tailored out or not performed? Which ones would you have suggested be applied, and how would you have tailored them?

5. *Communication of the tailoring decision-making process and the final tailoring is important to project success.*

   Considering what was learned in the chapter, and thinking about a project you are familiar with, what factors and risks would be included in a tailoring tool that you would use on a project? Develop a tailoring model and determine how you would tailor the project management and systems engineering processes on your project.

References

1. Wingate, L. M. (2016, October 27). *Project Management Tailoring Methods.* Technical Project Management Series, www.itmpi.org.
2. Kokatat. (n.d.). "Wake of the Finnmen." Accessed January 1, 2018, from https://kokatat.com/expeditions/wake-of-the-finnmen
3. Hicks, O. (n.d.). "The Greenland to Scotland Challenge." Accessed January 1, 2018, from http://www.ollyhicks.com/greenland-to-scotland-challenge/
4. Redbull TV. (n.d.). "Voyage of the Finnmen." Accessed January 1, 2018, from https://www.redbull.tv/video/AP-1NBYKYF9W1W11/voyage-of-the-finnmen
5. Wallace, J. (Rev.). (1688). *A Description of the Isles of Orkney,* written by the Rev. James Wallace, A.M., Minister of Kirkwall, about the year 1688.

Chapter 7

Methodology Synthesis and Application

The previous chapters all contributed to the overall understanding of how systems engineering is used within a project. Chapter 1 explained the background of the systems engineering discipline. Chapter 2 described how systems engineering evolved to become a set of processes used to help projects achieve the anticipated technical outcomes that are identified in a project scope. Chapters 3 and 4 introduced the Complex Systems Methodology complementary and unique processes from both project management and systems engineering, which are important for driving the project forward throughout the life cycle. Chapter 5 provided information as to how to use measurements to verify and validate performance to objectives. Finally, Chapter 6 described the methods for tailoring the processes so as to add value and not negatively affect the cost, schedule, and scope of the project.

In this chapter, all the processes associated with implementing the Complex Systems Methodology are described so that application of the processes is put into context and the direct application steps of the processes within a project is clear. The objectives of this chapter are to:

- put everything into context to understand the project and the environment within which it is operating
- describe how each of the steps will be applied

Each section below provides guidance on how to apply Complex Systems Methodology within the project. Section 7.1 describes the step needed to assess the project environment within which the project is being implemented. Organizational considerations, project life cycle, and environmental requirements are described, while Section 7.2 describes the processes to understand the project structure within which the systems engineering processes will be employed. Section 7.3 describes the step associated with applying the Complex Systems Methodology stakeholder-focused and solution-focused complementary and unique processes. The key processes are described, as well as the intended outcomes from using the processes.

Choosing measurements is an important activity that provides the evidence that the project is progressing as planned. The step to identify the appropriate measurements is described in Section 7.4. And Section 7.5 describes the step associated with tailoring the project. Finally, Section 7.6 provides the information on engaging the project and actively managing it throughout its life cycle.

As is consistent with the previous chapters, a case study is provided in Section 7.7 to demonstrate the use of systems engineering processes within a project. New concepts that are introduced in this chapter are presented in the Key Point Summary, Section 7.8. An Apply Now exercise is provided in Section 7.9 that provides the reader the opportunity to apply the processes learned in a comprehensive summary form.

Chapter Roadmap

Chapter 7 focuses on bringing all the processes previously learned in Chapters 1–6 into a comprehensive summary. This will allow the reader to understand how to apply the material learned into real projects. This chapter provides the step-by-step processes to:

- understand the environment in which the project will be implemented
- comprehend the project structure
- apply Complex Systems Methodology
- select meaningful measurements
- tailor the processes
- describes a case in order to assist the reader in application of theory
- provides a summary of the chapter concepts
- provides Apply Now exercises to practice the methods that have been learned

7.1 Step 1: Understand the Environment

In this section, the environment within which a project will be implemented is described. This environment can negatively or positively affect the project and the probability that it will achieve its intended outcomes. It is an important step to understand all of the environmental factors that can have an effect on the project and to address any environmental needs that must be fulfilled for the project to be successful.

The organizational environment that the project is in can greatly increase the potential for success or provide such challenges and risks that the project cannot be executed, leading to project failure. The impact should never be underestimated, and careful consideration of how project management and systems engineering are organizationally aligned should always occur, and risks associated with the structures should be mitigated early in the project life cycle. In particular, the following organizational structures should be reviewed:

- culture and alignment of the project management and systems engineering disciplines within the larger organization
- life cycle phase that the project is entering, in relation to the organizational norm
- project relationships with external organizations
- type of project management method(s) that will be used to manage the project
- life cycle phase(s) of the project that systems engineering is facilitating

7.1.1 Organizational Considerations

The important organizational considerations for a project include the structure of the organization and how the project structure fits into the overall organization, the culture that the project finds itself operating within, and the life cycle phase of the project in comparison to the overall organization. Section

7.1.1.1 describes the organization structure. Section 7.1.1.2 explains the cultural considerations, and Section 7.1.1.3 provides insight into the project life cycle phase implications.

7.1.1.1 Organization Structure

The way that the organization is structured can have a real and profound impact on the ability of project management and systems engineering processes to drive performance in a project and the systems engineer's ability to drive the project's technical development. Implementing both disciplines requires an organizational commitment to support the top-down approach, and both the project manager and systems engineer must be placed in a position with enough authority to drive change and elicit performance in the processes necessary to be successful.

Because systems engineering is an interdisciplinary approach, it requires authority to obtain work effort across all contributing organizational departments and divisions. Depending on how the organization is structured, functional groups may provide resources to a project, or the project can be self-contained and have all the labor, materials, and services under the direct supervision of the project. If any of the project's resources are provided outside of the project's control, negotiations must be conducted in order to secure participation and use of these resources.

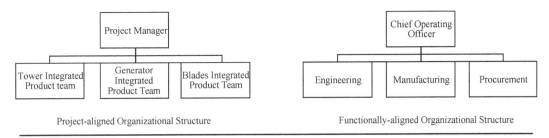

Figure 7.1 Two Organizational Structures

As shown in Figure 7.1, a project-aligned organization structure will support the project manager in obtaining all of the necessary resources, materials, and services needed to perform the scope of the project. The project manager will be responsible for ensuring that the appropriately skilled personnel are brought onto the project and perform as required. The career path progression, skills training, management activities, and all other personnel-related activities are the responsibility of the project manager. Materials and services, such as necessary computing resources, software, procurements, etc., are also the responsibility of the project manager.

Although the project manager typically secures all of the project resources and manages the project team as a whole, it is the systems engineer who is responsible for identifying the technical staff, effort, materials, and services that are required to perform the technical scope of the project. The systems engineer is responsible for providing the list of needed resources to the project manager in a timely manner, so that the commitment for these resources is secured early in the planning phase of the project. There will be a strong interaction between the project manager and the systems engineer throughout the project to manage the resources, addressing change requests along the way. In fact, the systems engineer will provide to the project manager performance, acquisition and procurement, risk reduction, and governance information that generally has resource implications. Something that may seem to have limited impact to the project baselines, such as the need to recruit and replace a key technical resource who leaves the program, could in fact drive a cost and schedule impact to the project that the project manager must address.

In the functionally aligned organization structure, the functional department or division is responsible for recruiting, training, managing, and retaining staff, providing the personnel with the skills and experience to support project work. This model aligns the responsibility of the staff management with the functional manager, while supporting project work through a negotiated internal agreement with the project managers. There are any numbers of permutations of these two models. The project manager will find each of them challenging.

The further away the responsibility lies for direct personnel management, the higher the risk to the project, as the project manager will have less control over labor hours devoted to the project tasks and schedule. In addition, if departments or divisions are managing their staff using a different management method—such as flexible software development—then these resources could be operating on different timelines and tempos. This may increase the risk associated with not directly managing resources by not being able to negotiate labor that matches the needs of the new project.

Finally, the organizational resources associated with materials and services may not be in alignment with the materials and services that a project requires. Examples of these include the lack of space for co-located staff, financial systems that cannot handle the unique reporting requirements of the project, or insufficient computing resources such as storage, network throughput, or required software. Because of these potential issues, an understanding of the organizational environment within which the project is encapsulated, and any gaps that may exist as a result, is an essential first step in understanding and mitigating the associated risks.

7.1.1.2 Organization Culture

An important consideration for implementing project management and systems engineering is in the organizational culture. Culture refers to the shared beliefs and behavior that are engrained in an organization and are demonstrated by personnel within the organization. The organizational culture generally reflects the senior leadership's words, actions, and the way performance is rewarded or punished. Rewards for poor performance, punishment for doing the right thing, as well as rewards for good performance are observed, and behavior that elicits positive personal results is emulated by the personnel throughout the organization.

An organization's culture can embrace or reject processes. Obviously, one that rejects processes will not be favorable for the implementation of either discipline. Generally speaking, the more the organization embraces process discipline, the less risk tolerant that organization will be. The organization that rejects process is typically very comfortable with risk and feels that it can achieve its outcomes in spite of the lack of consistent, defined, and repeatable processes. If an accepted organizational culture is well embedded or well established and it is not a process-supportive environment, the risk to the project will increase substantially. In some cases, it may make the ability to perform complex project management impossible. Understanding the culture within which the project will be implemented is an important consideration.

Often organizations evolve from one model to another. For example, a research and development organization that initially resists process may find that, as they move toward providing a producible design, the requirements for additional process to ensure safety, quality, repeatability within the manufacturing processes, etc. are needed, and they must move forward into a more rigorous model. Implementing additional processes into a culture that is unaccustomed to process discipline will be exceptionally difficult and risky. Overcoming this risk will require mitigation in the form of strong communication from senior leadership and ramifications for non-compliance to the newly implemented processes.

7.1.1.3 Project Life Cycle Phase

The life cycle phase of the project is important in relation to the organization's culture and structure if these are all not in phase and aligned. For example, in an organization that is established and in which strong processes are used, a research and development project with a lightweight tailored approach to processes may seem to be behaving outside of the norm. Conversely, in an organization that is focused on research and development, the implementation of a production project with a need for strong scope, schedule, and cost controls will be seen as overly burdensome from a process perspective. This incongruent perception will drive risk in the project that will be attempting to implement the appropriate disciplines in a counter-cultural manner. Both the project manager and the systems engineer should take under careful consideration the overall life cycle phase of the project and each work breakdown structure product life cycle phase so that the appropriate mitigations can be applied.

The project management life cycle phases as described originally in Chapter 2 are:

- project discovery and establishment
- preparation and planning
- realization
- governing
- completing

The full description of the systems engineering life cycle phases described in Chapter 1 that should be carefully considered include:

- new idea exploration
- definition
- concept development
- design and development
- production/construction
- operational maintenance and support
- deactivation/closure

7.1.2 Complex Project Organizational Considerations

Along with the internal project relationships that must be considered, any association with external organizations must be carefully evaluated for potential risk to the project. The risk increases as each contributing organization is added. This is due to each organization having its own structure, culture, and life cycle phase that may or may not be in synch with the project's internal structure, culture, and life cycle. The complexity increases not only for the project but also for each organization, which must manage the project activities within their own organizations as well as along all of the integration points. For example, an organization (A) that is co-developing software for a project in another organization (B) may need access to test, verification, and validation resources within the other organization (B). Due to company proprietary issues, this may not be allowed. Alternative solutions will need to have cost and schedule impacts evaluated and factored into the overall project budget and schedule to accommodate the needs of the other organization. Many organizations that are working in highly collaborative environments fail to address these issues until far along the path, when the impacts are felt more strongly.

The more organizations, and the more loosely coupled organizations (without binding contracts), that are involved in a project, the higher the project risk. The further from the norm of the project's organizational structure, culture, and life cycle phase that the external organizations are, the higher the project risk as well. The project manager and systems engineer should carefully consider these factors early in the project definition and establishment phase and implement the appropriate risk mitigation.

7.2   Step 2: Structure the Project

Once the project manager and systems engineer understand the organizational environment, the project must be structured according to its level of complexity and where the project elements fit within the life cycle phase of the project as a whole. The life cycle phase of each of the project products, as identified in the previous step, must also be evaluated.

The type of project management method(s) that will be used to manage the project must be identified at this time. As discussed in Chapter 2, the overall approach to the project may be the application of either a traditional waterfall method that will apply processes in a sequential mode, or a flexible method that applies project management processes in a cyclical and iterative mode. Different work breakdown life cycles may also require the application of different project management methods. This complicates the project and makes it difficult, but not impossible, to manage.

The project manager must approve project management methods within the project so that the outcomes from the different styles can be addressed and the overall project schedule can be maintained appropriately. The systems engineer should also be apprised of any decision as to project management methods to be used within the areas of technical product development.

A serious consideration arises when a project is being enacted to upgrade or enhance an existing operational capability. They often can lack clear information about the current state or status. These projects can be complicated (and possibly also complex), as there may or may not be configuration-controlled historical information and documentation associated with the operating capability. It can be difficult to approach these projects from a top-down systems engineering perspective, or to obtain a fully traceable path from the stakeholder requirements through the decomposed design. And the staff that understands the most about the system may not understand the need for systems engineering if they have developed and operated the current system without a previous application of the discipline.

In situations such as these, the project needs to be structured in such a way that the existing design is brought into the project system design and assessed against the stakeholder requirements. This matching of the existing capability to the stakeholder requirements is a necessary step that may not be understood by legacy knowledge-bearers, but it is a critical step to ensure that the new project's design captures the current design in the early stages of the project.

Another important consideration is how well the organization is poised to support the project. As described in Section 7.1, an organization may be structured to support the project with human resources, financial reporting, safety and security, communications technology, administrative services, etc. The project manager must make a determination as to what services it needs from the organization, and then ensure that those resources are brought to bear for the project. If projects are competing for resources (e.g., multiple projects require a new software capability, driving requirements or timing that does not meet the needs of the project), these must be resolved so that the project can be successful. Risk will be high until all organizational environmental resource requirements are identified by the project and an internal agreement is reached on the support that the project will obtain.

The easiest project to manage is a new stand-alone project that is organizationally aligned as described in Step 1. This type of project uses one type of project management method and has all the appropriate processes applied from the start. If a project follows a single project management methodology and appropriately integrates all processes from the beginning, then the complexity, and therefore

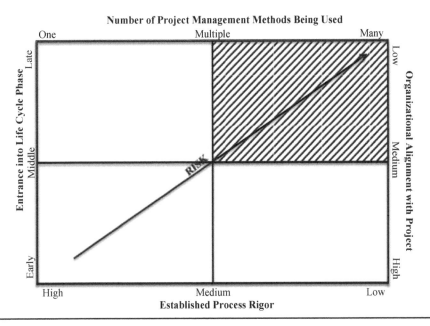

Figure 7.2   Project Structure Risk

risk, is low. If, however, a project is being set up in the middle or late into the life cycle development, organizational alignment is poor, the number of project management methods being used throughout the work breakdown structure is many, and the established process rigor is low, the complexity and risk for the overall project is high. This is demonstrated in Figure 7.2.

7.3   Step 3: Apply the Processes

Once the project is structured, the next step is to implement the processes that are needed by the project in a way that will support the project. As described in Chapter 2, project management is responsible and accountable for the project outcomes and provides the overarching structure for controlling cost, schedule, and scope. Project management processes support the overall mission of the project and incorporate work across all disciplines. Systems engineering processes help drive stakeholders' needs into a vision of the end state. That vision is then used to create a technical design that, through test, verification, and validation, will ensure that the system that is designed meets the stakeholder's needs. This is a holistic, top-down systems approach in which the major activity is decomposing the top-level vision down to the lowest-level part in the system and then maintaining control of that decomposition throughout evolutionary change, which is a natural process across the life cycle of the project. The stages of a life cycle provide structure and natural evaluation points to assess the readiness of the project to progress to the next phase.

Systems engineering processes provide the structure to integrate and link activities; ensure testing, verification, and validation to a single set of agreed-upon requirements with defined specifications; reduce risk; and ensure that changes are carefully and methodically designed into the system, all while providing visibility, configuration control of all parts, and full traceability so that change impacts can be understood before costly changes are implemented. Sound systems engineering practices reduce complexity and unanticipated emergent systems behavior. As described in Chapters 3 and 4, the Complex Systems Methodology has both complimentary and unique processes that will be applied to each project in a manner appropriate to the needs of the stakeholders.

Both of these processes provide the structure to address not only projects that enter at any phase of the life cycle, but also projects that require the use of different application of project management methods. As with flexible project management and systems engineering processes, the iterative loop is an integral part, and the practitioner must feel comfortable with progressive elaboration, as these activities add layer upon layer of new information as the processes progress. As long as continued forward momentum is achieved, then the project can expect a successful outcome. An experienced project manager and systems engineer will anticipate when the implementation of the next process is appropriate in the project, but these guidelines will help in determining an order of operations for implementation. As described in the following processes, the breadth—the number of processes used as appropriate to address the needs of the project—is explored. The depth—the amount of rigor and effort that is put into each process—is explained in Section 7.5.

7.3.1 Stakeholder-Focused Complementary and Unique Processes

In this section, the processes directly associated with stakeholders are described for new projects entering into any phase. These include processes associated with stakeholders, collaborations, communications, cost, schedule, and scope.

7.3.1.1 Identify all Stakeholders, Roles, and Engagement Requirements

This step captures the combined project management and systems engineering responsibilities of identifying and actively addressing stakeholder needs throughout the project. The project manager will manage all stakeholder engagement activities associated with the project as a whole, while the systems engineer addresses stakeholder engagement in the technical scope in order to provide a technically sound solution for their needs. The systems engineer works in tandem, but tightly in synch, with the project manager to identify all stakeholders and to actively manage the processes associated with stakeholder engagement.

The first implementation step is to identify all stakeholders, their roles and engagement, then document these in a stakeholder register. This process captures all of the necessary information that will be used throughout the project to engage with interested and key persons in a manner that leads to project success. All stakeholders that could impact or be impacted by the project are identified during this process. Throughout the life of the project, any stakeholder issues, concerns, suggestions, or requests are reviewed, documented, and resolved. Simple spreadsheets are maintained to track all stakeholder information, issues, or concerns and their resolutions (see Chapter 4, Figure 4.2, page 109). As with all the project management and systems engineering processes, this is an ongoing exercise, one which will include adding new stakeholders as they are identified and addressing their concerns, issues, and needs throughout the project so it is clear that, at the end of the project, they will know that their needs have been fulfilled.

7.3.1.2 Form Binding Relationships

Many projects, especially complex ones, rely on external organizations to perform parts of the project. This reliance on external activities that cannot be directly controlled is laden with risk, yet often leverages the strengths of external individuals and organizations to the benefit of the project. Once the stakeholders are known, agreements with external entities should be formalized, documented, and signed by the appropriate authorities so that binding relationships are created. This provides the authority and responsibility for organizations on both sides of the agreement. Fidelity around the work that is agreed upon and in what time frame it will be performed is essential, as is an understanding of communications requirements, processes for decision making and change, and the consequences of nonperformance. The formalization of these agreements is essential for risk reduction. How the formalization occurs is dependent on the organizations involved.

Complex projects often require collaboration across internal and external organizations to be successful. Working together to reach a common goal that could not be attained by an individual organization is commonplace. These negotiated partnerships can lead to formal contracts or other binding agreements. In many cases, internal collaboration is also required across organizational lines of authority and responsibility. Regardless of the boundaries that must be crossed to form and operate within these relationships, it must be appreciated that all partnering organizations are often indispensable and necessary for the success of the project. The difficulties associated with conflicting organizational objectives, different process methodologies, or differing perceptions must all be resolved so that risk does not arise from the relationships. Negotiation and compromise is needed, and success often comes down to the quality of the personal relationships that are formed.

7.3.1.3 Plan Communications

As both the project manager and systems engineer will be communicating with stakeholders throughout the life of the project, defining the communications and decision-making processes are an important step. Regardless of who is communicating with whom, the progress of the project must be communicated in a timely and effective manner, and any issues, concerns, or decisions must be addressed efficiently. The stakeholders' communications preferences are generally captured in the stakeholder register, as identified in Section 7.3.1.1. The more stakeholders there are for a project, the more challenging communications become. A communications plan describing exactly how the stakeholders will be engaged is a good way to mitigate the risk of poor communications, which can lead to a failed project.

Typical barriers to communications need to be addressed. Cultural and language barriers and discipline-specific terminology and colloquialisms that could cause confusion must be identified and a path forward developed to mitigate the risks associated with these communications barriers. How communications will be performed (e.g., memos, document exchanges, e-mail, web sites, social media, face-to-face discussions) must be specified along with any requirements for confidentiality, format, type of software, and versions of software that may impede the sharing of information between two or more organizations.

Just as the project manager has the responsibility to communicate with stakeholders, the systems engineers must have a plan as to how he or she will communicate and make decisions with the stakeholders that are engaged in the technical scope. Stakeholders are often involved in driving design decisions and therefore need to be in regular communications with the project team. The systems engineer must communicate about the progress of the technical scope of the work to the stakeholders, as well as with the project manager and the project team.

Communicating about the project on both management and technical topics can best be done through visualizations, such as summary or control charts, diagrams depicting status and health, or other graphics that show progress against the established plans. These types of charts can be used to convey information such as earned value, schedule, and budget, as well as technical information such as key performance indicators and technical performance measures. The project manager will determine how best to present the project information based on information from the key stakeholders and the organizations requesting information.

To enable the project team to communicate, the organizational infrastructure that is required to facilitate the communications must be identified and provided as a requirement to the organization. Project needs, such as the ability to communicate across geographic and organizational boundaries, must be addressed both technically and through process. Special consideration must be given for data management requirements that span organizations. Decisions need to be made regarding the location for data storage and access. Data integrity and data availability must be high. If multiple organizations require simultaneous access to the same data, some type of replication may be required. A data management plan will be a necessity to ensure that all of the challenges are addressed and that all organizations have the access they need to the project data.

7.3.1.4 Clarify Stakeholder Needs

Once all of the stakeholders are identified and a communications strategy is developed and implemented, the agreed-upon work that is to be accomplished by the project must be defined. This process is often iterative and overlapping to the previous steps. Obtaining an agreed-upon vision of the project outcomes is critical and can be much more difficult than expected to acquire. This is because most stakeholders will hold their own interpretation of what success means. It is imperative to make sure that the key stakeholders are in agreement as to the target that the project must aim for, and that the agreement is formally documented. That vision is generally reflected in a stable project charter. The documented agreement becomes the baseline for which all other project documentation derives.

Once the project manager has obtained an agreed-upon vision with the stakeholders, the systems engineer will analyze the technical needs of the key stakeholders. Understanding how the project outcomes are expected to work in operations is a critical first step. Time and effort should be applied to developing use cases and understanding the overall operations concept that is envisioned. These text-based stories form the guidance from which the solution will be developed. A spreadsheet is often used to uniquely identify each story, map it to the unique text documents, and visualize the sequencing and interactions of each. The spreadsheet also helps control the configuration evolution.

The systems engineer uses a casual, conversational style to elicit interaction specifics until it is clear what is within the project boundaries and what is outside of those bounds. All stories must be traceable to an expected project outcome and should express all assumptions, preconditions, and interactions dependencies and constraints, to the extent that they are known at that time. The impacts to scope associated with reliability, maintainability, safety, etc. must be captured and documented in clear language. These stories may evolve throughout the project as learning occurs and the project progresses.

7.3.1.5 Develop Baseline Cost and Schedule

Common practice is that when a project is initiated, an overall budget and schedule is recorded. These can sometimes change as the project undergoes initial structuring; additional funding may be needed, and/or it might be determined that the anticipated schedule must be modified based on the analysis of the stakeholder needs. However, once a project structure has settled down and the key stakeholders agree to the budget and overarching schedule, it is then time to determine the project management method or methods to apply. If a project is starting in any life cycle phase other than in an initiation phase, developing a schedule and budget that meets the needs of the project will be riskier and require careful consideration to manage that risk. This is because the project will not be completely straightforward, and there will be many unknowns that may emerge and drive cost and schedule impacts.

Determining the process method to use to address project-related cost and budget is an important step. Whatever method is used should be held consistently throughout the project life cycle, because changing a methodology midstream adds risk and causes significant issues in the interpretation of data and information from the use of measurements. Cost, budget, and forecasting methods are generally auditable and controlled by laws and regulations. The project manager will work with the appropriate financial experts and key stakeholders to determine which methods will meet the needs of the project.

Once the overarching method has been established, the project manager works with the systems engineer to allocate budget to the various work packages. The allocation of budget requires an understanding of what has been estimated to perform the work package, including labor, materials, and services, and how the basis-of-estimate was made. Procurement and acquisitions activities are identified and accounted for within the work packages. The compilation of costs need to only be at a level that provides useful information for making risk-mitigating decisions. Work breakdown structure chargeable account codes are then set up and communicated to the appropriate project team members.

Determining at what level to incorporate the work packages being provided by collaborators is also needed. If the cost and schedule information is required for roll-up into an integrated project view, the requirement for this must be specified in the binding agreements. Collaborators should also be provided with guidance on the following:

- specifications for compiling basis-of-estimate
- the format for the budgets that are submitted
- the format for the costs that are collected
- the inclusion or exclusion of subcontracted procurements and acquisitions
- the communications methods that will be used to convey the information securely
- the tempo that will be used to provide the information to the project manager

The formats and tempo must match the overall project construct so that all of the cost and schedule data is comparable. The treatment of monetary exchange rates and impacts must also be determined when they are appropriate to the project.

When the roll up of all of the work package budgets is done, the total budget, inclusive of any appropriate contingency and reserve, is established. The cost baseline is set and the budget will be managed over the life of the project per the agreed-upon processes. Using standard budget performance tracking or earned value management, the costs expended throughout the project are assessed regularly and compared against the budget. Deviations are addressed and contingency used in response to risk mitigation and realization.

The schedule development is tied into the cost development activities, in that the sequenced list of tasks that make up the project work breakdown structure is the driving force for the basis-of-estimates that feed into the budget. Developing the overall project schedule is the responsibility of the project manager. The systems engineer ensures that the technical effort is captured in the overarching project schedule. Both must ensure that labor effort is realistically and accurately reflected throughout the project schedule to minimize risks associated with a lack of availability of critical project skills.

Integrating schedules from flexible project management methods into a complex integrated schedule requires an agreement and commitment by participants as to the overarching schedule. It also requires an agreed-upon approach to handle flexible project management input into the schedule on an iterative basis, rather than all at once at the beginning of the project. Most often, this information is left outside of the integrated schedule due to the complexity of incorporating elements that are developed and evolved over time. But this defeats the purpose of the integrated schedule and increases the risk to the project, as the integration of key project deliverables will not be appropriately factored in.

Standard practice demonstrates that an integrated schedule, inclusive of all of the elements of a project, must be included. Processes to integrate the ordered list of work, the planned releases, and the associated integration testing must all be included in the overall project schedule. How that is accomplished will be based on the negotiations and agreements between the project manager and the responsible work breakdown structure management. However, all flexible methods include enough common activities to develop a reasonable approach. For example, a common approach called Agile can generate a reasonable and reliable velocity per cycle time, and a prediction as to how long a project will take. It can also adapt to earned value reporting using finished or completed work packages. A product roadmap (visual representation of the overall project) showing the anticipated sequence of deliverables over time and all integration points can assist in the communications of the integrated plan.

Both the project budget and schedule constitute project baselines that must be managed throughout the project. There must always be an up-to-date and accurate accounting of how much cost has accrued and been committed, and how much work activity has been accomplished. These are compared to the plan so that adjustments can be made in a timely manner. Change that naturally occurs in the course of a project is assessed for impact to the budget and schedule, and if there is an impact, it must be addressed with budget and schedule contingencies that have been established.

7.3.1.6 Baseline the Project Scope

All of the work that will be done during the project to meet the needs of the stakeholders is captured in the project scope. At the highest level, the project scope reflects the use cases and operations concepts that have been developed, as defined in Section 7.3.1.4. During the scope definition, the project work breakdown structure is defined. This informs the schedule and budget activities in an iterative manner, as described in Section 7.3.1.5.

Both the project manager and the systems engineer engage in processes that could affect the scope baseline and therefore have the responsibility to ensure that the scope baseline is carefully managed. Establishing a project scope baseline is an important early step in the project. The documentation, acceptance of the documentation and content as the baseline, along with the change control activities required to control the impact of change associated with scope evolution on the cost and schedule, are critical activities that both the project manager and the systems engineer will engage in. Management of the baseline throughout the project is an imperative if the project is to achieve its objectives within the schedule and budget that have been allocated.

Through progressive elaboration, the systems engineer will evolve the technical solution that meets the needs of the stakeholders. It is common during the highly iterative and recursive phases that follow for the scope to evolve in ways that could impact the overall cost and schedule of the project. It is therefore imperative that through design elaboration, the cost and schedule impacts are well understood and taken into consideration.

7.3.2 Solution-Focused Complementary and Unique Processes

Depending on the phase of the life cycle that a project enters, and the life cycle phases within the project, development of the solution set that meets the stakeholder needs can be straightforward and mature (such as for a production project) or can occur in a highly iterative and recursive manner (such as for a design and development project). Projects can also be made up of combinations of these activities. As described in Step 1 (Section 7.1.1.3), the life cycle phase(s) will affect the stability of the requirements and the level of expected changes to the project scope. Once the project manager and systems engineer know the project environment, structure, and scope, as defined in Section 7.3.1, the technical definition phase of the project can begin in earnest.

In this section, the processes directly associated with developing the project solution are described. These include processes associated with the technical solution, resources, architecture, specifications, integration, risk, governance, and quality.

7.3.2.1 Formalize Technical Requirements

The definition of the technical solution will start with the project scope baseline. This is the time in the project when the systems engineer, in collaboration with all of the project contributors, defines the technical description of how the needs of the stakeholders will be met. Up until this time the stakeholders have been specifying what they needed and why, how they expected it to perform, who would be working with the system once operational, when the system would be available, and where they needed the system to operate. The project systems engineering management plan is developed at this time. This document will describe all the processes that will be used to support the systems engineering approach throughout the project life cycle.

Any part of the project that requires design and development will go through iterations of decomposition, progressive elaboration, refinement, and evolution. In Step 3, a simple project that is based in the production of a known item will be very straightforward. The development of the work breakdown structure will have been informed by project requirements and the resulting work packages, but there will not be significant ambiguity or confusion regarding which process is driving another. In complex projects, particularly those that are developing some combination of hardware and software in various levels of product maturity, the work breakdown structure will be increasingly difficult to place into a linear process timeline. Experienced practitioners of systems engineering become comfortable with the interlacing and iterative nature of the systems engineering processes, particularly as relates to project scope development. Practitioners that are not as well seasoned in the art of systems engineering may find it confusing and frustrating to understand which processes feed other processes.

To help alleviate some of that confusion, a spreadsheet can be used to identify stakeholder requirements and link them to the associated technical requirements that derive from those. The spreadsheet can be used to further decompose the technical requirements to specifically address detailed design tasks and attributes associated with the system, product, subsystem, assembly, subassembly, and parts of the proposed technical solution. This traceability matrix is used to visualize the requirements to facilitate tracking the verification of performance to the requirements and to note stakeholder validation as it occurs throughout the project life cycle. A specification tree is also developed at this time to demonstrate the decomposition of the system.

7.3.2.2 Obtain Resources Required

As the planning of the technical scope is being completed and the work breakdown structure and project schedule are being developed, all of the necessary resources (as known at the time) are identified. These include:

- enterprise resources
- labor resources, including specific skills required, allocations of hours
- materials, equipment, and services
- procurement

These categories of resources affect the cost, schedule, and scope of the project and must be considered and planned. These resources often come from the organization, as described in Step 1 (Section 7.1). Labor might be allocated from an organizational department; enterprise financial software or purchasing processes, materials, equipment, and services may be supplied from the organization rather than directly obtained by the project. It is the project manager's responsibility to ensure that all required resources are available to the project. And it is the systems engineers' responsibility to ensure that the project manager knows what resources are needed to implement the project technical scope.

As was described in Section 7.3.2.1, this process is interlaced and integrated with many of the other processes, with progressive iterations driving the final solution. The systems engineer must identify the technically driven needs appropriately and provide the information in a timely manner to the project manager. Each project is different and, depending on the needs of the project and the organizational environment within which it is embedded, the project manager will engage to secure the appropriate resources. A prudent method to address the needs of the project is to put in place a resource management plan, as described in Chapter 3, Section 3.2.1.1.1 (page 77). The plan is typically a text document that describes the needs of the project and the processes that will be used to actively manage resources throughout the project life cycle.

During this step, the project manager, with support from the systems engineer, must identify, recruit, train, manage, and retain all of the project staff. It is possible that some labor resources will not be needed until a future date, and so the appropriate timing for hiring staff is a key consideration. In reality, throughout most projects as they evolve along the life cycle, the project labor resources require constant rebalancing. The addition, change, and reduction of skill sets to meet the requirements of the project in each phase of the life cycle are required. Labor, by named individual or job title and specific allocated hours, is added to the integrated schedule using the framework of the work breakdown structure that was created during the scope development step.

If any organization-sponsored enabling systems are required to support the project, those requirements must be identified and conveyed to the project manager and to the organization. In some cases, after negotiation between the project manager and the organizational departments or divisions that own the resources, a decision is made that the capability that is needed will not be provided by the organization, but will became an item that must be purchased directly by the project. In this situation, the cost of the procured enabling system must come from the project budget. In some cases, this becomes a direct impact to the contingency. In other cases, the scope of the work must be reduced in order to allocate budget for the unforeseen purchase.

In order to ensure that the overall needs are addressed and initiated in a timely manner, an acquisition and procurement plan (reviewed in Chapter 3, Section 3.2.3.2, page 86) is developed and provided to the project manager for action. The systems engineer specifies what, how, and when these purchases are needed. Any contractual arrangements, including acquisitions, procurements, memoranda of agreement, and other binding agreements identified in the scope of the project may require action at different times in the project life cycle. The plan identifies how these purchases will be integrated into the project. All pertinent information is included; specifications, tolerances, dimensions, need dates, acceptance criteria, reporting needs, and other requirements. As the project progresses, new requirements for acquisition and procurement may emerge and need to be dispositioned. They are added into the plan and scheduled appropriately. Before entrance to another project phase, a review of the performance associated with contracted activities is completed.

7.3.2.3 Define the Functional Architecture

The functional architecture serves the purposes of demonstration as to how the system is expected to function. The boundaries of the system, the expected flows, as well as the interfaces are identified in the graphical representation. The design objective of the functional architecture is to progress from a representation of the stakeholder requirement to derived performance requirements that flow from that function. A specification tree and allocations flow down to a logical depiction that can be used to demonstrate how the system fits together.

All parts of the system and the relationships are identified for consideration during the technical design. It does not require technical expertise to develop a functional architecture; however having an understanding of process flows is helpful. This top-level architecture provides confirmation from the stakeholders that the design requirements are well understood before time and effort is spent on developing the associated physical architecture.

It is important to remember that the definition of the functional architecture is an art. The designer will have limited information and will basically be using the vision and stories, and the requirements or needs of the stakeholder, to develop a visualization of an intended system. There is no one right answer or one right way to draw out the functional diagram. There may be a significant number of iterations to the diagram until the stakeholders confirm that it matches their vision of how the system will work. This process should not be shortchanged, as it forms the basis for the technical architecture.

7.3.2.4 Design the Physical Architecture

Design of the physical architecture requires the participation of subject matter experts with knowledge in the technological areas that are included. The physical solution is designed to address all major elements of the system that support the function, as described in Section 7.3.2.3. Simply stated, if there is a box and a line on the functional architecture diagram, there is a corresponding technical approach as defined in the hardware, software, network capabilities, and all physical connections on the physical architecture. As described in Chapter 3, the physical architecture defines form, fit, function, and interfaces.

Physical interface requirements are expected to emerge throughout the design and development phase of the project, as they are generally dependent on the technical solutions that are included. As described in Chapter 4, Section 4.2.1.1.1 (page 112), the development of an interface management plan will assure that all the processes necessary to identify, capture, and obtain owning organization acceptances for all interfaces are established. Interface identification is a highly iterative process that starts at an abstract level and continues throughout decomposition until all interfaces at all levels have been identified. Interface requirements, performance specifications, and performance parameters for each interface are documented in an interface control document and put under configuration control. The connectivity required in specific technical terms is identified in the interface control documents. Tracking all interface control documents in a spreadsheet will assist in the test, verification, and validation activities.

As the physical design evolves and matures, the documentation must keep up with the changes so that it is always possible to follow a path from the highest-level requirement (i.e., the system) to the lowest-level requirement (i.e., the part). The need for this becomes obvious when a change or trade is being considered and the impacts resulting from that change to the overarching system are being scrutinized. The physical architecture can be considered complete when it can serve as a schematic for the system. That level of maturity is required for production/construction.

7.3.2.5 Establish Quality Activities

Project quality activities are the responsibility of both the project manager and the systems engineer. Overall project quality is the responsibility of the project manager, who continuously assesses project quality and validates the quality with the customer throughout the project and at the culmination of the project. The project manager is also responsible for measuring process quality, or integrity, within the project, including metrics and measures that demonstrate performance to plan. The processes for validating quality are established upon project initiation and managed throughout the project life cycle. Stakeholders validate the quality of the system using measures of effectiveness, key performance parameters, and validation activities to determine whether the system, as delivered, has met their needs as stated in the requirements and as addressed through approved change requests.

Technical quality is the responsibility of the systems engineer. Quality is measured quantitatively (as a specification or performance parameter) and qualitatively (as a value). Measures of quality come from the establishment and tracking of the integrated set of measurements, such as measures of effectiveness, measures of performance, technical performance measures, and key performance parameters, as described in Chapter 2, Section 2.1.4.1.4 (page 39) and during test and verification activities that confirm performance expectations. Verification is a process that confirms that every system element designed, developed, produced, or purchased has the performance characteristics that adhere to the specifications and that perform as anticipated when integrated into the system.

In most projects, in order to track quality performance, a requirements verification matrix is created to track the testable, non-ambiguous requirements in clear, simple statements of intent. Each requirement has a specification and intended verification activities that can be tested and proven, as well as validation criteria that has been established by the stakeholders. A spreadsheet can be used to capture and track the quality activities. The use of flexible project management methods may require the use of different measures to assess quality performance, such as visualization boards to show quality through movement within the continuous flow of activities.

7.3.2.6 Establish Risk Strategies

Risk to a project can come in the form of impacts that decrease the probability of successful outcomes. It can come from opportunities missed for cost savings, schedule compression, or delivering better-than-expected scope to the stakeholders. Risk can also come from a known situation in which the operations model, the current technology, or the standard processes will inhibit the project from achieving the stated requirements and specifications that are needed to realize a fully operational system as envisioned by the stakeholders. Both the project manager and the systems engineer have a responsibility to reduce risk and seize opportunities on the project. How they go about doing so requires different strategies. To establish risk and opportunity strategies on a project, the project manager needs to implement risk management processes, and the systems engineer needs to implement technical risk reduction.

7.3.2.6.1 Project Risk and Opportunity Management

The project manager sees risk as threats to the constrained cost, schedule, and scope environment and therefore performs project risk management to mitigate those risks. The project manager typically develops an overarching project risk management plan and risk register to capture, prioritize, and mitigate risks and capture and address opportunities. The risk management plan identifies the methods for addressing risks and opportunities and provides guidance for managing draws on cost and schedule contingency. As described in Chapter 3, Section 3.2.4 (page 87), the risk register is generally developed using spreadsheet software and is actively managed throughout the project life cycle. Project managers base their risk decisions on a combination of experienced judgments about potential impacts or rewards and quantitative analysis results.

7.3.2.6.2 Technical Risk Reduction

Systems engineers understand that technical risk is a common and natural occurrence associated with the process of design and development. As part of the plan to reduce technical risk, each work breakdown structure work package is reviewed, and those that contain technologically immature products, or where mature elements are required to perform within vastly different specifications (e.g., reduced costs, faster throughput, higher frequencies, etc.), can be targeted for risk reduction activities.

A plan to reduce technical risk through the use of trade studies, modeling and simulation, prototyping, and testing can provide clarity on what is being attempted. Each of these techniques provides information on alternatives. Technical risk reduction is a highly iterative process and, as designs mature, opportunities and risks that might have been unforeseen earlier in the phase may present themselves and require resolution. The objective is to evolve the risky elements of the project to a stable and risk-free version that meets defined specifications and can be reproduced dependably.

Technical risks and project risks are often interrelated. As technical risks and opportunities are explored, additional resources may be required, and schedules and scope may be negatively affected. Alignment of the project management and systems engineering risk management processes through the application of Complex Systems Methodology will help to ensure that actions taken to reduce technical risk do not negatively impact the overall project constraints. The use of measures of effectiveness, measures of performance, key performance parameters, and technical performance measures as measurements that are assessed regularly within standard project management reviews ensure that technical risk is reviewed and dispositioned at the project level, and that impacts due to technical risk reduction are addressed through formal project change control.

7.3.2.7 Establish Governance

Project governance can be described as the controlled management of change. It is a structured process, which provides guidance and careful insertion of accepted change into the project. The process of controlling change provides the basis for deliberate decision making. The project manager is responsible for overall project change control associated with the baselines of scope, schedule, or budget. The project manager is also responsible for establishing the configuration management methodology for project documents. The methodical approaches that will be used to synthesize, integrate, test, verify, validate, and review the project throughout its life cycle are defined and established in this step as well.

The systems engineer is responsible for development and implementation of the technical configuration management processes. Always knowing the current configuration of the design and the produced or constructed version of a system, products, subsystems, components, and parts is critical. Configuration control is an imperative throughout the life cycle of the project. The interaction with the change management process is developed at this time, and a process that provides templates for engineering change requests and notices, which have impacts on the project's configuration, must be put in place. In large or complex projects, a configuration status accounting or project control role will be established to manage the processes.

7.4 Step 4: Choose Measurements

The choice of measurements, which are indicators of progress, is an important next step in implementing a project. Measurements must provide early warning of issues, verify that the product(s) meet the required specifications, and validate that the outcomes meet stakeholder needs. Measurements provide a mechanism for inducing forward movement.

A knowledgeable and experienced project manager and systems engineer should be involved in the choice of measurements. These experts can help select and validate useful and relevant measures and metrics. Invalid measurements can result in poor decision making and lead to project failure. Diligent measuring of activities that only give the appearance of progress but do not actually measure progress in meaningful ways wastes resources and impedes visibility into the true status. This heightens the risk to the project and prevents the opportunity to address real issues.

Measurements should be chosen to minimize the negative effects on the project and provide timely decision-making visibility, as well as providing confirmation that the project is progressing as planned. Measurements are made through the use of measures (quantitative) and metrics (qualitative). They can be subjective (based on opinion), or objective (observable and repeatable). The full description of the measurement processes is available in Chapter 5. Section 7.4.1, Essential Measurements, provides a summary of the activities that the project manager and systems engineer should select from.

7.4.1 Essential Measurements

Essential measurements provide the necessary information for control of a project. Use of the measurements provides visibility into key performance and life cycle phase progression readiness. They include baseline, resources, risk, change, and configuration controls, as well as programmatic and gate reviews. Implementation of these measurements requires a planned approach. As with all of the other processes, the choice of measurements is an integral part of the previous steps and not a stand-alone selection. They need to sit on a firm foundation that provides for the ability to collect, analyze, and react to information that emerges. Essential measurements are fully explained in Chapter 5, Section 5.4 (page 149).

7.4.1.1 Baseline Control

Schedule, cost, and scope baseline control is the primary objective of the project manager. Measurements that provide the visibility to assess the level of conformity in planned versus executed project activities are essential.

7.4.1.1.1 Schedule Control

Control of the project schedule requires an understanding of all the tasks needed to complete the project, how long it is anticipated to take for each task, which tasks are expected to take the longest, and where there is float (slack) in the schedule to address schedule risk. Control also requires constant vigilance applied to a dynamic and evolving set of parameters. A project schedule baseline is compared against progress at timeframes determined by the project manager during the tailoring process (Section 7.5).

Schedule control uses measurements that are presented from reporting tools such as spreadsheets or Gantt charts, or through earned value management reports. Either method will provide comparisons between planned versus actual charges across all work breakdown structures. Processes to audit results for accuracy, to resolve blockers to schedule progress, to address schedule contingency needs, and to manage change must also be established.

7.4.1.1.2 Cost Control

Control of project costs is dependent on many complex factors outside of the immediate control of the project manager. There can be a high level of unknowns in the areas of acquisition, procurement, collaborations, research and development, etc. Organizational norms associated with financial processes and procedures, and the cultural effect on the quality, accuracy, and level of detail of the collected cost information, can add or subtract from the overall risk. These all make cost control a risky endeavor, requiring a careful approach while setting up the project and diligence throughout the project.

Having a clear understanding of the risk profile of the project, the project manager chooses a cost management methodology for which to manage the overall project budget. Chapter 4 describes the full cost management methods that are available. The project manager sets the budget contingency and reserve. Then the budget is decomposed throughout the work breakdown structure and control accounts established so that costs can be reviewed at each work package level and in summary.

Cost control uses measurements that are presented from reporting tools, such as financial performance reports, or through earned value management reports. Each method provides comparisons between planned versus actual spending across all work breakdown structures. The earned value cost controls are combined with the schedule controls to provide integrated views of performance. Processes to audit results for accuracy, to address budget contingency needs, and to manage budget change requests are also established.

7.4.1.1.3 Scope Control

Controlling scope is the responsibility of both the project manager and the systems engineer. The scope baseline provides the basis to track work progress across the work breakdown structure. It also provides the ability to understand the potential impact of change requests on the planned system and to assess ramifications on the cost and schedule before the change is made. This is an imperative, as cost and schedule overruns are often caused by uncontrolled scope.

Every project has a defined scope with a solution definition that is expected to meet stakeholder requirements. There is also an expectation that throughout the design, development, and integration phases of the project, the scope will evolve and mature. Measurements that are used to understand the evolution and its impacts are essential. These include activities such as establishing intermediate milestones and gates at which technical experts can convey their successes and the project can assess progress of the project scope. Scope measurement should be made on all levels of the project, including components, parts, subsystems, interfaces, product, and system, according to the appropriate life cycle phases of each. Scope performance is often measured with performance metrics. These are reviewed in Section 7.4.2.

7.4.1.2 Resource Controls

Maintaining control of resources is a necessity, but can be challenging. Since resources include everything that the project needs (labor, material, equipment, and services managed by, produced by, or obtained by the project), it is identified as an essential measurement. Without resources, the project cannot go forward.

Resource controls are generally report based. Organizational systems typically provide data on labor hours worked and receipt of materials, equipment, and services. The project may have the ability to automatically compare and contrast the organizational data with project plans to analyze the comparisons. Differences in labor hours charged versus planned, actual delivery of materials and equipment versus schedule, or any other deviations are evaluated and risk mitigated by the project manager and systems engineer as appropriate.

7.4.1.3 Risk Control

Controlling project risk can be accomplished through active management. Project risk and opportunity management, and technical risk reduction can be measured at the tempo required by the project. The risk register provides the control with which to identify and assess the priority, impact, and disposition of all risk and opportunity items. Technical risk controls are implemented to ensure that the results of trade studies, prototyping, and modeling and simulation are understood and that the decisions made based on those results are integrated into the project plan as appropriate. Essential measurements associated with both of these risk processes are through graphical representations of status, reviewed regularly by the project team during management and technical reviews.

7.4.1.4 Change Control

The importance of change control cannot be overstated. Project managers use change control to ensure that consideration is given with regard to the impacts of the proposed change across the project. It is not used to inhibit change from occurring. Generally, there is a cross-disciplinary change board established that reviews all proposed changes and makes a recommendation to the project manager and/or the systems engineer, who have the final say. The process for controlling change is implemented by the project manager at project initiation and managed throughout the project life cycle. Contingency is applied to approved changes as required. Essential measures include a change control log (spreadsheet), which is maintained for historical and audit purposes. The outcome of the project is reflected in the baseline plus the approved changes.

7.4.1.5 Configuration Control

The systems engineer is responsible for establishing configuration control methods, which the project team adheres to. Configuration responsibilities and activities are typically distributed across an integrated team. Configuration control measurements can be performed at all stages of the project in a proactive manner (recordkeeping validation, active version control, review of documents prior to acceptance into a system of record, etc.), or in a passive manner (audits). The choice of the methods for controlling configuration is up to the systems engineer and the project manager.

7.4.1.6 Reviews

Establishing the number, type, and timing of project management reviews is the responsibility of the project manager. The systems engineer is responsible for the technical reviews. Both of these types of reviews serve multiple purposes, although in many cases the programmatic reviews are used as communications venues, and the technical reviews are used as control gates. Reviews can be used to confirm status, critique progress to plans, or act as gates to ensure that the project does not progress until it has been assessed and deemed ready. Chapter 5 described a standard set of technical design review gates, such as systems requirement review, conceptual design review, preliminary design review, critical design review, etc., that can serve as measurements associated with progress. The use of a formal set of defined gates provides a robust construct to evaluate and measure technical evolution and ensure continued forward movement.

7.4.2 Performance Metrics

Which technical performance metrics are chosen is dependent upon the project. Often, the more technically challenging or complex the project is, the more performance metrics will be used to track progress. The systems engineer, in collaboration with the stakeholders and the project manager, is responsible for choosing the appropriate technical metrics for the project. In most cases where earned value management is being used, technical performance metrics reflecting the performance to plan for scope will be required as well.

As part of technical scope management, metrics such as measures of effectiveness and key performance parameters (establishing stakeholder views of performance) and measures of performance and technical performance measures (establishing solution provider views on performance) are most commonly used. These metrics are described in detail in Chapter 2. Key performance indicators are also used to identify the most important performance areas for the project, as described in Chapter 5. Key performance indicators and technical performance measures can be used together to identify the highest technical risk areas that are an imperative for the project (key performance indicators), and then track progress associated with the risk reduction activities (technical performance measures). The systems engineer establishes the specifications, thresholds, and limits with key stakeholders and works with the project manager to develop the appropriate reporting mechanism and tempo for review at the project management level.

7.5 Step 5: Tailor the Processes

After understanding the external environment of the project and structuring the project accordingly, Complex Systems Methodology processes are scrutinized for use on the project. Some processes will not be used because there is no requirement for them. This happens in small simple projects—for example, a project that has no acquisition or purchasing requirements will not need a procurement plan. However, there are very few processes that can be left out completely from a project.

Before the project manager eliminates a process, careful evaluation should be conducted to identify the downstream interactions that rely on that process's outputs. An understanding of the essential processes, as reviewed in Chapter 6, is an imperative. Errors made by eliminating processes are often realized late in the project, when it is difficult to rectify the omission, and they often come with unforeseen and unfortunate consequences. Any decisions to eliminate a process must be formally documented and approved by the project manager.

The processes identified in Section 7.3 Step 3 above serve the purpose of guiding a project throughout the life cycle. When applied appropriately, these processes increase the probability of project success and stakeholder satisfaction. Tailoring of these processes is an art which requires a careful approach. The benefits of proper tailoring will be realized in the achievement of successful outcomes without undue burden from the cost of the processes. Experienced project managers and systems engineers use the knowledge of the processes, an accurate understanding of the needs and life cycle of the current project, an appreciation of the project management methods that are being used (flexible and/or traditional), along with previous experience and lessons learned to compare and contrast the needs of the current project and make qualitative decisions regarding the level of rigor that should be applied to develop and manage the processes.

A common misperception held by inexperienced practitioners is that the processes are effort intensive and costly without providing quantifiable results. In an effort to minimize effort, processes are eliminated or applied without rigor—with far-reaching consequences to the project. In actuality, implementing the appropriate level of rigor in the processes ensures project success. Tailoring decisions are based on risk avoidance through the proper application of process. How much rigor is applied to the processes needs to be considered in comparison to the risks associated with improper tailoring effects.

Tailoring can be applied in an upward (additive) or downward (subtractive) manner. Tailoring up refers to starting at a minimal level of rigor and adding rigor as the project progresses through the life cycle. This would be appropriate for a small research project, for example. A sentence, a paragraph, or a page of text could describe the activities that will be performed on a project to ensure that project outcomes are achieved. The simplest of management activities will be used. Rigor is added as appropriate over the life cycle of the project.

Tailoring down refers to both starting with an expectation that all processes will be applied with rigor and minimizing the rigor on those processes that are understood not to be required. This would be appropriate for a large, complex project. It starts with separate stand-alone plans, which describe in detail the processes that are to be used to actively manage the project. All project management and systems engineering processes across the project life cycle, along with all associated management activities, will be implemented and managed. Decisions to reduce the rigor of each plan are made on a case-by-case basis by the project manager in collaboration with the systems engineer.

To facilitate this decision making, a tailoring model (as described in Chapter 6) can be used to assist in the tailoring processes. The use of a tool or tailoring model to assist with tailoring decisions provides a repeatable methodology that can demonstrate the appropriateness of the tailoring decisions. It can also provide background on risk-related inputs that drove the decisions. A tailoring model can also help display information in a way that makes it easier to communicate. Through the communications processes, stakeholders are provided notice of tailoring decisions, and any questions they may have should be addressed. The model is used to document the association between the risks and the activities on the project that will reduce the risk. It also provides a statement of record for how the decisions were made.

All tailoring decisions are documented and approved by the project manager in consultation with the systems engineer and other subject matter experts. Communicating the tailoring decisions to the project team is necessary so that there is common understanding as to what processes will be implemented and how they will be implemented. Changes to the processes are generally stable throughout the course of the early life cycle phase, but often rigor will be added or processes that were tailored out (e.g., procurement or quality) are added back as the project moves from one phase to the next.

7.6 Step 6: Implement the Project

All of the process have been reviewed and selected, tailored, and applied. Measurements and metrics have been chosen and put in place. It is now time to execute and control the project using the documentation that has been developed, along with everything that has been learned in Chapters 1–6.

7.7 Case Study: Carolina Soap Market

All case study quotes are from interviews held with the individual presenting the background. In this next case, the interview was held with Lisa Burnham Bakowski, and all quotes are attributed to her.

Carolina Soap Market is the brainchild of Lisa Burnham Bakowski, entrepreneur and owner. Located in Raleigh, North Carolina, a hub of innovative and creative businesses, Carolina Soap Market is a perfect fit for the area. This prospering business was developed by Bakowski to feed her own creative drive. As she explained, "So many products on the market today subject us to different chemicals, such as deodorants, body lotions, hair products, and of course soap." As a matter of course, she closely monitors the content of products that are part of her and her family's daily life, careful to make ecologically sound and conscientious choices to limit their exposure to harsh chemicals and preservatives.

Two things make Carolina Soap Market products unique. Bakowski decided to work with a North Carolina theme, with soaps that would reflect sights and scents common to the state. The other hallmark of the brand is the inclusion of beer in the soap. With the emergence of craft beers, including beer in artisan soaps seemed to Bakowski to be a natural differentiator. "You can tell your friends that you bathe with beer! And they also lather and bubble up nicely due to the natural sugars in the beer," she points out.

Always one to do crafts, she started making her own lip balm and body butters and soon found that making natural, environment-friendly products had the potential to be much more than just a personal quest and hobby. She also discovered that it was a passion that could be converted into a business that would appeal to like-minded individuals. Making soap the way she wanted, with the ingredients that she could stand behind, appealed to her, and the idea to start the business was formed. "I realized it was something that I enjoyed and could do well," said Bakowski.

Carolina Soap Market is a business that offers handcrafted artisan soaps, lip balms, and assorted other products related to skin care. "The soap is crafted in small batches and individually cut, which makes it one of a kind. We use only skin-loving oils and quality ingredients that are gentle to the skin." As Bakowski pointed out, "The handcrafted soap I create is formulated to be less cleansing and milder to the skin. It is different from store-bought soap, [which is] generally made with synthetic products (resulting in a detergent), which can strip the skin of its natural oils. Carolina Soap Market soaps clean just as well and also leave the natural oil barrier in place." This makes for a superior product that fulfills a need in the market of quality, natural skin care products.

7.7.1 Project Charter and Plan

As Bakowski began to research an approach to providing her high-quality artisan soaps to the public market, she also continued to experiment with her products. She explained that, "I spent an incredible amount of time researching the market space, soap-making processes, and materials that would be needed. I would apply what I learned through experimentation. I would not be willing to go forward with a product until I was truly satisfied that it was of the appropriate quality. It was also critical that I could describe the product and how it was made in sufficient detail to anyone who would ask, so as to inspire confidence in what I was offering." The more she learned, the more she felt inspired and confident in the business idea.

As she researched ideas over the course of a year, she would document them in the notebook. "I realized that the more I learned about the topic, the more I was enjoying it. When I get interested in something I generally go all in, but this was different. In addition to being thoroughly engaged in researching the idea, I was becoming more committed to the business idea, and pieces naturally and easily started falling into place," Bakowski said. Without any formal business training, the notebook became the place where an informal business plan was developed. She started sketching out ideas for the name of the business first. As she explained, "I knew what I wanted. I had an idea for what the name should be, how it would sound, what it should look like, the branding, fonts, and colors. I wanted the name to be simple, easy to remember, and draw an image for buyers that reflected well on the region that would provide the natural ingredients that would be used."

7.7.2 Stakeholder Engagement

The Carolina Soap Market key stakeholder is typified in a specific demographic: women between the ages of 35 and 55, single or married, with or without children. This target group may or may not be career oriented. They most likely will be cognizant of product content and be environment conscious. In addition, they may lean toward purchasing organic and natural products. Most importantly, they are willing to spend more money (as a gift to themselves or for a friend) on artisan soap when a store-bought alternative is available for less. In order to find this demographic, engage with them and entice them to the products, secure sales, ensure positive affirmation in the social media, and win return customers, a communications strategy (see Section 7.7.3) would be employed.

7.7.3 Communications Management

In order to effectively communicate with the key stakeholders, Bakowski decided to focus marketing through social media. Her requirements were that social media had to support highly visual representations (photos and video), enable readers to provide feedback or comments, and make it easy for readers to follow the company on social media, thereby obtaining a direct marketing path to interested stakeholders. Allowing the option for interested stakeholders to follow the company provides a way to easily keep them informed about the company and also provides a straightforward way to promote new offerings and advertise events in the local North Carolina areas where the soap would be available for purchase.

Photographs of the products had to be high quality and provide a visually appealing image that would entice a feeling. Each of the soaps is named after the North Carolina genre represented by the ingredients, colors, and scents that are included. Carefully wrapped with a cardboard sleeve, the soap's "personality" is described in a title and ingredients list. For example, Outer Banks Breeze, with its soft blue, white, and green colors, calls up an image of a day at the beach, or, as Bakowski says, "cool breezes and memories of North Carolina beaches." This particular soap is scented with spearmint and eucalyptus essential oils. Videos that are provided through social media show different phases of the manufacturing processes.

As a first step, a basic website was built, which described the business and the products for sale. Initially Bakowski did not want to invest in a website shopping cart, where buyers could purchase directly from Carolina Soap Market, until it was clear that the business could support the annual costs of these website features. Instead, Bakowski decided to use a third-party online e-commerce marketplace as a storefront. This was the least risky approach, as it allowed her to pay in small increments and without long-term contractual agreements. This strategy was followed for a year, and, upon doing well for one full quarter, she decided to upgrade the website to add the shopping cart feature so that purchases could be made on the website directly.

7.7.4 Scope Management

In order to move forward on the business, Bakowski needed to answer several questions about the business scope, such as what products would be offered, how many types of soap would be made, and how much product would be produced at a given time. In addition, the production equipment would need to be purchased, so she needed to consider what was needed, in what quantity, and how often it would need to be maintained or replaced.

Bakowski also needed to determine how many ingredients and of what quantity had to be purchased, how much would need to be retained at any given time, and how often resupply would be required. This information was highly dependent on the formulas she was using and was therefore closely tied to the research and development activities she was engaging in to determine which products and types of soaps she wanted to include.

In order to answer these questions, Bakowski went back to her research. Her approach to materials when she made soap as a hobby was different than what would be needed to run the business. For soap making as a hobby, she only purchased the bare essentials. For the business, however, she needed to establish a different level of supplies. As she explains, "I used an iterative process of developing, testing, learning, documenting, creating a formula, testing more broadly, revising, documenting, and revising the formula. The results of this stream of activities allowed me to determine the final formula or recipe that would be put into production. It would also help me develop my list of ingredients, materials, and equipment to purchase."

A method of tracking status information in real time would also be needed to ensure she would not run out of critical supplies. Bakowski purchased software that specifically supported the industry and provided all the appropriate tracking. It allowed her to input her formulas and recipes and to identify the number of products sold for an automatic calculation of the diminished materials. As her supplies run low, a notification would be made so that she could reorder. This software provided a level of automation that made the business component run with less effort on her part, allowing her to focus on the production and marketing aspects of the business. Each of the major processes associated with business planning are described below, including architectural design (Section 7.7.4.1), requirements management (Section 7.7.4.2), and research, development, and design synthesis (Section 7.7.4.3).

7.7.4.1 Architectural Design

Bakowski wanted to offer something unique. She also wanted to keep things simple and straightforward, while having the ability to change the offerings if they did not sell well. Her research led her to conclude that offering specific limited offerings is more effective in sales than providing too great an assortment of offerings, and that people often become overwhelmed with too many choices. This seemed to be particularly true when scents were involved. Moving from one to the next causes a sensory overload, and often results in a lost purchase as the person moves on to other products. Therefore, she decided to offer a core group of soaps and then add some seasonal and special-edition soaps, where she could also market-test new ideas. The architecture of the business is therefore designed around five strategic offerings. These include handcrafted artisanal soaps, lip balms, and solid lotion bars made from natural ingredients. Holiday gift sets and soap savers are available for purchase as well. The offerings follow the guidelines that they must reflect popular attractions of the geographic region, be made with high-quality ingredients, and have a unique twist, such as provided through the incorporation of local regional beers into the soap.

With the state name of Carolina in the business name, she wanted to align the themes of the products with natural locations within the region. "I spent time researching locations within the Carolina area that evoked the images that I wanted the soaps to convey," she explained. For example, NC Gardens soap includes scents commonly found in the botanical gardens and arboretums of North Carolina, such as mild lavender, which she embellishes with calendula petals that are suspended inside and on the top of the soap bar. Special attention is paid to the colors and designs of the soaps, so that they also convey the feel of the local area represented by the soap. Madison County soap, with a fruity scent of black raspberries with a hint of vanilla, is the color of creamy vanilla, with a swirl of black and red, and topped with poppy seeds. Even the packaging text is carefully crafted to compliment the soap color.

In addition to the focus on Carolina locations, the explosion of craft beer merchandisers opened a prime opportunity for Bakowski to tie her products into an established and growing popular market. By developing a process for incorporating beer into soap, she is able to provide a product that appeals to the same stakeholder group that is willing to purchase higher-quality, higher-priced products that are unique in the marketplace.

7.7.4.2 Requirements Management

The requirements associated with all activities of the Carolina Soap Market are inclusive of the following:

- Business requirements
  o business system (computer and software for accounting, production and sales support)

> o processes (for marketing, sales)
- Production equipment and supplies
  - o processes (for production)
  - o production equipment
  - o safety gear
  - o ingredients
  - o packaging and shipping
  - o other products such as seasonal soaps, kits, and packages

7.7.4.2.1 Business Systems and Processes

Having an industry-specific, commercial, off-the-shelf software package available for small business owners was extremely useful to Bakowski. There were many available to choose from, but she chose one that would allow her to track all her recipes and ingredients, automatically deduct from her inventory based on the soaps that she makes, and maintain an inventory of raw materials and finished products. Then when she makes a sale, the software updates the ingredients lists based on the soaps that have sold and provides notification for those ingredients that are getting low.

The software also helped her determine the price point for her product. As Bakowski described, "I did not want to undercut myself in the marketplace; however, since I did not have a strong business background, I wasn't sure what the proper price was." She researched competitive offerings to determine what level of cost would be acceptable to the target stakeholders across geographical boundaries. The software helped her consider all overhead costs that needed to be recovered. With that information in hand, she was confident that her products would sell at a fair, competitive price.

In addition to processes associated with running the business, Bakowski also needed to determine the processes she would need for marketing and selling the products. She feels comfortable trying new things and making changes as required to find and reach her target market. Social media and face-to-face selling seems to be the best approach for this business. Eventually she will document easy-to-follow standard operating procedures for the business so that as the business grows, employees will be able to perform the activities in an established manner.

7.7.4.2.2 Production Activities

Production activities include the processes, equipment, safety gear, and ingredients needed to make the product. In addition, the packaging and shipping, as well as other products that are offered, are described in the following sections.

7.7.4.2.2.1 Production Processes

All soap undergoes the process of mixing the wet ingredients into a batter and pouring this into a soap mold. This is called *cold process* and includes combining scent and color with sodium hydroxide and common oils to trigger a chemical reaction process called *saponification*—converting the ingredients into soap. Once in the mold, Bakowski leaves it to harden for approximately 24 hours, then removes it from the mold and allows it to sit for another 24 hours prior to being sliced, like a loaf of bread, into one-inch bars. After curing on a rack for approximately five weeks, enough water has evaporated to make the bar hard enough so that it will last longer and be gentle on the skin.

Although soap-making processes are well established, Bakowski had to design her own formulas and recipes, which required modifications to standard processes to address the complexities of adding an alcohol-based ingredient into the soap as well as to achieve the color, consistency, and scents that would become the signature of Carolina Soap Market products.

7.7.4.2.2.2 Production Equipment

The production equipment needed to provide Carolina Soap Market products includes soap molds, stainless steel mixing bowls, stainless steel measuring spoons, a scale that measures grams and ounces, and cleaning materials. The fragrances and oils used drive the need for high-grade stainless steel tools. A soap cutter was also purchased to assist in slicing the soap at precise widths.

7.7.4.2.2.3 Safety Gear

The ingredients that are used, as well as the process itself of creating soap, can create a significant safety hazard that had to be addressed and mitigated. Sodium hydroxide is a caustic chemical that can cause serious damage if it comes in contact with the skin, can cause blindness if it gets into the eyes, and can cause death if ingested. In addition, the use of any ingredient with alcohol, such as in beer, adds complexity and safety concerns, as the liquid needs to be heated to evaporate the alcohol content, which is flammable. In order to protect against these safety hazards, Bakowski purchased and makes use of personal protective gear, including goggles to protect her eyes, work gloves, and clothing that provides adequate coverage.

7.7.4.2.2.4 Ingredients

The Carolina Soap Market products include ingredients that fall into six primary categories: oils, fragrances, and colors, along with three base product materials. Some of these are:

- Oils: olive, coconut, rice bran, castor, tallow, and organic sustainable palm
- Essential oils and phthalate/paraben-free fragrances: orange essential oil, essential oil blend (cedarwood, orange), Litsea Cubeba essential oil, lemongrass essential oil, spearmint essential oil, eucalyptus essential oil, phthalate-free fragrance, craft beers
- Natural colors: kaolin clay, ultramarine, kaolin clay oxide, mica, Moroccan red clay, activated charcoal, yellow oxide
- Base products: distilled water, sodium hydroxide, sodium lactate

Additional ingredients may be added or quantities changed, depending on the effect desired. In two popular soaps, poppy seeds or calendula petals have been added both to the soap and as a finishing touch on the outside of the soap.

7.7.4.2.2.5 Packaging and Shipping

The packaging of the Carolina Soap Market products was an important consideration for Bakowski. As she explains, "I wanted my packaging to provide a sense of quality. Of course, I would follow labeling requirements for weight, but I also wanted to list all of the ingredients, so that if anyone had an allergy they would be aware of exactly what was in the product. In addition, I didn't want the soap to be delivered in plastic. It was important to me that the packaging was ecologically friendly with minimal waste. The simple paper band seemed the appropriate fit." Perhaps even more importantly, she wanted the packaging to evoke a feeling of the buyer receiving a gift. "I wanted my customer to feel as if they were getting a treat; something quite special," she said.

Shipping soap can be expensive and tricky. It must get to the destination safely and cannot be damaged through rough handling. For these reasons, all products are shipped with recycled bubble wrap and paper shred to protect the inner contents of the packages. The products are shipped at the lowest cost possible by using first class and priority mail. Shipping costs depend on the weight of the product being shipped and the location to which it is being shipped. Because of these challenges, Bakowski opted not to ship internationally for now. Dealing with international tariffs, customs, and shipping is an additional layer of complexity that was not something that she wanted to deal with at this time, but might be part of the future growth plan for the business.

7.7.4.2.2.6 Other Products

Bakowski offers several additional products: Soap pouches and holders to preserve the life of the soap, seasonal soaps, lip balms, lotion bars, and gift packages are all available for purchase.

7.7.4.3 Research, Development, and Design Synthesis

The development of the Carolina Soap Market business required a significant amount of research, development, and design synthesis to bring together products that were differentiated in the heavily saturated and mature marketplace. Bakowski spent a considerable amount of time completing internet searches, reviewing soap-making groups on social media, and reviewing comparative business sites for information that she needed to plan the business. Often questions that were asked by others were things that she had not thought of yet or experienced, reducing the risk of her running into that problem in the future.

Most time consuming was the development and design synthesis activities. Once Bakowski felt that she had enough information to produce soap, she then needed to work out all of the details of the formulas and ultimately create the recipes that she would use to produce the soap in quantity. "I did not want to make too many mistakes trying to get things right, and I felt more prepared after the research," she said. "Designing soap requires precision. The ingredients must be carefully weighed. Oils get heated to a high temperature. Some of the ingredients are dangerous, and you must use protection." When she describes her process of working with beer, she explains, "You cannot just open a beer and pour it in. You need to boil it down to remove some of the alcohol." This process of heating and cooling must be done just right. A proper cool temperature needs to be maintained during the soap-making process or it will be too thick to add color and scent. The final touches that would be made to the soap, such as color distribution, embellishments, and decorations, were decided on in this phase as well.

Once the formula and recipe were acceptable to Bakowski, she would document them formally so that they could be repeated in production. Then it would be time to assess the desirability of the design through testing, verification, and validation (see Section 7.7.8). If the testing was successful, the recipe would move into the production queue for large-scale testing and then full-scale production.

7.7.5 Schedule Management

The Carolina Soap Market was in the research and conceptual design and development process for a full year. In year two, a website was set up with a pointer to a third-party online e-commerce marketplace. Social media was also launched. After a solid quarter of earnings, the website shopping cart was launched and the third-party e-commerce connection terminated. Each production run of 36 bars takes up to two hours. The business is currently running in the operations phase, with continual, ongoing new product design and development.

7.7.6 Acquisition Strategy and Integration

Bakowski's acquisition strategy for her requirements followed a minimalistic approach. Only purchasing the equipment and supplies that were absolutely needed, and in the smallest quantities, allowed for a slow and carefully orchestrated low-risk growth path. She would buy from wholesale markets geared specifically toward soap makers, in bulk (gallons or by the pound) or in small quantities (grams and ounces) when appropriate. She adjusted her purchases to address the popularity of the products.

Not one to make purchases without careful thought, Bakowski followed her strategy of slow diversification, offering new products in small amounts and only after diligent market testing. This kept her costs down as well, since, for example, a new scent would not be purchased in bulk until her research confirmed that it would become a desirable product.

Integrating the craft beers required close collaboration with the breweries. It was important to have the connection between the beer and the soap, so name use was important. Bakowski would obtain permission to use the name of the beer for the soap, which provided a marketing and advertising benefit to the brewery as well. These types of close collaborations between craft product manufacturers are common, since they usually result in name recognition for both of the brands.

7.7.7 Risk and Opportunity Management

Bakowski identified a series of risks that she felt were necessary to mitigate. The highest-priority risk was the financial one. After doing the initial research associated with the supplies she would need to purchase, she was not sure how much she wanted to get into the business. Her concerns were around investing too much in a website, in buying too much of the required ingredients, and then not having the sales to justify the purchases. As she clarifies, "I wanted to start small, but knew I had to initially purchase additional equipment and ingredients to supplement what I already had acquired while making soap as a hobby. But I did not want to invest too heavily until I knew what sales would actually look like. Even once I started the business, a week would go by without an order and my stress level would go up. Then I would get orders, and feel that I made the right decisions. I tried not to stress and worry, but that was not always possible."

Another risk that needed to be mitigated was in not using products that might cause harm to her customers. Ingredients that were known to trigger reactions, such as those containing nuts, were eliminated. No nut oils or products that might contain nuts are used in any of the products.

Another risk was in advertising her products on social media and websites. It was difficult obtaining high-quality product photographs. After trying several solutions herself, such as photographing inside, photographing outside, and creating a lightbox that could be used to photograph small objects, Bakowski was still dissatisfied with the results. She finally reached out to a professional photographer to assist her and has been satisfied with the quality of the product photographs.

Finally, she identified a risk having to do with the quality and hardness of the soap over its life cycle. Handcrafted soaps require a drying-out period between uses to retain their hardness. They need to be stored away from water, or they decompose and become soft, unappealing, and sometimes unusable. Customers are advised how to care for their soaps through verbal instructions in person, and for online orders, a paper copy is included with each order detailing how to care for handcrafted soap. Carolina

Soap Market also provides soap pouches and holders that assist in providing an environment that allows the soap to dry between uses.

7.7.8 Quality, Test, Verification, Validation

As with any craft product, quality is of prime importance. The quality of the product affects the reputation of the brand. With craft soap, features such as the hardness of the soap and the lather, conditioning, and cleansing ranges are all important. The colors in the soap must be strongly appealing. And most importantly, the scent must evoke the proper images and memories.

Bakowski started off making small batches and thoroughly market testing the product before committing to additional supplies. Then she would test them herself, making sure that the quality was as she envisioned. She would provide samples to a small group to test, then obtain their feedback and make adjustments accordingly. This iterative process reduced her technical risk as well as her financial risk. If a product did not pass these quality tests, the recipe would not be used further.

The packaging also had to be of good quality. Originally Bakowski used thicker card stock to wrap around the soap, but due to the cost, she wanted to try a thinner stock. As it needs to hold the soap tightly, be able to withstand printing of the brand information, and protect the soap contents, she soon found that the thinner card stock was not working, so she returned to the original design.

7.7.9 Governance

Bakowski governs the business through review of statistical measurements and careful change and configuration controls. She regularly reviews the statistics on who is buying the products and which products are selling well. Social media is used heavily to promote the products, specials, and locations where the products will be sold on site. Occasionally, targeted advertisements will be put out for an event to draw additional interest. Followers are a good indicator of interest in the brand.

Configuration control is critical. Bakowski manages this process through attention to detail. Carefully weighing and measuring the ingredients for the recipes, using high-quality stainless steel equipment for mixing and production, and following established processes maintains the quality of the products.

Change happens regularly in the business, so flexibility is a necessity. Change processes are managed through the evaluation of product quality, product success in sales, new product offerings, and opportunity management. Bakowski says, "I only offer a small core group of products at a time. If something is not selling well, I replace it. When I get feedback about a product or an idea for a product, I will carefully evaluate, test, and verify the product before I introduce it into the brand."

7.7.10 Outcomes

The Carolina Soap Market has achieved more than Bakowski had envisioned from when she started the business to the current time. Her brand is growing, and she receives positive feedback from her customers regularly. Her target demographics appear to be generally correct; however, in some cases she is reaching a younger population than expected.

She is actively involved in marketing and sales, attending local pop-up events, street fairs, festivals, handcraft marketplaces, shows, and other places where arts and crafts vendors sell. She is also pursuing additional collaborative partnerships, which should help expand the business. In addition to producing a large number of her own craft soaps, she has also used a recipe designed to compliment a partnering collaborator's business initiative. Bakowski has also provided soap gift sets for non-profit fund-raising activities and has provided soaps to compliment special events held at local breweries and coffee shops. She is currently planning for scaling up based on the opportunities that are emerging.

7.7.11 Lessons Learned

Bakowski experienced many lessons learned before and during her start-up of the business. And although it has been challenging, it has also been greatly rewarding. The more research she did, the more she felt that it was the right thing for her to do. When she came up with the name, the colors, the scents, and the packaging, it felt right. "It felt good and right every time I made a decision about the business, and that was a driving force that kept me going," she said. Her husband's support and confirmation that it was a venture worth trying and that the risks were worth taking provided the grounding that she needed to press on through all of her doubts.

A key lesson that she learned was on using the storefront prior to going live with her own website shopping cart. She felt that, "As scary as it was, I should have just gone with the investment." However, being so early on in the development of the product offerings made the move seem too risky at the time. Now that her product offerings are consistent, and new product offerings are inventive and unique, she is finding more confidence in how she evolves the business.

In fact, when considering the evolution of the business model, she mentions that she wishes she had done some additional work on learning some of the business side. "It would have been helpful if I would have had more of an outline of the game plan from the start. I had to learn on my own based on experiences, which made it more challenging. On the other hand, had I considered all of the business activities that needed to be done, I might have been overwhelmed!" she explained.

7.7.12 Case Analysis

This case demonstrates the evolution of a business focused on highly iterative and flexible research and development activities. The project of developing the business was approached as would be expected of a project in a research and development life cycle phase. Using a notebook to develop the plans is a standard approach. Stakeholder engagement was high, and the stakeholders remained engaged throughout the business implementation, weighing in on the product designs and providing suggestions and recommendations to help evolve the products.

Communications management has evolved and is measured as being effective, and scope management has been used to develop the products in a controlled environment. Effective processes have been used to schedule production work and acquisition resupplies. An approach utilizing the concept of scarcity is highly successful in this market and is working well for the brand. Risks are being effectively managed in priority order. Quality, test, verification, and validation processes were all used to reduce technical risk and to provide a solid, stable set of requirements, in the form of recipes, into the production phase.

The business has achieved the original goals that were set and continues to experience growth and expansion. The outcomes are positive. New products will continue to be inserted into the brand using the same careful processes, which have been shown to be successful to date. As the business continues to grow, additional business processes will be needed.

7.8 Key Point Summary

This chapter brings together in summary form all of the information reviewed in Chapters 1 through 6. Steps are described on how to apply the processes to any project. Each step provides context and summary points that should be considered. The environment that the project will operating in, the organization structure and culture, the project's life cycle phase, and the complexity of the project itself impact the way the project is managed. Process discipline provides the construct that enables the project to proceed in an expected and controlled manner. The step to identify the appropriate measurements to ensure that the project stays on course, both managerially and technically, is also provided. Once the decision is made on the appropriate processes to apply, the step of tailoring is described. Applying the appropriate rigor to the processes that are applied to the project ensures that the effort required to perform the processes is appropriate to the expected outcomes.

7.8.1 Key Concepts

- The project environment can negatively or positively affect the project and the probability that it will achieve its intended outcomes. All of the environmental factors that can have an effect on the project need to be addressed to mitigate the risk.
- Systems engineering processes provide multi-disciplinary projects the structure needed to be successful.
- How and where the disciplines of project management and systems engineering are located and operate within the organization are critical enablers.
- As a system evolves, it often requires the application of more process rigor.
- A project must be structured according to its level of complexity and life cycle phase, as well as the life cycle phase of each of the project products.
- Systems engineering integrates and links activities and ensures testing, verification, and validation to a single set of agreed-upon requirements with defined specifications.
- Systems engineering processes are designed to reduce risk, ensure changes are carefully and methodically designed into the system, provide visibility, control parts evolution through the use of configuration management, and provide full traceability so that change impacts can be understood before costly changes are implemented.
- Errors made by eliminating processes are often realized late in the project, often coming with unforeseen consequences; therefore, the decisions to eliminate a process must be carefully assessed.
- Measurements provide early warning of issues, verification that the product meets the specifications, and validation that the outcomes meet stakeholder needs. They also provide a mechanism for inducing forward movement.
- The systems engineer and the key stakeholders establish the specifications, thresholds, and limits of the technical solution.
- The systems engineer works with the project manager to develop the appropriate reporting mechanism and tempo for technical reviews at the programmatic level.
- Tailoring can be applied in an upward (additive) or downward (subtractive) manner. Tailoring up starts at a minimal level of rigor and adds as the project progresses through the life cycle. Tailoring down uses rigor in all the processes and minimizes rigor on those processes minimally required.

Table 7.1 Apply Now Checklist

COMPLEX SYSTEMS METHODOLOGY™ SM (CSM™ SM)
APPLICATION FOR ALL PROJECTS

Questions to Ask	Responses		
STEP 1. Understand the Project Environment			
How does the project fit into the organizational environment?	Project-aligned organizational structure	Functionally-aligned organizational structure	
What level does the organizational culture support process discipline?	Fully embrace process discipline	Does not embrace process discipline	
Is the project life cycle in alignment with the organization core business?	Directly in line	Mostly in line	Not in line
What part of the life cycle is the overall project in?	Construction/production	Design & development	Research & development
How many external organizations are involved in the project?	None	One	Two or more
What percent of binding contracts are in place for all external project participants?	100%	Less than 75%	Less than 50%
How complex is the project?	Not complex	Complicated but not complex	Complex
STEP 2. Structure the Project			
What percentage of the project is in a different life cycle phase than the overall project?	None to less than 25%	Less than 50%	More than 50%

(continues on next page)

Table 7.1 Apply Now Checklist (cont.)

COMPLEX SYSTEMS METHODOLOGY™ SM (CSM™ SM) APPLICATION FOR ALL PROJECTS

Questions to Ask	Responses		
Where is the project entering into the overarching life cycle phase of the system?	Beginning or early on	Middle	Late in a life cycle
How many project management methods are in use?	One	Two or more	
Level of required organizational environmental support in place (e.g., financial systems, manufacturing systems, etc.)	High	Medium	Low

STEP 3. Apply the Processes

Are the stakeholder-focused documents/processes in place?	Stakeholder	Collaborations	Communications	Needs analysis	Cost baseline	Schedule baseline	Scope baseline	WBS
Are the solution-focused documents/processes in place?	Requirements	Resources	Functional architecture	Physical architecture	Quality	Risk	Change control	Configuration control

STEP 4. Choose Measurements

Which measurements will be used?	Cost & schedule baseline	Scope baseline	Resource controls	Risk control	Change control	Configuration control	Reviews	MOE, KPP, KPI, MOP, TPM

STEP 5. Tailor the Process

How significant is the project tailoring?	None to some project tailoring	Heavy project tailoring

7.9 Apply Now

Table 7.1 provides a checklist of the detailed steps that have been reviewed in the chapter. Using the checklist, the reader can expand on his or her own personal experiences as applied to a project of interest to gain a full understanding of the approaches described throughout the chapter.

To explain the use of the table, for each row, identify the specific information related to your envisioned project. In Step 1, identify how your project fits into the organizational environment, how the organization's culture addresses processes, etc. For Step 2, address the process questions related to structuring the project. In Steps 1 and 2, the farther to the right of the table, in general, the higher the project risk will be. Step 3 helps establish the completeness of the process implementations. And Step 4 assists in the identification of the optimal measurements for the project. Step 5 identifies the tailoring level to be used. After all this information is input into the table, a firm foundation for project implementation is established and can be implemented.

Once the table has been completed, the following questions provide the opportunity for the reader to assess their understanding of the material.

1. *An organization's environment, within which the project will operate, is an important consideration.*
   Considering a new project implementation, describe which organizational structure you prefer the project operate within.
2. *The organization's culture often supports or impedes a project's success.*
   Describe the culture that you would find most supportive of the project you envision. Explain how you would address a culture that is not supportive of the project.
3. *The project life cycle phase, and the life cycle phase of the products within the project, must be addressed in the project management and systems engineering approaches.*
   Describe how you assessed the life cycle phase of the project. Explain any differences in approach for the project based on the life cycle phases of the products within the project.
4. *Project complexity drives different organizational considerations.*
   Provide a description of the organizational considerations you have identified for your project. Explain how these organizational requirements will be addressed.
5. *Structuring the project ensures that the project management methods address all constraints that have been identified in the environment.*
   Identify the project management method(s) that will be used in the project. Explain how these will be integrated into a single project view.
6. *All project management and systems engineering processes must be considered, but not all may be needed for each project.*
   Describe which project management and systems engineering processes may be considered for exclusion from your project. Explain the reasoning behind the choices. Provide evidence that the downstream impacts of the elimination of a process will not impact the project.
7. *Measurements provide early warning of issues, allow verification of specifications, validate that outcomes meet stakeholder needs, and provide a mechanism for inducing forward momentum.*
   Describe the type of project management and systems engineering measurements you would use for your project. Explain the reasons for your choices and the outcomes you expect they will facilitate. Provide the characteristics of the metrics and measures that will be used. Describe how often, or at what tempo, you will assess these measurements and why you determined that tempo was optimal for achieving successful outcomes for the project.
8. *Tailoring activities require a solid understanding of the life cycle implementation of all of the processes associated with project management and systems engineering.*

Provide a description of how the project management and systems engineering processes would be tailored for your project. Explain the detailed thinking behind the choices. Account for any decisions that result in a tailored approach to a process.

References

1. Agile Alliance, (2017). *Agile Practice Guide.* PMI Global Standard. Newtown, PA: Project Management Institute.
2. Carolina Soap Market. (n.d.). "Carolina Soap Market." Accessed January 12, 2018, from https://www.carolinasoapmarket.com
3. Ibid.
4. Hindy, S. (n.d.). "The Craft Beer Revolution: How a Band of Microbrewers Is Transforming the World's Favorite Drink." *The Economist.* Accessed January 12, 2018, from https://www.economist.com/news/business-books-quarterly/21600664-master-microbrewer-analyses-revolution-hops-and-dreams

Chapter 8

The Future of Systems Engineering

This chapter explores a vision of the world that is yet to come. Ideas for how technology advancements may affect different disciplines are explored, and suggestions for ways that the discipline of systems engineering can positively impact those projects so that they are able to achieve the positive outcomes that will change the world are presented.

Chapters 1–6 contributed to the understanding of systems engineering as a discipline and described how it effectively operates within the project management discipline and methods, such as with traditional and flexible approaches. Each chapter explored the Complex Systems Methodology, which provides a structure and framework using standard project management and systems engineering processes and tools in an integrated way to increase the potential and probability for achieving positive project outcomes. Complex Systems Methodology is a particularly useful approach with complex and/or highly creative projects. Chapter 7 brings together all the concepts and puts them into context so that the reader understands how to directly apply Complex Systems Methodology to projects.

The objectives of this chapter include:

- setting the stage for discussing the future application of systems engineering
- describing a view of projects that will be found in the changing world
- discussing how Complex Systems Methodology can be a positive force in achieving success on projects in the future

In this chapter, Section 8.1 provides a brief review of the origins and evolution of systems engineering through to the early 21st century. A summary of the background of the systems engineering discipline and the evolution between the systems engineering and project management disciplines, as well as other standards-generating organizations that led to complementary and intersecting processes, is described.

Section 8.2 is meant to be a think piece, and most likely will not address many topics that should be included. However, it includes some main themes that will provide enough information to generate inspired thought. It takes a fresh look at the changing world and provides a view on societal responsibilities, human sustainment, interaction, and accomplishment needs. These needs drive projects that will address welfare and prosperity, technology, communications, and others. Then, Section 8.3 provides a comprehensive view of the process discipline evolution and reiterates the coherent approach that will ensure that the highly complex projects of the future are attainable.

This chapter concludes with Section 8.4, Key Point Summary, and 8.5, Apply Now. These sections provide key concepts associated with this chapter and provide exercises that will assist the reader in applying the key lessons within this chapter.

Chapter Roadmap

In Chapter 8, the current systems engineering discipline as has evolved over the last decade provides the backdrop to assess how the world is changing and adapting. It also provides information regarding the challenges that will demand a coherent integrated process approach as projects become more complicated and complex. This chapter provides information that allows the reader to:

- understand the current stage of systems engineering in projects
- comprehend the evolving world
- understand the nature of the multidisciplinary environment
- appreciate the processes that will be most important for each of the focus areas
- review a summary of the concepts presented in the chapter
- practice Apply Now exercises to ensure understanding of the material provided in the book

8.1 Systems Engineering Origins

In this section, the origins of the systems engineering discipline are outlined and the specific processes that make up the discipline are summarized in preparation for the discussion of how these can be applied to address the complexity challenges of the future. The key points will include a description of how systems engineering:

- emerged from a need to achieve positive outcomes
- provided a method to address the production of products that were integrated across different engineering disciplines
- provided a holistic and synergistic perspective of the product
- has been accepted as a formal discipline with associated standards
- has been successfully implemented in organizations around the world to manage complex projects

Systems engineering first emerged as a method that could be applied to interesting inventions that could be mass produced and widely distributed. Activities that were done as part of a product invention were differentiated from the engineering practices that would be used to move that invention into a producible product. This separation was driven by the move to digital technology from mechanical and analog electronics. With smaller, more cost-effective components, products could be designed, produced, and made available to a wide population at reasonable costs. This shift in focus brought about the processes that would make up the discipline of systems engineering. In fact, the words "systems engineering" were first used in the 1940s by Bell Telephone Laboratories, where primary activities of systems engineers—optimization modeling, queuing theory, and probability theory—were being used to solve problems associated with broad military systems during World War II.[1] The RAND Corporation had also been developing analogous processes associated with systems analysis that would also tie directly into the systems engineering discipline.[2]

As the digital revolution drove incredible growth in industry, complicated and complex systems were developed, produced, and brought to market. Methods were created to carefully capture the stakeholders' requirements and to ensure that the incorporation of all disciplines (including software development) were put into place and incorporated into the processes, so that the final product or service had the highest probability of meeting the stakeholders' expectations. The United States Department of Defense and the National Aeronautics and Space Administration were both early adopters of this systematic, integrated approach to the product life cycle. The complexity of the systems that they were developing for both war fighting and space exploration required a methodical approach that would provide the rigor needed throughout the design, development, production, and operations of these highly integrated systems so that they could be operated safely, securely, and with the intended results.

By the later part of the century, the need for an integrated systems approach was clear. Systems built on heavy software–hardware interactions and interfaces had emerged. Competition from commercial enterprises in the global marketplace pushed organizations to provide high-quality, high-performance products, while also meeting profitability targets. In addition, cross-discipline systems were becoming the norm, along with information-dense products associated with command and control activities and enterprise systems that were distributed across organizations and into external organizations.

To address this need for engineers that could apply the appropriate processes to ensure that the system development would achieve all expected targets, universities in the United States began offering courses that would ultimately produce a pipeline of engineers who were familiar with and could employ the systems engineering concepts that had been formalized through military standards, such as the MIL-STD-499A.[3] Over time, organizations formed to set standards for the systems engineering discipline and provide forums for individuals performing the systems engineering tasks through which they could gather and support one another. Papers and books outlining systems engineering practices were published and made available in the mainstream. The formal job title of Systems Engineer was used for recruitment and hiring, although the job descriptions varied dramatically. By the early 21st century, the systems engineering discipline had developed into a degree-awarding formal discipline at both the undergraduate and graduate levels. Certificates were being granted for those who demonstrated their knowledge and experience in systems engineering, and training and conferences were held around the world. Discipline-specific terminology and standards were developed and communicated through the standards-setting organizations.

The emergence of the systems engineering top-down approach was fundamentally different from the common design-build-test engineering processes that were in use at the time. The top-down approach required an understanding of the stakeholders' vision, in terms that could be decomposed into requirements. These requirements would be converted into functional and physical architectures depicting how the technical solution would meet the needs of the stakeholders. From there the technical work would be decomposed into products. As the products were developed and produced, test, verification, and validation would be performed at the product level and in the system once it was cohesively assembled. The product teams were typically integrated, and all the needed disciplines for that product would work together to ensure that a quality product was delivered. This multi-discipline, holistic approach was a mainstay of the systems engineering discipline. The ability to use these processes to minimize technical risk and increase the probability of successful outcomes was persuasive.

To ensure that the process was well understand in and across multi-disciplinary teams and to limit the amount of wasted effort in creating processes, systems engineering standards employed on a project helped communicate the requirements of performance for the discipline. Common and accepted terms were adopted that conveyed the intended roles of the project systems engineers. These included terms that described not only the systems engineering processes, but also the systems engineers' perspective of the system and its component parts, including such things as qualities, properties, characteristics, functions, behavior, and performance. These terms referred to the systems-level attributes, understood to be defined by all of the decomposed parts of the system, as well as the interconnections between the parts of the system, which often generated emergent behavior.

As the systems engineering discipline evolved, the processes being used began to intersect with the project management processes associated with control of project cost, schedule, and scope. Organizationally, each process discipline was often reporting to different lines of authority, causing some level of conflict or divergent activities. The activities that overlapped generally were associated with stakeholder engagement, risk, and change management, as decisions made in the technical areas have a direct effect on the cost and schedule of the overall project. In some organizations the risk of contradictory guidance, opposition, incompatibility, or incongruence is easily solved through the alignment of the systems engineer (responsible for the *technical* process discipline) directly under the supervision of the project manager (responsible for the *overall* project process discipline). Other organizational alignments may work, depending on the personal abilities of the individuals in each position and the level to which they are in agreement as to areas of responsibility and authority.

Standards are the established methods for accomplishing tasks within both the project management and the systems engineering disciplines. They are the measure against which quality or compliance can be assessed. Implementation of systems engineering standards reduces complexity, provides a methodical approach to forward progression in a project, and is meant to minimize unexpected emergent behaviors from a system. As systems engineering evolved from invention and engineering disciplines, their standards are not entirely unique to the systems engineering discipline. There is significant overlap in standards with disciplines that routinely interact with the project. This causes some confusion about the necessity and value of a set of systems engineering standards, particularly in areas that touch on the same topics. This sometimes consistent, but often conflicting, overlapping, and incompatible guidance causes risks in systems projects if the common terminology and intended use of the process is left to individual interpretation. Identification of the standards-setting organizations is available in Chapter 1. Solving this process conundrum within a project is often left to the systems engineer, sometimes working in concert with the project manager. Providing a solution across the standards-generating organizations has only just begun. The evolution of standards that impact systems engineering is shown in Figure 8.1.

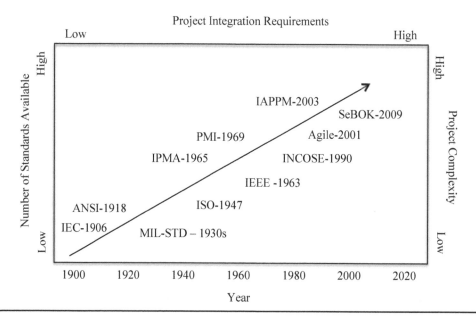

Figure 8.1 The Evolution of Standards

Figure 8.1 identifies some of the most significant standards that affect project management and systems engineering. The year of introduction is shown along the *x*-axis (bottom). The *y*-axis (left vertical) shows the number of standards that were available for both disciplines, from low to high. The top axis shows a scale from low to high of the project integration requirements, and the right vertical axis provides a scale from low to high of project complexity. What this demonstrates is a continuing development of standards from various organizations that impact both project management and systems engineering. Without a doubt, standards will continue to evolve to address the changing nature of projects.

8.2 Adapting to the Changing World

This section focuses on exploring the changing world and the anticipated impact on project implementation. Project complexity is expected to continue to increase as a globally connected society continues to incorporate more interconnected technology into the mainstream. The four categories explored in this section are societal responsibilities, human sustainment, human interactions, and human accomplishments. These four categories can be described as follows:

- Societal responsibilities call attention to projects that are driven by the needs of the society. This is often influenced by culture and environment.
- Human sustainment refers to projects associated with welfare and prosperity—the activities that humans engage in as part of their unique existence on the Earth.
- Human interaction focuses on projects that address the needs of humans to interact with their surroundings and other humans.
- Human accomplishments provide insight into the human need to pursue creativity, physical excellence, and additional capability, knowledge, and capacity.

As shown in Figure 8.2, each of the four major categories further breaks down into subcategories that typically drive technology projects. These will be discussed in the following sections. This figure represents a system of activities that are often measured by high-level indicators to assess overall systemic health and vitality of the world, countries, cities, and populations. There is a vast array of measurements,

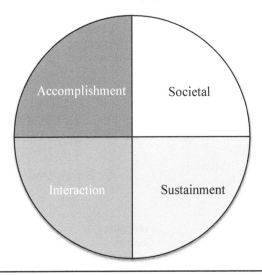

Figure 8.2   Impact Areas

metrics, and indicators available, but two of the more common indicators include economic development and growth. Economic development refers to the activities associated with improvements in living standards, ultimately affecting life expectancy, infant mortality, and literacy levels of the population. Economic growth is the measure of incomes and expenditures, including trades that a country is supporting. These indicators are key measurements that influence decision making at the highest levels. Those decisions will trickle down to each of the areas described in the following sections. Political and economic policy decisions can impact the support and funding levels needed to drive social investments. Societal projects that support the adoption of new technologies, projects that improve general living conditions, projects that enhance mobility and transportation, and projects that support human accomplishment rely on strongly supported policies.

Sections 8.2.1 through 8.2.4 investigate the activities in each of these subcategories that systems engineering will address in future projects. The potential risks and impacts to these projects if the appropriate processes are not implemented will have far-reaching effects, considering the complexity and interconnectedness that is anticipated. Therefore, it is important to investigate the makeup of these projects and to determine how systems engineering processes can be employed to ensure optimal outcomes.

8.2.1 Societal Responsibilities

The societal responsibilities section describes projects that are driven by the needs of society. Projects associated with society take into account the culture and environment, and yet provide basic services common to all societies, such as transportation, power, waste management, public protection, etc. (see Figure 8.3). They also provide the core safety and security services for the society. Although not all-inclusive, these highlighted activities form the basic structure within which individuals can be sustained and accomplish things that would otherwise be unattainable. Projects that address societal needs generally have a high requirement for accessibility, safety, and stability. Careful planning can also make these assets extensible, so that as the population grows, the system can accommodate that growth without a complete replacement.

Societal responsibilities meet core human needs through culture, environment, government, and society at large. Depending on the stability and maturity of the environment that the individual is

Figure 8.3 Societal Responsibilities

embedded in, the behavior of the society will affect his or her ability to feel physiologically safe and secure and be provided opportunities to socialize and develop his or her potential. The following sections (listed in alphabetical order) describe the areas of social responsibility that will likely have the highest technological growth and evolution in the coming decade and that will require a careful application of systems engineering:

- disaster mitigation and recovery
- ecology and conservation
- energy and power
- information security
- mobility and transportation
- social services and defense
- waste management

Each of these topics is explored in the following sections.

8.2.1.1 Disaster Mitigation and Recovery

The amount of information, analysis, statistics, and forecasting that is in current use to predict disasters, provide early warning, and then deal with the aftermaths is better than it has ever been. In a multi-organization collaboration, the International Disaster Database has compiled annual statistics since 1973.[4] "EM-DAT contains essential core data on the occurrence and effects of over 22,000 mass disasters in the world from 1900 to the present day, conforming to at least one of the following criteria:

- 10 or more people dead
- 100 or more people affected
- The declaration of a state of emergency
- A call for international assistance"[5]

Categories of natural disasters according to the EM-DAT include:

- "Geophysical: earthquake, (dry) mass movement, volcanic activity
- Hydrological: flood, landslide, wave action
- Meteorological: storm, extreme temperature, fog
- Climatological: drought, glacial lake outburst, wildfire
- Biological: animal accident, epidemic, insect infestation
- Extra-terrestrial: impact, space weather"[6]

These natural disasters have a tremendous impact on society, and because of this, organizations that model weather patterns and make predictions are relied on to help those in positions of authority provide early warning to the general public. The ability to provide reliable predictions has evolved through the use of high-performance computing and advanced modeling and simulation capabilities. Communications, processes, and methods to provide assistance by governments following an event have also evolved dramatically. More people are able to seek aid immediately after and well into the following months/years, which affects the statistics. The number of people killed from disasters is often reduced by the mitigations (such as early warning) that are provided. "In 2016, the number of people killed by disasters (8,733) was the second lowest since 2006, far below the 2006–2015 annual average of 69,827 deaths."[7]

The potential success of mitigations must be considered in combination with other factors that are not controllable, such as the type of disaster, the location that sustains the damage, the socio-economic conditions at the site, etc. For example, a significant share of the deaths from natural disasters have occurred in economically impoverished societies, whereas natural disasters occurring in higher income economies that sustained damage report higher economic losses but fewer deaths overall. The total number of individuals affected and the estimated economic losses caused by disasters have shown increases as populations have expanded into areas prone to natural disasters such as earthquakes and floods.

In addition, there are non-natural disasters, such as explosions, loss of ships at sea, acts of war, exposure to pollutants, building collapses, transportation accidents, technology-triggered ecological disasters (e.g., tsunami affects electrical grids, which shuts down pumping systems, causing mass flooding), structural fires, etc. Significant impacts are seen in loss of life, impacts to quality of life, and economic and infrastructure damages. As the impacts from natural and non-natural disasters are significant to both individuals (sustainment and interactions) and the societies they live in (ecologically and economically), the list of potential weather prediction and disaster mitigation and recovery projects that will need to be executed soars.

The probability that projects in the next decade will continue to address these topics is high. These projects will inevitably include integrated technologies that can address the multidimensional, interrelated components of the systems that are or might be affected. A heavy emphasis on requirements will be needed during the project definition phase, as the underlying disciplines associated with the fields of disaster planning, emergency response, etc. are still evolving and influenced by most recent events and experiences. Mitigation and recovery look different to the stakeholders, and perceptions about the quality and quantity of repair, as relates to the pre-disaster status, often emerge. These positions need to be carefully weighed against the expected financial burden and recovery schedule of the mitigation. Careful attention to stakeholder identification and needs will be required, as the vulnerabilities that will need mitigation will undoubtedly be unique.

Both the planning for and the actual execution of projects associated with disaster mitigation and recovery can benefit from the additional application of systems engineering processes that can add value to these types of projects. As these projects generally enter at the early life-cycle stage, tailoring of the processes can be done appropriately. They should, however, always include the processes associated with communications and decision support, risk management, modeling and simulation, and trade studies.

8.2.1.2 Ecology and Conservation

Each of the world's societies is bound to responsibly use, and ultimately preserve, the world's resources for future generations. Every individual can contribute to the initiatives set forth by society to this end. Over the last decade, a focus on ecology and conservation has increased as the damage to the environment from previous generations' actions becomes apparent and requires action to resolve. The changes in the earth's biosphere is also driving an awareness of declining and deteriorating natural systems that have been treated as permanent features. Species extinctions, reduction in the ice sheets, glacial melting, air and water pollution, deforestation, reduction of viable rainforests, and changing weather patterns are just a few examples of this evolving environment.

Ecology refers to the interdisciplinary field of systems, particularly living organisms and the environment of the Earth. Conservation is the protection, management, and restoration of natural environments. Projects that address societal needs associated with ecology and conservation have a high need for stakeholder management. Knowing what the population considers valuable and what changes they are willing to adopt to reverse damage or conserve resources is critical. These types of projects are often funded by grants and endowments, enter in the early project phase, and are often carefully managed.

Outcomes are widely published in an effort to affect governmental policy, behavior of industry, and economic decision making. Garnering support from stakeholders through consistent and rigorous research is an imperative. Providing strong communications to the larger communities to enact change is critical. Although all systems engineering processes may be applicable, depending on the size and scope of the projects, tailoring of the processes is appropriate for these projects.

8.2.1.3 Energy and Power

Society relies on energy and power and provides the infrastructure necessary to collect and store enough energy to provide power to its populations so that all of the needs are met. These needs include activities such as those related to mobility and transportation; lighting; cooking; powering electronics, tools, and equipment; providing heating and air conditioning; etc. There are renewable (e.g., wind, thermal, hydro-electric power) and non-renewable (i.e., fossil fuel) sources of energy, and optional sources may continue to emerge over the next decade.

Power will remain a critical need with a growing demand. If environmental and access concerns continue to influence the evolution of energy and power, there will be a shifting of draws on electric sources and may evolve to a model in which fossil fuels are no longer needed to power transportation, and energy is collected from multiple sources (both renewable and non-renewable), stored, and distributed from the perimeter of the population areas. Individuals may one day only need to be close to electric power for devices to be charged and may be unaware of how the energy is collected and stored. The overall impact on the environment from these changes remains to be seen, but this trend does indicate a need to think about projects that will evolve, upgrade, and modify existing homes, businesses, and means of transportation.

Energy includes all materials and processes that can be accumulated predictively, collected and stored, delivered as a product, and provide a source to draw from. Energy is generally measured in joules or watt/hour. The amount of energy available to power products is limited to the amount of energy that is stored. Energy can be collected and stored from sources such as fossil fuel, the sun (solar), wind, wave, thermal, or nuclear. Power is calculated in joules per second or watts[8] and represents the measurement of how much energy is used per unit of time, or the rate at which the energy is delivered, transferred, or used. Power is delivered in a constant quantity and cannot be stored.

Any activity that requires a power source is dependent on this critical resource. How a society obtains its energy and distributes is power is dependent on many factors, such as the availability of sources of energy; the facilities in place to collect, store, and secure the energy until used; and the infrastructure available to transport the energy to the location of need through the distribution of power. Expectations from the population of users are high. They expect easily attainable, readily available, extensible and surge capable, safe, and redundant access to power of consistent quality for all of their needs. Projects such as these must carefully consider all of these stakeholder needs.

Without the ability to light, move, energize, or efficiently and safely provide energy and power to the population, modern civilization does not exist. However, the collection, storage, and distribution of these resources come with significant environmental impacts. The next generation of technical capability in energy and power will affect all current systems. Entrance into the systems will be along the entire life cycle. Some new capability will be implemented. Many upgrade projects, utilizing existing operating systems, will be actualized. Therefore, it will be necessary to understand the processes to manage project technical scope that enters at different locations along the life cycle phases. This is an area of systems engineering process that is not mature and will need evolution (see Section 8.3 for more detail). A rigorous application of systems engineering processes—including specialty-engineering activities that consider human factors, reliability, resiliency, availability, maintainability, safety, security, and quality —must be incorporated.

8.2.1.4 Information Security

Without a doubt, information security is an important focus area for the next decade. The increase in technically connected devices and embedded technology will continue to increase throughout the next decade. The strong push to host personal information on shared servers on a wide network for almost every transaction across all disciplines (financial, health, purchase history, etc.) requires complementary assurances as to the accessibility, quality, and security of the information.

There may be debate about where the responsibility lies for providing information security; however, it is clear that individuals providing the required information can only hope that the organizations collecting the information have put in place the appropriately rigorous security systems to protect this personal information. Individuals must also rely on a higher authority to stop the theft of personal information, much as they do for the theft of personal property. Just as individuals do what they can within their circumstances and control to protect their physical belongings, they must do so to protect their personal information, while relying on the holder of the information to safeguard the information that is outside the individuals' control.

As new technology is introduced during the next decade that shifts the control of personal information from the guardianship of the individual to that of multiple organizations, while also collecting more personal information about the individual and integrating it into the collective databases, safeguards must be in place to assure that information is safe and secure and remains accessible to the individual and only those others that have been approved for access. Personal information must also be secured against unintended uses, such as when an organization collecting the information has an incentive to sell it for profit. The hosting organizations may be operating within their legal purview to take actions that may not be in the best interest of the individual, even when the impact to the individual may be profound. And those individuals may not have any recourse (legal or behaviorally—in other words, there may be only one provider of the service). In addition, the integrated nature of the next generation information environment may actually inadvertently combine personal data into ways that make a breach that is exceptionally damaging to the individual due to the compilation factor (compiled information can become more sensitive than the data parts that it is made up from).

Unlike the previous decade's practice that the buyer should beware and not invest in products or services without fully checking them out, this next generation of products and services will be complicated, complex, and exclusive to the point that the average buyer will not understand the potential impacts or may have limited to no options but to sign away their rights. With the incorporation of biometrics, the risk increases even more. If biometric data were to be stolen and somehow replicated, what would recovery look like for the individual? Each person also may not see the compilation of their data being done in the background, so that when it is put together into a cohesive story, it actually provides a high-risk picture of who they are, what they do, and where they go. For a systems engineer (or a project manager), there are ethical boundaries that may be tested in these types of projects. Although these practitioners will also find themselves vulnerable to the same impacts as the average person, they will be in a position to influence the direction these projects take. The awareness of the potential impacts can help inform the designs.

There are many benefits to the use of technology that is currently in existence, and even more are expected in the next decade. But just as with physical threats, proactive defensive postures by both individuals and organizations are but a single line of defense. Additional measures must be enacted to provide ways and means to find and prosecute offenders and to resolve information security breaches along with all the resulting individual impacts, which generally falls outside the purview of the organization(s) providing the technical collection, storage, and access. This is why this particular category falls under societal responsibilities. When technological solutions become so embedded in the environment that the individuals are required to submit personal information into systems outside of their own control, the responsibility shifts to the organizations hosting the information, and the ramifications to the individual can only be minimized, not completely mitigated. The appropriate societal organizations must take responsibility for implementing projects that address assessment and mitigation associated with information security.

Projects that plan and execute information collecting, access, and hosting services with which to provide security solutions and resolve known security gaps will increase notably in the next decade. These projects would be expected to enter, on average, in an early phase of the project. Considering the complexity and significant personal and organizational risks that are associated with information security, systems engineering processes associated with test, verification, and validation will be extremely important. Communications and decision support, technical risk reduction, quality, measurements, and governance will be critical processes to employ.

8.2.1.5 Mobility and Transportation

One responsibility of society is to provide an infrastructure that supports mobility and transportation of both humans and goods. Mobility refers to the ability of a person or thing to freely move from one location to another without regard to how the movement occurs, whereas transportation refers to the systems that are in place to enable the transfer of persons or objects from one location to another. The quality of transportation and the mobility offered by the transportation options often lead to those societies' level of perceived prosperity and desirability and hence draw more of the population to those locations. The expectations are that there will be accessible, efficient, and safe options that can be counted on to deliver the anticipated performance. The ability of the infrastructure to gracefully handle surge capacity is important as well. A society will always have projects that are associated with mobility and transportation, because the developments of these assets provide value in opportunities for economic prosperity and growth. Conversely, lack of options in this area can limit an economy. A vibrant society will have mobility and transportation options that fit their inhabitant's needs.

Projects that support improvements in air (airports, aircraft, drones), waterway (ships, channels, ports), and ground (roads, highways, automobiles, bridges, tunnels, rail) and in the interactions between these (e.g., communications, tracking) can therefore be expected throughout all established societies in the future, but at differing levels of investment. Mass transportation, modular systems, ultra-high-speed trains, self-driving or driverless cars may be entering or increasing their footprint in the market. Modifications to the way products are transported and delivered, such as automated delivery with drones or replacement of trucks on the road with tube transport systems, could change the very nature of logistics and impact mobility.

Population migrations and homestead decision making will be heavily swayed by the implementation of projects such as these. Individuals using these resources have an expectation that they will have the ability to move from place to place and to have what they want, when and where they want it. They expect to be kept safe and secure while using the resources. A level of redundancy in the systems, as well as a number of options if one system becomes unavailable so that there is minimal disruption to the individual, is also expected. Providing the solution that meets these requirements means that systems must be mature and stable, effective, predictable, safe, and secure. Requirements such as the need for reliability, maintainability, accessibility, safety, stability, and extensibility need to be designed in, tested, verified, and validated. This is where systems engineering can provide tremendous value.

Countries and cultures are unique enough that it is difficult to provide a one-size-fits-all model for addressing these types of projects, which range from highly technical to heavy-lifting construction projects. The next generation of technical capability in mobility and transportation will affect all current systems. Entrance into the systems will be along the entire life cycle. Some new capability will be implemented. Many upgrade projects, utilizing existing operating systems, will be actualized, and the processes to manage project technical scope that enters at different locations along the life cycle phases will be required. The rigorous application of systems engineering processes—including specialty-engineering activities that consider safety, security, human factors, reliability, availability, maintainability, and the environment—must be applied to ensure optimal outcomes associated with these projects.

8.2.1.6 Public Protection

The category of public protection includes activities that protect and serve the public, such as emergency services and defense. The category of public protection is a societal responsibility but may be seconded to organizations that are established to perform the duties on behalf of the population's governing bodies. This occurs with fire and rescue services and law enforcement. Comparatively, defense contracts are generally contracted to established organizations that are able to handle the sheer size and scope of the often mega-million, multi-decade military defense projects.

In many locations, firefighting stations are stand-alone operations. Some are volunteer run. Future technology projects associated with firefighting most likely will address the most common results of fires, including civilian fatalities and injuries, property losses, and firefighter injuries incurred while fighting a fire. They may also aim to integrate communications between all firefighting organizations so that they can be called up for large-scale incidents (natural or unnatural disasters or fires that span traditional geographic boundaries). The fire protection system (including sensing, suppression, and extinguishing elements) will become more embedded in buildings in autonomous, yet connected ways. For example, a building may have fire and smoke detectors and suppression systems, and these may be connected to a network that triggers an alarm at the local fire station.

Projects of the future associated with firefighting may be executed to integrate sensors and control systems with the equipment. Robotics could play a large role, as could situational awareness technologies. Another potential area of project focus may be on investigating and optimizing firefighting plans of attack to minimize risk for the firefighter. Projects of these types would benefit from the application of systems engineering processes associated with test, verification, validation, quality, safety, measurements, interface management, technical risk reduction, and modeling and simulation.

It is difficult to imagine a scenario in which technology evolution will not impact the activities of law enforcement and defense. Technological advancements will affect all areas of communications, assessment, predictions, recordings, tracking, and protection. Cross-industry sharing of private data from organizations to law enforcement and defense is inevitable. Projects of the future will include any number of new and unique products and services, including autonomous or unmanned vehicles, advanced clothing, bionics, robotics, and various methods to identify, incapacitate, and subdue targets. Technology that helps minimize collateral damage will be desirable. Spending on these projects will be limited only by the amount of funding that can be allocated and constrained only by the limits of the imagination. These types of projects generally enter in the early phase, are heavily weighted in research and development, and then progress all the way through production and operations. The application of the appropriately tailored full suite of systems engineering processes is appropriate for these projects.

8.2.1.7 Waste Management

The final societal responsibility category that will be addressed in this chapter is waste management. As humans, we generate an enormous amount of waste that must be processed in ways so as not to pollute the environment. Some waste is biodegradable; some is toxic for generations to come. Efforts at recycling, reuse, and waste reduction; commitments to reduce food waste; cleanup of toxic areas; implementation of safe disposal methods; dealing with greenhouse gas emissions and ocean pollution; and addressing electronic components waste (along with the stored data) are problems that society is attempting to address. However, rethinking how humans treat waste requires innovation and radical change. If current trends in thinking about these topics continue into the future, efforts will continue to find a way to return as much product back into the manufacturing environment as possible to be recast as something new, and then to turn the remaining waste into energy. This will create a wealth of new projects in advanced thermal and biotechnologies, projects to develop separation capabilities for sorting waste, and process improvement projects to revise standard ways of treating waste at both the individual and organizational levels.

Systems engineering processes that are focused on stakeholder engagement and communication will be key. These projects most likely will come into existence both in the early phase and mid-phase of the project—for example, where an existing operational capability is upgraded to handle a new waste-processing tool. Additional systems engineering focus areas should include interface management, technical risk reduction, modeling and simulation, trade studies, quality, measurements, and test, verification, and validation.

8.2.2 Human Sustainment

Section 8.2.1 focused on meeting core needs of society through culture, environment, government, etc. Comparatively, this section focuses on the needs of the individual operating within that larger society. Projects that impact the ability of humans to sustain themselves generally require elements of both welfare and wellbeing (see Figure 8.4). Welfare refers to individuals' ability to sustain their lives, while wellbeing considers the individual's quality of life. These are the things that are an imperative to every individual human, regardless of the larger environment they live in. In Section 8.2.2.1, the category of welfare explores future technology projects associated with things that are imperative to sustaining life, such as shelter and sustenance (food and water). Section 8.2.2.2 explores future technology-related projects that impact the areas of employment, health, learning, and prosperity.

Although originally proposed as a theory in 1943,[9] Abraham Maslow's Hierarchy of Needs Theory is still a valid and respected theory that helps categorize and consider five basic needs that motivate a person's actions. These are categorized as:

- **Physiological.** Shelter, sustenance, health
- **Safety and security.** Financial worth, health and well being
- **Social.** Human interactions
- **Esteem.** Contribution to society, work
- **Self-Actualization.** Achieving one's full potential, contributing to a higher purpose

Figure 8.4 Human Sustainment

Although Maslow proposed that one level had to be achieved before one would be motivated to focus on or pursue the next level, he also realized their interrelatedness. As with most processes (including systems engineering), there seems to be a natural tendency toward an iterative, recursive, and cyclical nature in its application rather than a strict linear application of the processes. Since systems engineers are comfortable with this non-linear approach, they are well qualified to contribute to projects that address requirements fulfillment in this manner.

Diverting from the Maslow theory by categorizing activities differently, this section captures current trends in technology in alignment to current literature trends. For example, much of current literature aligns activities such as employment, health, learning, and prosperity under what could be categorized as wellbeing. These categories reflect activities that, depending on their quality, can vastly affect an individual's *well being* versus one's *welfare*. Due to the structure of most societies, these are not necessary features of obtaining the basic necessities of life, such as are captured in the category of welfare (shelter and sustenance). Ultimately, projects that are associated with human sustainment provide benefits that support, enhance, simplify, or in some way make things better for the individual. There is not a hard division between these, and a case could be made to categorize each of these under different headings.

Using Section 8.2.1 as the basis for discussion, this section focuses on technology advancements associated with an individual's needs for shelter and sustenance (physiological, safety, security). Section 8.2.3 will address technology associated with human interactions (social, esteem); and Section 8.2.4 will explore the area of technological development associated with human accomplishments (self-actualization).

8.2.2.1 Welfare

Humans must have shelter, food, and water to survive. Technology projects that address these musthaves generally address changes in the societies that require different approaches based on demographic trends (e.g., urban migration patterns) or that provide opportunities to optimize how things are done. Understanding the requirements that are driving projects in the area of welfare, as well as the quality, safety, and security requirements that are associated with them, is an imperative. In this section, technological projects addressing shelter and sustenance (both food and water) are discussed.

8.2.2.1.1 Shelter

Current trends seem to indicate that urbanization—the movement of employment opportunities to cities and immediate outlying areas, with the resulting decline of rural living—will continue into the future. Consistent with the past, it is likely that the development of housing will continue to be tightly tied to employment locations and transportation options, as they are highly integrated. "Increased population growth and urbanization impose new challenges on transportation, health, and other modern infrastructures, while at the same time, systems solutions and technology itself can adversely impact air and water quality. There are clearly many other examples of these interdependencies, both positive and negative."[10]

The complexity of implementing technology into housing is dependent on several key considerations. These include requirements associated with where and how the technology:

- is being implemented into existing structures
- is being built new
- will be resourced
  - o funding
  - o availability of materials and services
  - o location (urban, suburban, rural, wilderness)

Trends of the populations themselves, unique to each society, will determine how readily technology will be integrated into housing. Ubiquitous technology insertion into housing itself, as well as the installed appliances, will not be a matter of choice by the individual, but will be become part of an individuals' life upon entering a new dwelling or upon replacement of existing appliances. This section focuses on the potential technology projects associated with integrated systems that might be found inside these structures.

Housing technology includes temperature regulation, lighting, media, security systems, and access to utilities. Almost all associated technology of this type will be ubiquitous. Moving into this type of structure assumes acceptance of these enhancements, which will include sensors and interactive systems, most likely embedded with artificial intelligence algorithms that can learn from the occupant's behavior and adjust the temperature, lights, or utilities without requiring interaction with the occupant regarding their opinions on the matter.

Sensors will be able to detect both benign events, such as when the heat should be increased, and harmful events, such as water leakages, smoke, fire, and excessive carbon dioxide levels. The migration to voice commands will eliminate the need for the multitude of remote controls. Smart lighting, which will adjust by voice-activated programming and commands, as well as by sensing the environment, will eliminate the need for touch points, such as switches, dials, and connections. The use of cameras will become commonplace to assist in security monitoring, including facial recognition capabilities for keyless entry to the structure itself and any room within the house. Monitoring systems capable of providing visualization (temperature sensing, status of all systems), locator services that track movement throughout the structure, as well as emergency notifications to all remote devices will be the norm.

Next-generation technologies appear to be evolving to make physical touch points and electronic access points obsolete, eliminating the requirement to physically position a switch or button or to connect at specific pre-determined locations within the house. This would obviate the need for much of the current infrastructure supporting those systems, including switches, cables, plugs, wires, through-the-wall charging, touch screens, specialized devices required for network, video, and audio, etc. If this were to happen, it would create radical change within the construction, production, and sales industries and would also create a potentially massive technical waste issue that will need to be addressed, preferably before widespread implementation.

Other ubiquitous implementation of technology will come in the form of sensing, monitoring, and perhaps controlling the use of the utilities. The use of water and power will be prime targets for monitoring and control, especially if these resources are scarce. Real-time notification or understanding of usage often curtails waste. The technology of the future will not only track and notify the user of the amount of the resources that are being used (most likely in comparison to the surrounding occupants in similar households), but will help users develop a plan on how to eliminate waste and optimize how they are currently using those resources. In extreme cases of limited resources, technology could be used to control the amount of resources available to residents.

For appliances, technology trends are interconnecting these in ways that will allow automation throughout the logistical support structure. Commonly referred to as "smart" appliances, these washers, dryers, refrigerators, and other task-specific machines (e.g., vacuum cleaners, toasters, ovens, dishwashers) will use universal software and connectivity to assess the user's preferences, schedule activity to meet those needs, and communicate with the logistics chain to obtain upgrades, repairs, maintenance, supplies, or even replacement of the unit itself. These can be expected to self-assess energy efficiency and provide analysis and suggestions as to what actions should be taken to resolve unsatisfactory performance. One can imagine that the user will be prompted to make new purchases as new, more energy-efficient products become available, perhaps even providing an analysis (from a financial interface) of when the user's budget will accommodate the purchase, arranging for the delivery and the disposition of the current unit to a recycling center.

Technical projects that address these changes must consider security, stability, dependability, and redundancy as part of the systems engineering processes that will be applied. Interface management will be the most challenging aspect of these projects, although approaching these projects in various life cycle phases to address existing infrastructure that is in various stages of maturity will be a close second. This will drive a need for robust test, verification, and validation. As the impacts to other organizations and processes throughout the supply chain will be significant, a strategic view to stakeholder engagement will be required.

8.2.2.1.2 Sustenance

Future technological changes associated with the supply of food and clean water, as they are life sustaining, can be highly controversial. In these next sections, the most promising evolutionary paths for each topic will be explored separately. The world continues to struggle with uneven distribution of these critical resources, causing hunger, thirst, water-borne illness, and malnutrition. The quality and safety of food and water, as well as generation and treatment of waste, are topics that are high on every society's priority list. The nature of agriculture (inclusive of crops and animals) is evolving quickly. Product characteristics, consumption patterns, geographic dispersion of the end points, and the availability of options drive the often-tight integration of the entire production-to-delivery supply chain. Each of the sections below describes the future technology that may influence these two critical markets.

8.2.2.1.2.1 Food

As societies become more aware of the interactions among food, water, and wellbeing, the choices they make symbiotically affect the offerings that are provided. Food traceability provides the view of the whole food chain and the impact of food on an individual. This knowledge drives decision making for both the providers and the consumers. When enough emphasis is placed on one area, or removed from another, the supply chain must react. Conversely, the behavior of the supply chain can influence consumer behavior by providing food that is readily available and does not have a high monetary cost. The providers also have a responsibility to the consumer to provide nutritious, quality, safe foods. Information concerning food impact on health may cause consumers to modify their consumption based on that knowledge, thereby putting counter-pressure on the supply chain to revise their offerings. This cycle is readily apparent in all societies.

It is anticipated that future technology will optimize the food supply chain so that widely dispersed or heavily congregated population distributions can be provided for, while providing visibility and traceability from the originating location through the processing, storage, logistics, and retail of the products. In addition, future technological solutions are expected to ensure the quality and safety of the food supply. Interconnected communications will allow systems to relay information about any food within the supply chain, both good and bad, instantaneously. This will enable companies to see the entire supply chain in near real time and both react quickly to situations that might impact the safety, quality, or distribution within the food supply chain and take any necessary mitigating action.

The current approach to food safety often centers on technical interventions or the after effect of contamination that occurs between the food source and the consumer. The evolution of agriculture will impact the direction that future technology will take. As agriculture is influenced by factors such as population density, societal acceptance, the cost of implementing new technology, and climate, technology may also evolve from an intervention approach to a proactive approach.

Proactive approaches may include urban agriculture that is housed vertically in compact infrastructures. "Utilizing advances in digitization and automation, automated vertical farming has proven to be 100× more productive per acre than traditional farming."[11] It also provides a controlled environment that has the potential to reduce the issues that require interventions. It is also expected that water usage in vertical agriculture will be reduced compared to traditional farming methods by reclaiming and recirculating the water, thereby limiting its environmental impact. Increasing the resiliency of the food supply through genetic modifications or protecting it from pests and blight by growing in a carefully controlled environment are advances that are either already in place or are currently under investigation and could significantly change the way agriculture is handled in the future.

Methods such as aquaculture (fish farming) and hydroponics (soil-free agriculture), although heavily reliant on water initially, over time minimize water consumption, eliminate environmental damage from land-based agriculture, and potentially eliminate the need for agrochemicals. Methods to replace living animals with manufactured or cultured animal-based or insect-based protein are potential solutions to address food availability, safety, access, and environmental concerns. Having the ability to utilize vertical agriculture facilities and cultured protein factories that are operated by robots, monitored by sensors, and provided specialized and customized care made possible by real-time analysis could increase crop yields and sustain a larger population with less environmental impact.

Printed food on demand is another technology that could see a surge in the future. "Imagine a world where consumers can print their own foods on-demand as needed and only as much as they need, essentially eliminating food waste and food loss!"[12] These future capabilities will be enabled by technology. Sensors, tracking systems, testing and analysis could potentially minimize the need for human involvement in the food supply processes. The potential for decreasing food waste through smart harvesting (right time, enabled by robotics), optimized shipping, distribution, and inventory management is significant.

Food waste is a distressing byproduct of current agriculture processes. The potential to mitigate this risk through technological advancement is encouraging. Options to reduce spoilage and increase the quality and safety of food are currently being investigated. The technology evolution that could enhance the interventions that are in place today include activities that reduce or eliminate pathogens in the food processing area of the food supply chain or extend the shelf life of the product. These include thermal (hot intervention), nonthermal, and chemical (sanitation) methods. Future advances in thermal methods may use radio frequency, microwave, advanced thermal (electric current), and infrared technologies to heat and destroy bacteria, while nonthermal methods may use any one or a combination of high-pressure pulsed electric fields, radio-frequency electric, ultraviolet light, irradiation, ionizing radiation, cold plasma, and nonthermal plasma (high-voltage electricity) to destroy anything that might generate illness. In addition to these, chemical sanitation methods are also being investigated.

Projects that deal with the food supply chain must be considerate of strategic shifts and unplanned changes that can be carefully monitored through stakeholder engagement activities. Risk management and risk reduction will be important processes to follow, as will test, verification, validation, quality, and measurement processes. These projects would generally enter early in the life cycle phases, usually with heavy research and development activities, but they can also span throughout all life cycle phases into operations.

8.2.2.1.2.2 Water

Having available, plentiful, and clean water is essential to societies. Human life cannot be sustained without daily water intake. Organizations that supply water must manage water levels (availability), quality, and safety. The scarcer the water, the more costly it becomes to the individual and to the society that must procure water from non-local sources. This situation usually leads directly to new research opportunities to secure additional ways to obtain water. Wastewater management is generally tightly integrated with water supply management, as the amount of water that must be reclaimed, filtered, purified, and returned to the water source is significant and must be carefully managed. Each society has unique needs and pursues its own solutions.

New technology is being developed to address the needs for obtaining this critical resource in the future. Particularly interesting is harvesting water from the air using natural processes and the sun or facilitating the extraction of water using artificial leaves. Some technology solutions for obtaining clean water or reclaiming water for clean water needs are currently available, such as desalinating ocean water (reverse osmosis), reclamation and recycling of wastewater, and to a lesser extent, extraction of water from air. However, future technology will make these methods more efficient and cost effective by introducing advancements in filtering, purifying, reclaiming, and recycling, which are expected to reduce the amount of energy (and associated costs) required to extract clean, potable water.

Future filtering technology may see the introduction of high velocity, self-cleaning water purification filters. Removing salt from water has always been an expensive endeavor, but with new technology and materials, options are emerging to create strong ultra-thin membranes or perforated filters that can handle the water pressure required for standard desalination with much less energy for extraction. Advances in purification technology are coming from the biotechnology industry in the form of bio-digestors. And reclamation and recycling technology includes both new renditions of plumbing and the use of standard sinks, showers, bathtubs, and toilets, incorporating on-site reuse, rainwater diversions, and other methods to collect and reuse this great source of wastewater. Finally, new sensing technologies that provide information for quality and timely decision making include portable systems using artificial intelligence algorithms to detect pollutants in real time and satellite-based sensors that can be used to detect spectral signatures and quantity of substances in large water sources.

Projects that address the water supply must implement safety, risk management, risk reduction, test, verification, validate, quality, and measurement processes. Projects addressing water management, research and development, or implementation would benefit from the appropriately tailored systems engineering processes wherever they enter in the life cycle phase.

8.2.2.2 Wellbeing

When the basic necessities of shelter and sustenance are taken care of, other needs must be met to create an atmosphere of wellbeing. In the next sections, the expectations of how future technology will affect employment, health, learning, and prosperity will be investigated.

8.2.2.2.1 Employment

The number and types of jobs that are available to populations are highly dependent on what is happening within that society. If the changes that have been described in Sections 8.2.1 and 8.2.2 are realized, a massive number of disruptions in the workforce will affect many organizations, and consequently, the types and number of employment opportunities that are available. Organizations will have to change, adapt, or be replaced by others that will meet a growing need for new or repurposed products and services.

Opportunities provided by technological advancements will allow (and encourage) broad collaboration and risk taking. Socially funded and supported businesses that provide products and services that are not based on needs expressed by society but are instead driven by creative individuals will likely continue to proliferate.

Trends seem to indicate that a significant amount of automation and efficiency improvements are to be expected. In addition, enormous amounts of data and information will be collected in unfiltered, partially filtered, and fully filtered forms. Recommendations will be generated from highly integrated, yet widely disparate, sources, calculated by artificial intelligence using complicated algorithms. Trusting these recommendations will be necessary, as fact checking the analysis and resulting recommendations would be extremely difficult. "Intelligent" personal assistants that help make decisions across all personal and work-related boundaries will be our constant companions.

The skills needed to address the future technology may be hard to find, and societies that are lagging in their technology implementations will find their populations falling further behind. New skills that are, as of now, unknown will most likely emerge, whereas skills that are critical today may no longer be necessary in the future. Organizations will need to adopt a learning culture that matches the overall changes in the way people are investing in new knowledge. With the technology of today, the way people learn is fundamentally changing to a highly compartmentalized, visual, on-demand, and constantly available learning mode. In the future, this trend is likely to continue to include having extremely flexible, on-demand, and dynamic training environment that is tailor-made to an individual based on their knowledge and experience with the subject.

For the recruiter trying to fill positions, being able to utilize learning algorithms to identify the best candidates in a broad field of highly diverse individuals will be an imperative. The future of technology that will be used to search for talent to fill roles within organizations may include the proliferation of digital talent platforms that promote the brand of each individual and that include powerful artificial-intelligence–driven search algorithms to help match employers and potential candidates.

Although it is difficult to say which skills will be most valued in the next decade, it is apparent that in a world of technological change such as predicted, the focus would be on technological expertise. In addition, however, skills that address systems-related issues—such as being able to reduce chaos, spot patterns, identify and analyze trends, and manage complex activities—will be needed. Fortunately, this is a skill set that systems engineers have.

8.2.2.2.2 Health

Another important aspect of wellbeing is health. In this section, technology associated with health monitoring, detection of health issues, and medical procedures and interventions are discussed.

The current technology trends have evolved from an era of preventative medicine provided through established health care systems, to one of having more options to self-monitor and assess personal health status in real time. This can be observed in the proliferation of physiological (biometric) surveillance tools such as step counters and heart rate, sleep, and oxygen level monitors. Although some of these have been popular throughout the last decade, what makes the current and future versions of these tools different is the way that they are integrated into other systems. From specialty devices to integrated wrist-worn wearables, the trend of developing devices that are the least invasive, are impervious and rugged, and are highly personalizable will continue.

In the future, these detection tools will be further integrated into consumer products such as clothing, shoes, undergarments, glasses, headphones, jewelry, and furniture or fixtures to further provide individuals the opportunity to monitor their health. For example, integrated health assessment systems located in sinks, toilets, and showers could become alert systems for colon cancer, diabetes, and pregnancy, notifying the user upon use of their current health status. These could also integrate with systems to provide real-time access to medical records and historical data, providing a comprehensive view for the individual as to their overall condition. As artificial intelligence is incorporated, alerts would be raised if a threshold were crossed, perhaps offering options as to a course of action. Individuals could be better informed about their overall health status and be alerted to potential actions they should take. This holistic view of an individual's health status would then provide unprecedented real-time (presumably) accurate information to medical staff, should the need arise. In addition, real-time interactions with health care professionals through the internet may become the preferred method of communications regarding one's health.

There are significant risks to these systems as they evolve, including information security and data quality, which could lead to misinterpretations or poor advice, perhaps opening the door to nefarious intent. For example, it's possible that the responsibility for assessment of an individual's health may shift away from the health care professional and onto an artificial source; predictive health indicators could be used to deny services; or insurance companies could obtain the right to access individual files to determine their insurability or determine the level of treatment. These will need to be carefully mitigated through rigorous risk-management processes.

The technology associated with the medical profession is also expected to undergo sweeping changes in the next decade. The trend toward trying to make procedures as non-invasive as possible will continue, both for the comfort of the patient as well as the associated recovery times. Trends will continue toward augmented reality visualizations that help medical personnel develop treatment options, robotics-empowered surgery, gene therapy, and personalized drug dosing. All of these capabilities are enabled by a full systems architecture, including processes such as signal processing, interfaces, hardware, software, and sensors, both mobile and stationary.

An example of the radical changes that could revolutionize the medical industry, as well as many interfacing industries, is the success of using three-dimensional (3D) printing for pharmaceutical drugs. Although the technology of 3D printing has matured during the previous decade from manufacturing into foods and other products, the process of stereolithography, which uses a layering technique to form a physical design, has now been identified as a potential method for developing ingestible, tailor-made drugs. A system such as this, integrated with the health information compiled by the technology as discussed previously, would radically change the requirements for medicines. Optimal dosing, drug combinations to optimize performance, and micro-dosing would impact the lives of the patients who would only have to take one small dose, specifically designed for their genetic makeup. Artificial intelligence could add value by searching for solutions that would meet very specific criteria for that individual. Other relevant examples are miniaturized "pass-thru" or implantable devices that would perform diagnosis, or tailored gene therapy that would transform activities within the body.

Projects that enter into this field usually radically displace other products and services and therefore start at the early life cycle phases with heavy research and development. All projects of this type would benefit from the appropriately tailored systems engineering processes. The amount of systems engineering processes that are necessary to ensure that all interactions, interconnections, safety and security, test, verification, and validation needs are met is extremely high in this focus area.

8.2.2.2.3 Learning

Both the processes and the technologies that individuals use to learn and become educated are rapidly changing. Over the past decade, the market for internet-based courses has increased dramatically, and the textbook market has been moving to electronic versions for purchase and rental. Individual learners are interacting in different ways with content, desiring a more diverse, creative, and adaptive environment. And although traditional education—and the anticipated degree obtained with the education—is still an essential factor in recruitment, in the next decade other educational models may emerge to displace or replace some current standards.

Degrees cost money. The reason students pay is that they feel it will lead to solid, reasonably well-paid positions of employment. As new technology insertion causes shifts in the businesses that meet the needs of the society, skills that are needed will also change. Although it is difficult to say how and when they will change, there may come a time when employers realize more value for money spent on training than for education tuition reimbursement. Training can have a more direct line impact and can assist in the technical transfer understanding associated with proprietary processes and technology.

Some of the emerging trends in the field of education and training include:

- individual on-demand coursework or self-directed learning
- spoken language and interaction versus typed and written style
- game structures, providing interesting and exciting interactions, immediate feedback
- delivering content to learners in small, digestible packets

- combination of media such as virtual, 3D, augmented reality, or immersive visualizations (such as with virtual reality headsets and integrated systems that provide the sensation of being in the alternate reality)

All of these ideas are to some level in place today; however, what makes the future potentially different is the way that the next generation of learners will want to absorb new information. If they embrace and pursue these, then popularity may drive more advancement, wider proliferation of the technology, and acceptance of the implementation styles into the mainstay of organizations. If there becomes a way to comparably assess training of this type to the job requirements, and if standards are adopted that lead to certifications that become more valuable than formal education, then that will drive the organizational staffing recruitment and retention processes of organizations in the future.

Technology will reach from the development and delivery of content to interconnections with other learners and will provide remote access to information. The varieties of ways that material will be presented to assist learning will be significant. This will include real-time analytics to address evolving needs on the fly and reinforced learning over time that incorporates different learning modalities to increase retention and applicability of the material. It is anticipated that technology embedded in courses will eventually be able to assess individuals' existing knowledge based on their responses, and be able to tailor the learning to those individuals, only advancing according to their proved ability to comprehend and apply what they have learned. It potentially could, with the right algorithms, add learning modules and collaborations to assist students in gaining experience with their knowledge and proving that they are qualified to advance to the next level.

Projects that support these disciplines will always need to consider that all of this real-time and connected education and training must have the infrastructure in place to support it. These projects will also tend to be transformative and may replace existing technology but may also enter at various phases of the life cycle. Important processes to consider will include test, verification, and validation, and quality.

8.2.2.2.4 Prosperity

The idea of prosperity as a part of wellbeing encompasses how individuals maintain an average (or better) standard of living in the society within which they reside. Technology-based societies rely on exchanges for goods and services. This can take the form of bartering (goods/services for goods/services) or currency exchanges (physical and electronics).

In this section, the future technologies that will be used to manage these exchanges are explored. As with all the other sections, this is not an all-inclusive list but investigates the most relevant technologies that may evolve.

All areas of exchanges have undergone radical change in the last decade, and it is expected that this evolution will continue. How individuals invest, purchase items, and store money have all changed. The banking industry—as an established, highly regulated, infrastructure- and process-heavy institution—is being challenged by financial services technologies providing alternative storage and investment services, and digital currency (cryptocurrency).

A move away from the handling of physical currency to electronic currency has fundamentally changed reliance on established banks and the way that currency is used between organizations. For instance, lending practices, once the domain of the banking institutions and credit unions, have morphed to technology-enabled crowdfunding. In the crowdfunding model, new ideas for products and services are marketed through the internet, and individuals can invest in these ideas in exchange for anything from basic information, to discounts, to the product or service once it is produced. This has been a massive change for entrepreneurs, who can obtain funding for their ventures without the costs or the overheads associated with working with the established banking institutions.

Another scenario that will continue to push the electronic currency trend is in the need to find viable options to address cross-border and foreign exchanges. This is most impactful to business-to-business transactions. The cost of these transfers in the traditional banking environment can be significant. Moving to cryptocurrency could revolutionize the way that businesses collaborate and fund projects. The implementation of electronic currencies could significantly decrease the cost of international projects.

How individuals invest is expected to continue its evolution into the future. Significant access to information on the internet, alternative investment websites, and algorithm-driven investment assistance will continue to provide unique opportunities that will broaden individuals' interpretation, understanding, and ability to invest in ways that will suit their needs.

Technology is rapidly evolving, increasingly sophisticated, and involves complex technologies. Technology-driven payment processing and the availability of internet-based purchases will continue to inspire organizations to further enable these functions or risk going out of business. There is expected to be a radical impact in the future for point-of-sale systems and processes as more and more automated, "intelligent" systems come on-line to assist the buyer in selecting just the right item from anywhere in the world. Increased use of mobile, highly integrated technologies that can scan the environment for internet-based transactions will offer suggestions tailored to an individual's needs (based on social media activity) and within the price they can afford (based on their currency exchange information). The use of existing technology in new ways and the proliferation of new technology will be a constant threat to established businesses that are unable or unwilling to adapt.

As these technology projects are implemented, the systems engineer must focus on security as a primary outcome. The risk-management and technical-risk–reduction requirements will be significant. Systems engineering processes associated with test, verification, validation, quality, communications, stakeholder engagement, and managing interfaces will all need consideration.

8.2.3 Human Interactions

Human interaction projects address the needs of humans to interact with their surroundings and with other humans. The two primary focus areas associated with human interactions that will be addressed include entertainment and communications (Figure 8.5). Technology exerts a significant influence in how, when, where, and why humans interact with each other and with the world at large around them. It also provides opportunities to interact with the world in unique technological ways across the spectrum of text, audio, video, and virtual and augmented realities. Projects associated with human interaction generally have a quality focus, but they often are also driven by research and development and the need to be first

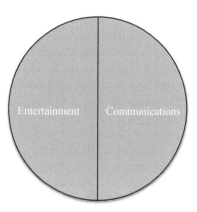

Figure 8.5 Human Interactions

to market. The benefit provided by these projects is the ability to more quickly and easily access other humans and to have an enhanced experience that showcases each person's uniqueness and personality.

8.2.3.1 Communications

Communications technology includes all technology that is used for informal to formal conveyance of information from one individual, organization, or society to another. Communication drives behavior and crosses all life areas. It is how individuals obtain information, share stories, keep in touch, learn, and collaborate. It is how people get to know each other. Sharing information about ourselves—both intentionally through postings on social media and unintentionally through the digital signature of each person associated with their activities on the internet—provides a breadth and depth of information that presents a picture of who that person is, what they believe in, how they behave, and what they hold valuable. These same persons may not ever intentionally divulge this information to a stranger during common conversation. However, the technology is in place now, and will become more pervasive in the future, to provide a digital image of each person that can be accessed by anyone with an interest and the means. In this sense, personal branding may increase in importance, so that what an individual wants to portray is the image that actually is portrayed.

Along these same lines, the concept of privacy will continue to be debated, with solutions that remove all privacy, offer a risky privacy (may be accessed), or promote complete privacy offered in competing solutions. There may be more of a shift to less formal, yet more private, types of communications—for example, those that are not stored over time or those that incorporate end-to-end encryption capability. On the other end of the spectrum, organizations may choose to employ technology that closely monitors workplace communication, retains and reads all traffic, and analyzes statistics on the frequency, parties involved, and times of conversations. Depending on the desires of the organizations providing the solutions, and on the demands from individuals and organizations, the tools and infrastructure that will be developed and implemented will evolve heavily over the coming decade. The quest for obtaining easy access to all technology-based resources while minimizing the requirements for revalidating sign in, yet keeping everything secure, will continue. Options for the future may include technology that interprets:

- facial features
- body motion—personal and unique gestures and motions
- body recognition
- voice recognition/commands
- brain waves

The future technology will require exceedingly interconnected devices capable of behind-the-scenes collecting of information, performing data analysis, synthesizing massive amounts of intelligence, assessing needs, tailoring recommendations to personal tastes derived by knowledge obtained throughout the internet and social media, and then storing that information for future use.

Virtual and augmented reality may provide advanced communications options in the future. The appearance of someone being in the same room, through augmented reality, holographic images, or 3D renditions, will vastly improve virtual communications channels. Translators that can convert the speakers' written and spoken language on the fly to spoken and written versions matching the listener's language will be a world-wide game changer. And the move to eliminating physically connected devices will open up a new world of mobile options.

Systems engineering challenges associated with communications technology will include embedded and/or integrated technology that is complex and enabled by software, heuristics, algorithms, and artificial intelligence. Integration challenges will be significant, with power, signal, and connectivity requirements being strong drivers that must be carefully managed. Security and data integrity, stakeholder engagement, requirements management, test, verification, validation, traceability, cross-discipline integration and communications, agreed-to protocols and standards, and technical risk reduction processes are necessary to ensure positive outcomes.

8.2.3.2 Entertainment

The field of entertainment in both production and use has dramatically changed in the last decade, based on the rapid propagation of new entertainment-focused technology. As technology continues to evolve, the future technology associated with entertainment is expected to blur the lines between virtual and reality. Access will continue to expand, with different levels of immersion available, depending on the location where it is used. For example, a complete immersion system will include full surrounding experience and augmented visualizations, while a mobile experience may only have audio and limited visualizations without any sensory components. Integrated systems will provide instant access to digital entertainment, and all entertainment will be available on demand.

Access to movies or other video entertainment will continue to evolve both on the production and viewing sides. Today, productions incorporate many more collaborative partners. More independent movies are making it into the mainstream, and there are more options for accessing performances from any source. The availability of high-quality production equipment available to anyone at a reasonable cost, and locations from which to share the results, is fundamentally changing the nature of this type of entertainment.

Instant access to any venue through 360-degree visualization technologies that provide a realistic feel to an experience are available in some forms now. However, future versions will include enhanced auditory and sensory algorithms, sensitive to human body mechanics and actions, that provide an experience as close as possible to "being there." The ability to travel and experience activities that are beyond the reach of many will be increased. With the enhanced reality environment, watching or participating in sports and adventures, games, concerts, and travel will become available to individuals who may not otherwise have been exposed to that world.

The ability to actively participate in the entertainment, whether alone or with known, unknown, or virtual players, will also enhance the experience. Active participation may become a more preferred entertainment option. The ability to choose how scenarios play out will provide a simulated reality in which an individual can direct the experience and experiment with different outcomes.

With the advent of short-range wireless technology, the move away from wired connectivity began. In the future, this trend is expected to continue and will provide options for wearable miniature devices that pick up voice commands and provide ultra-high quality sound through personal sound delivery systems.

The requirements on the infrastructure and networks to provide this type of fast, integrated, and highly audio/visual experiences is extreme. Systems engineering can provide enough structure into these highly creative projects to ensure that the requirements for the infrastructure and networks are defined and can be met, the interfaces are considered in the design, and the quality is as the stakeholders expect.

8.2.4 Human Accomplishments

Human accomplishment projects address the needs of humans to make a difference in the world. They encompass the human need to pursue creativity and physical excellence and add to existing capability, knowledge, or capacity. Technology enables these projects.

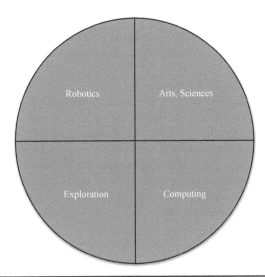

Figure 8.6   Human Accomplishments

Projects associated with human accomplishment (see Figure 8.6) are often state of the art and based on research and development. Often the vision of the outcome is not well defined. These projects push the boundaries of experiences and knowledge. Therefore, the processes that are used to direct them are focused on those that measure, test, verify, and validate outcomes and keep the project moving forward to whatever achievement is desired. Although there are many categories of human accomplishment, the ones that are identified in this section are representative of those that will generate technological projects in the next decade. These include:

- arts and sciences
- computing
- exploration
- robotics

8.2.4.1 Arts and Sciences

The technology of the arts and sciences is too broad to be defined in detail in this chapter, but the impact of technology on both of these highly creative and expansive categories is such that they must be at least referenced here. Both categories will inform and influence all other areas of human existence and will be limited only by the bounds of imagination. Each discipline will experience technological growth and advancement, which will impact the capabilities and achievements.

The advancing worldwide view of the Earth and the universe will drive creativity to new levels. Artists will be able to incorporate media of all types. More individuals will be able to investigate their creative abilities, share with a vast audience, and expand their brand. The uniqueness of each individual and how he or she sees the world will be expressed through technological capabilities that give easy access to resources and ideas and will spark creative imagination.

Systems engineering processes for art and science projects will enter throughout the life cycle phases. These highly creative projects can have appropriately tailored systems engineering processes applied that provide structure to ensure that the project outcomes meet the intent of the artist, scientist, and other stakeholders.

8.2.4.2 Computing

Computing technology drives many of the advances that have been identified in this chapter. Future technology advances in computing focus on a broad range of topics, such as:

- ease of access—single sign on, biometrics
- ability to access from any location or source—across platforms (cars, house)
- ease of use—voice activated, no hard wires
- consistency of access—excellent performance, streaming, internet
- serve their intended purpose—provide interesting content, tailored ads, access to desired media, a venue for personal expression, real-time analysis for decision making, a broad spectrum of quality research and information on demand—and enable transfers associated with exchanges (of all types: products, services)
- automate things we do not want to do

It is clear in today's environment that security of the systems that provide the products and services associated with computing are a critical concern. The lack of interoperability between devices, differing power requirements, different access methods, and the complexity of hardware/software interfaces will continue to complicate and add risk to projects that rely on the view of a highly interconnected world. In computing projects, all systems engineering processes should be applied so that the full complexity is understood and addressed.

8.2.4.3 Exploration

As humans, our curiosity demands exploration. Technology enables and emboldens exploration. Although different individuals can view exploration uniquely, two frontiers of technology-enabled exploration appear to be focused on space and underwater environments. In either scenario, ventures to extreme outposts for scientific investigation and tourism are expected to increase in the future.

There is a difference in these types of projects from the others already explored, in that individuals that pursue these projects often have a high risk tolerance and are willing to sacrifice safety for the promise of experiences and learning.

Projects associated with exploration must consider safety, risk management, and risk reduction, particularly in relationship with the environment and with individuals or communities that are not directly related to the project or that could be negatively impacted by the project. For example, a project that is funded to build an underwater observatory would need to consider the risks associated with the environmental impact or the risk to individuals that visit. The risk that a systems engineer might assign when thinking about these types of unintended risks may vary dramatically from the risks that the individuals in charge of the exploration might be willing to personally accept. The systems engineer will need to be able to assess the risks associated with the entire project scope, as well as the stakeholders' views, and take the appropriate mitigation steps.

Exploration projects are most often generated in early phases, include heavy research and development activities, and may not ever enter a production or construction phase (e.g., one-off projects with no follow on), which leads to a limited need for repeatability within the design. There will also be an expectation that there will be a significant test and verification need, prototyping, and a highly evolving design.

All processes that are employed should help drive the design to successful outcomes within the risk parameters that have been established.

8.2.4.4 Robotics

The advancements in robotics will continue to evolve and expand in all areas of society. Future technology will continue driving new automated and/or autonomous robotic solutions to assist and entertain humans. Robots can take on undesirable or tedious work. They can perform longer, operate more efficiently, perform tasks that are not possible for humans, and exist and operate in hostile environments. Without the intelligence and emotion that accompanies human behavior, robots do not need to be convinced to perform an activity and can be activated without consideration to other needs. This makes development of robotic capabilities highly desirable.

Advancements in robotics focus on those capabilities that:

- replace human activities with those that are more efficient or effective
- perform a service to humans
- provide knowledge and capabilities beyond human capacities

As an example of the efficiencies that will be gained in the next decade, massive changes will occur as robotics are embedded and integrated into the entire logistics stream. From design, to production, to delivery, robotics capabilities will disrupt all types of current capabilities and replace even more. Visually guided autonomous mobile robots (e.g., drones, self-driving delivery vehicles), cryptocurrencies, and agile services enabled by aligning needs with available service providers in real time will transform the logistics value stream.

Robotics will interact and provide assistance to help humans. Robotic technology embedded in outerwear, through the use of exoskeletons or through technically enhanced prosthetics, may provide assistance to humans through replication of lost capabilities (e.g., overcoming limitations brought on by illness or injury) or may add capabilities (e.g., heavy lifting, enhanced vision or hearing). These types of enrichments may lead to robotics that can increase performance in sports or other adventurous activities by adding strength, balance, and micro-effects that counter any negative performance results brought on by inexperience. In addition to physical life enhancements that the inclusion of robotics may bring, they may also provide us with entertainment, intellectual stimulation, and companionship.

Whether or not human-modeled robots or robots modeled after things that appeal to humans (animals, art, etc.) are designed and adopted into society depends on the society. Although a tremendous curiosity, robots that never move or perform actions that match the expectations of the observer may be more disturbing than comforting. And robots that can "out-think" humans by evaluating circumstances with an expanded knowledge base or through advanced learning algorithms may also cause concern that, without being able to understand the rationale behind certain actions, and without conscience to moderate behavior, they might pose an increased risk to humanity. These considerations will affect the successes of designs and will impact the future direction of technology development.

As with all other areas of human accomplishments, robotics projects will most likely be generated in an early phase and have a heavy research and development focus. Unlike explorations, however, repeatability will be a requirement. Successful models will be desired throughout societies once they are marketed and become popular. This means that production-ready versions have to be developed. Projects such as these require systems engineering process implementation throughout many life cycle phases, taking the design from research and development through to the operations phase.

8.3 Systems Engineering Process Evolution

The technology transformation that has been described in Section 8.2, if it comes to pass, will literally change the world and will profoundly impact the way that humans interact within that world. Each project of every type, duration, and complexity, in every discipline, will benefit from the structure of both project management and systems engineering process disciplines. As described in Section 8.1, these two disciplines have evolved along different and yet sometimes overlapping paths. It is currently challenging for a project that is dealing with different life cycle phases ongoing at one time, or is utilizing multiple project management disciplines within one project, to come to terms with how best to approach the project in order to increase the probability of achieving successful outcomes. Individual practitioners understand that there is currently an opportunity to integrate these complementary yet disparate methods into a more coherent holistic methodology, and that by doing so will create a framework that can be highly effective on future projects within all disciplines.

The Complex Systems Methodology, which blends the common and unique project management and systems engineering processes into a unified approach, is an effective and efficient way to approach projects in each of the future scenarios described in this chapter. As this methodology combines the most logical process activities across all life cycle phases into a cohesive, inclusive, highly adaptable, and tailorable framework, it naturally also reduces the constant debate as to which processes are applicable to each project. It allows the focus for these projects to be on the quality of the project outcomes versus on the processes themselves. The project managers and systems engineers that employ this approach can immediately attend to the value-added project content. Subject matter experts, along with project managers and systems engineers, must still keep informed about the discipline's evolutionary changes. However, as standards continue to shift, evolve, and merge within and between the various disciplines, the basic framework provided by the Complex Systems Methodology remains stable. The implementation of this methodology provides a strong framework to approach all future projects.

8.4 Key Point Summary

This final chapter provides a brief review of the evolution of systems engineering through the early 21st century. It provides the context for how project management and systems engineering processes developed and came to have complementarity and intersecting process areas. It also describes the additional standards-generating organizations and how they influenced project processes. The chapter then explores a vision of the future and identifies how systems engineering may engage with projects that span every area of civilization. It describes potential projects that will be implemented to address needs in society, human sustainment, interaction, and accomplishment.

The chapter concludes with an understanding that there is an opportunity to integrate these complementary methods into a coherent holistic methodology, thus creating a framework that can be highly effective on all future projects. The Complex Systems Methodology blends the common and unique project management and systems engineering processes into a unified approach, while still acknowledging standards evolution within the individual disciplines. This makes it a powerful tool for efficiently and effectively mastering complex projects across all disciplines. It combines the most logical process activities across all life cycle phases and provides an inclusive, adaptable, and tailorable framework that mitigates the risk of unproductive focus on processes instead of value-added project context.

This book provides the tools, techniques, and framework to successfully facilitate the creative project management of projects entering in any life cycle phase and throughout all disciplines.

8.4.1 Key Concepts

Systems engineering emerged from a need to achieve predictable positive outcomes. It provides a method to address the production of products that crossed engineering disciplines and need to be integrated. And it provides a holistic and synergistic view of the product.

Systems engineering is accepted as a formal discipline with associated standards. It is complementary and in some cases intertwined with other discipline processes, such as project management. Other standards-generating organizations provide standards that are used by both systems engineering and project management. These disciplines have been implemented around the world and have directly led to the successful completion of complex projects.

Future projects will touch almost every area of civilization. Societal projects are often influenced by culture and the environment. Human sustainment projects will address needs in welfare and prosperity. Human interaction projects focus on communications and entertainment. And human accomplishment projects will address creativity, physical excellence, knowledge, and other capability and capacity needs.

The Complex Systems Methodology, which blends the common and unique project management and systems engineering processes into a unified approach, is an effective and efficient way to approach projects in each of the future scenarios. It combines the most logical process activities across all life cycle phases into a cohesive, inclusive, highly adaptable, and tailorable framework. Most importantly, it allows the focus for these projects to be on the quality of the project outcomes versus on the processes themselves.

As standards continue to shift, evolve, and merge within the disciplines that generate them, the basic framework provided by the Complex Systems Methodology remains stable and provides a strong framework to approach all future projects.

8.5 Apply Now

The following questions provide the opportunity for readers to assess their understanding of the material presented in Chapter 8.

1. *Projects that emerge in the area of societal responsibilities will require processes that focus on safety and security.*
   Considering these categories of new projects that might emerge, identify an area to which you would be most interested in providing systems engineering support and describe why.
2. *Future projects that emerge in the area of human sustainment will require processes that focus on safety and quality.*
   Identify a human sustainment project that you would like to be involved in and explain which areas of process you feel would be most important and why.
3. *Human interaction projects include those that address communications and entertainment activities.*
   Explain how you would approach a project of interest in the areas of communications and entertainment and describe the most important processes that you feel must be implemented. Describe why you feel that way.
4. *Projects that focus on human accomplishments have a vast range of needs, from a small creative project to a space mission.*
   Explain how you would approach a project in each of the extremes. Describe which you would find more appealing and why.
5. *Complex Systems Methodology provides a framework from which to approach all projects, including complex projects.*
   Compare and contrast the value that the framework provides for a small project versus a large complex project.

References

1. Liu, D. (2016). *Systems Engineering: Design Principles and Models.* Boca Raton, FL: CRC Press/Taylor & Francis Group/CRC Press.

2. Ibid.
3. Ibid.
4. EM-DAT: The International Disaster Database. (n.d.). Accessed December 31, 2017, from http://www. emdat.be
5. Ibid.
6. Ibid.
7. Ibid.
8. International System of Units. (n.d.). Accessed December 29, 2017, from https://www.britannica.com/ science/International-System-of-Units
9. Maslow, A. H. (1943). "A Theory of Human Motivation." *Psychological Review, 50*(4): 370–396. Accessed January 12, 2018. doi:10.1037/h0054346–via psychclassics.yorku.ca
10. INCOSE. (2014). "A World in Motion: Systems Engineering Vision 2025." International Council on Systems Engineering. Accessed December 17, 2017, from https://www.incose.org/docs/default-source/ aboutse/se-vision-2025.pdf
11. Desjardins, J. (2017, December 6). "Next Generation Food Systems," *Visual Capitalist*. Accessed December 31, 2017, from http://www.visualcapitalist.com/future-of-food/
12. Fisher, W. (2017, June 6). "The Future of Food Traceability." *FoodSafety Magazine*. Accessed January 1, 2018, from https://www.foodsafetymagazine.com/enewsletter/the-future-of-food-traceability/

Acronyms

Chapter noted within [ ]

AC	Actual Cost [4]
ANSI	American National Standards Institute [1]
APM	Association for Project Management [1]
BAC	Budget at Completion [5]
BOE	Basis-of-Estimate [4]
CDR	Critical Design Review [2]
CI	Configuration Items [2]
CoDR	Conceptual Design Review [3]
CPI	Cost Performance Indicator [5]
CSM	Complex Systems Methodology [2]
CV	Cost Variance [5]
DoD	Department of Defense [1]
EAC	Estimate at Complete
ECR	Engineering Change Request [2]
EIA	Electronic Industries Alliance [1]
ECN	Engineering Change Notices [2]
ES	Earned Schedule [5]
EV	Earned Value [4]
EVM	Earned Value Management [4]
EVO	Evolutionary Project Management [2]
FDD	Feature-Driven Development [2]
FIRST®	For Inspiration and Recognition of Science and Technology [2]
GPS	Global Positioning Systems [1]
IAPPM	International Association of Project and Program Management [1]
ICD	Interface Control Documents [2]

IEC	International Electrotechnical Commission [1]
IEEE	Institute of Electrical and Electronic Engineers [1]
IEEE-CS	Institute of Electrical and Electronics Engineering Computer Society [1]
IEEE-SW	Institute of Electrical and Electronics Engineering Software Engineering [1]
INCOSE	International Council on Systems Engineering [1]
IPMA®	International Project Management Association [1]
ISO	International Organization for Standardization [1]
KPI	Key Performance Indicators [5]
KPP	Key Performance Parameters [2]
MBSE	Model-based systems engineering [4]
MIL_SND	Military Standard [1]
MOA	Memorandum of Agreement [2]
MOE	Measures of Effectiveness [2]
MOP	Measures of Performance [1]
NASA	National Aeronautics and Space Administration [1]
OMG	Object Management Group [2]
ORR	Operations Readiness Review [3]
PDR	Preliminary Design Review [3]
PMI	Project Management Institute [1]
PRR/CRR	Production/construction Readiness Review [3]
PV	Planned Value [4]
R&D	Research and Development [1]
RAD	Rapid Application Development [2]
SAE	Society of Automotive Engineers [1]
SEBoK	Systems Engineering Body of Knowledge [1]
SERC	Systems Engineering Research Center [1]
SEMP	Systems Engineering Management Plan [2]
SEMS	Systems Engineering Master Schedule [2]
SIR	System Integration Review [3]
SPI	Schedule Performance Index [5]
SRR	System Requirement Review [2]
SV	Schedule Variance [4]
TCPI	To-Complete Performance Index [5]
TEMP	Test and Evaluation Master Plan [2]
TPM	Technical Performance Measures [2]
TRL	Technical Readiness Levels [5]
WBS	Work Breakdown Structure [2]

Glossary

Chapter noted in [x]

A

Active Management/Engagement. The act of being continually engaged with all processes being performed on a project. [2]

Actual Cost (AC). In the Earned Value (EV) management method, the cost that has actually been recorded for the work performed during a certain time period. [4]

Agile. A project management approach most often used in software development that structures work in iterations and focuses on incorporating the voice of the customer. [2]

Allocated Budget. The amount of the budget that has been assigned to each WBS level or work package. [3]

Artificial Intelligence (AI). Software capability that includes learning and adaptation algorithms. [8]

Audit. A review activity performed by an authority or subject matter expert external to the project to evaluate compliance to a prescribed process. [2]

B

Baseline. A defined level, time frame, or, set of attributes that serve as a control and comparison point. [2]

Baseline Execution Index. Measures the percentage of tasks completed relative to the baseline task percentage anticipated to be complete at a particular moment in time. [5]

Basic Research. Generally involving theoretical or fundamental research, with an objective of acquiring new knowledge. [5]

Basis of Estimate (BOE). A detailed background describing how project costs were estimated and how the budget was built. [4]

Blocker. Something that is holding back expected progress. [5]

Bottom-Up Engineering. A design, build, test method that aims to enhance a system from the lowest levels up to the system level. [6]

Boundary. A demarcation the serves as a separator from one thing to another. [3]

Budget. A monetary commitment to an activity, such as a project, which is used to fund labor, materials, services, infrastructure, and other requirements. [2]

Budget at Completion (BAC). The calculation of the anticipated budget at the completion of the project. [5]

C

Change Control. The process of actively controlling evolution through an

275

Change Control, cont.

interdisciplinary, broad review of the change request prior to initiating change into a project. [2]

Change Control Board (CCB). A group of stakeholders who are responsible for assessing and approving change requests. [2]

Change Management. A controlled approach to assess change requests associated with the baseline cost, schedule, and scope of a project. [1]

Charter. An originating document that describes the anticipated outcomes of the project. [2]

Collaboration. An activity that refers to binding agreements that support project interactions to leverage the strengths of individuals, organizational entities, and organizations. [1]

Complementarity. The ability to gain benefits from two different and independent process sets, which add value due to the use of both together. [2]

Complicated. Typically meaning a product or services with many integrated parts, elements, or activities. [1]

Complex. Includes some elements that have operational and/or managerial independence from the main system and, when integrated with the main system, may behave in emergent ways. [1]

Complex Systems Methodology™ SM (CSM™ SM). A framework that compiles logical process activities from the disciplines for project management and systems engineering across all life cycle phases of a system into a cohesive, inclusive, and highly adaptable and tailorable framework. [2]

Compliance. Refers to the cooperation expressed that is associated with a directive. [2]

Composite Risk Index. A risk management calculation based on multiplying probability and impact assessments. [3]

Conceptual Design Review (CoDR). A review to confirm that the design, as envisioned, is reasonable, realistic, and attainable. [3]

Configuration. The record of authority for the current status of any item (e.g., hardware, software, system). [1]

Configuration Status Accounting (CSA). A role that is responsible for recording and reporting on the status of configuration control items. [3]

Configuration Item (CI). A unique item that is tracked as it evolves and is configuration controlled. [2]

Configuration Management (CM). The process of maintaining records of the version of the technical design. [1]

Conservation. The protection, management, and restoration of natural environments. [8]

Contingency. Schedule or funding set aside to cover unexpected costs or delays in a project. [2]

Cost. The amount of an expenditure used to gain something, in this case, an expenditure to fund a project. [1]

Cost Performance Indicator (CPI). In Earned Value Management (EVM), this index provides insight into cost performance through the ratio of earned value to actual cost. [5]

Cost Variance. In EVM, the difference between the budget and the amount committed at any given point in time. [5]

Critical Design Review (CDR). A quality activity that assesses the system design for readiness to enter into the production/construction phase of the project. [2]

Critical Path. An assessment of which scheduled tasks are planned to take the longest to complete. [2]

Critical Path Methodology (CPM). A process for assessing the time impact of critical and noncritical tasks along a time horizon. [5]

Cross-Disciplinary. Activities that are performed independently but in association with other disciplines. [5]

Crowdfunding. The collection of personal funds across unrelated supporters to initiate a start-up business or provide funding for other activities.[8]

Cryptocurrency. Digital currency that is used in exchanges of products and services. [8]

D

Design Synthesis. The process of mapping the functional and physical architectures to one another. [6]

E

Early Warning. The process of being notified in enough time to make conscious decisions about an appropriate way forward. [5]

Earned Schedule (ES). An EVM technique that reflects the cumulative value of the work completed versus planned. [5]

Earned Value (EV). In the EV management method, the calculation of the percentage of work that has been performed versus the amount of the allocated budget that has been assigned for that time period. [4]

Earned Value Management (EVM). A systematic approach to compare percentage of work that has been performed versus the amount of the allocated schedule and cost budget that has been assigned over a defined time period. Provides methods for compiling useful indices. [4]

Ecology. The interdisciplinary field associated with living organisms and the environment of the Earth. [8]

Economic Development. Activities that affect improvements in living standards. [8]

Economic Growth. Measurement of incomes and expenditures of a society. [8]

Energy. A force generally measured in joules or watt-hours; can be accumulated predictively, collected and stored, and delivered as a product; and can provide a source to draw power from. [8]

Engineering Change Notice (ECN). A documented decision on the authorization of a change to a technical design. [2]

Engineering Change Request (ECR). A notice that is used to identify a change that is required and initiates the change control process.

Environment. Surroundings in which a project operates, including the organizational culture, alignment, life cycle phases, interfacing organizations, and internal entities. [1]

Estimate at Complete (EAC). In Earned Value Management (EVM), a calculation that reflects an anticipated final cost estimate for the project at completion. [5]

Evolutionary Project Management. A project management approach that works with compartmentalized development deliveries. [2]

F

Factors. Circumstances and influences that have an effect on a process. [1]

Feature-Driven Development (FDD). A project management method that focuses on top-down architecturally driven functional views. [2]

Fit. Tolerance levels for which the form "fits" into the associated space of the design. [3]

Flexible Method. In project management, an approach to manage projects that does not follow the traditional waterfall method of project management. [2]

Float. The amount of slippage that a schedule can endure before impacting the critical path of a project. [2]

Force Multiplier. Refers to something that adds something more than the sum of the activities. [6]

Form. Refers to the specific tangible structure, shape, dimensions, or configuration of what is being described (e.g., square box). [3]

Formal Review. A planned event that follows strict rules and conventions. [2]

Foundation. A supporting, underlying element that other activities build on. [2]

Function. Function refers to the ability of what is being described to do what is intended in a predictable and stable manner. [3]

Functional Architecture. An abstract, yet detailed, graphical representation of all of the parts of the system and their interrelationships. [1]

G

Governance. Planned activities that are designed to control progress and manage change in a calculated and methodical manner. [1]

Gross National Product (GNP). Measurement of the total value of goods produced and services provided by a country. [8]

H

Holistic. A viewpoint of an entire interconnected entity. [1]

Hypothesis. A prediction that is tested to substantiate a statement. [2]

I

Indicator. A measurement that indicates a trend or position. [2]

Information Reviews. A review that is held to convey information rather than make decisions. [7]

Innovation. The ideation of something completely new. [2]

Interdisciplinary. Activities that rely on the participation of more than one discipline for success. [1]

Interface. The ways and means that two unrelated items communicate or are connected. [3]

Interface Charter. Document that identifies the stakeholders that will be responsible for identifying, baselining, and controlling change associated with interfaces. [4]

Interface Control Documents (ICDs). Documents that describe how one object, process, or other item is connected to another. [2]

Interface Management. Controlling actions associated with cross-boundary interactions and connections between system elements using tools such as Interface Control Documents. [1]

Invention. An original creation. [1]

K

Key. Describing something crucial, indispensable, or fundamental. [1]

Key Performance Indicators (KPIs). The measures that represent activities that are critical for the success of a project. [5]

Key Performance Parameters (KPPs). Quantitative measures that identify minimally acceptable thresholds and targets for performance. [2]

L

Leadership. A position of authority that also inspires and compels desired behavior. [1]

Life Cycle. The systemic end-to-end view (from conception to demise) of how an activity evolves over time. [1]

Life Cycle Phase. A categorization that captures the part of the systemic end-to-end view (from conception to demise) that a project is in a precise moment in time. [1]

M

Management Review. A review performed by the project manager to assess performance to plan. [2]

Matrix Management. Where supervisory responsibilities for project staff are maintained in another division, department, or organization, and the day-to-day direction of the staff is the responsibility of the project manager. [4]

Measurements. The combination of measures and metrics that are used by project managers and systems engineers to assess progress to baselines and other standard processes. [1]

Measures. Quantitative and objective values used to compare planned to actual outputs. [1]

Measures of Effectiveness (MOE). Defined by the stakeholders, values that constrain the technical requirements anticipated within the operational environment when the system is delivered. [2]

Measures of Performance (MOP). Used by the technical disciplines to compare the actual performance of the technical solution to the specifications. [1]

Memorandum of Agreement (MOA). Agreement made between two or more parties to participate in an activity or to provide products or services, generally without the exchange of funds between the parties. [3]

Metrics. Can be both qualitative and quantitative data that are used to measure the progress of an activity against a defined starting point. [1]

Mobility. Refers to the ability of a person or thing to freely move from one location to another without regard to how the movement occurs. [8]

Modeling and Simulation (M&S). Method of using both representative model and activity simulations to demonstrate the behavior of a system or elements of that system. [2]

Modularity. The ability to have stand-alone units that can also work together.

N

N^2 **chart.** A matrix used to identify and tabulate both functional and physical interfaces. [4]

O

Objective. Observable and repeatable assessment. [1]

Operational Readiness Review (ORR). A review that confirms a system's readiness to be delivered into the operational environment. [2]

Organizational Culture. A shared system of beliefs that drive the behavior of individuals in an organization. [7]

Organizational Structure. The hierarchical arrangement of an organization that shows reporting relationships. [7]

Outcome. The results, or expected ending state, of an activity. [1]

P

Performance Specifications. A set of derived performance requirements that flow from a function. [1]

Physical Architecture. A design that includes all elements, such as hardware, software, and human actions, and addresses form, fit, and function. [1]

Planned Value (PV). In the EV management method, the calculated value (cost and schedule) authorized and expected to have been applied during a specific timeframe. [4]

Power. Calculated in joules per second or watts; represents the measurement of how much energy is used per unit of time. [8]

Preliminary Design Review (PDR). A review that confirms that the preliminary design is consistent with specifications and may move to the next phase. [2]

Process. A defined series of actions that, if followed, will lead to an expected outcome. [1]

Production/Construction Readiness Review (PRR/CRR). A review that confirms that the design is ready for production/construction and that production planning is adequate to produce the design. [2]

Project. An activity with a defined scope or goal, to be accomplished within a specific timeframe, with a dedicated budget. [1]

Project Management. A discipline that provides a defined set of activities used to manage the performance of a project particularly focused around managing the parameters of cost, schedule, and scope. [1]

Prototype. A representation of an idea in a physical or electronic form used to visualize the outcomes of a project. [1]

Q

Qualitative. An assessment of data in non-numeric form, such as in words or images. [1]

Quality. Refers to a quantitative and qualitative assessment of the value of the outcomes, including conformance of the design to the requirements and specifications and to the value, as seen by the stakeholders. [1]

Quantitative. An assessment of data in numeric values. [1]

Quantitative. Refers to a quantity or number of something. [5]

R

Rapid Application Development (RAD). A method of flexible object-oriented project management that provides quick and iterative prototype development cycles. [2]

Rebaseline. The act of revising the scope of a project to fit into the budget and schedule. [5]

Requirements. Definition of the exact functionality that is desired to meet stakeholder needs. [1]

Requirements Traceability Matrix. A form used to track requirements through to verification and validation of performance. [2]

Research and Development (R&D). The activity of identifying an idea and attempting to prove or disprove the theories, then evolving the idea until it is a workable solution or product. [1]

Rigor. The measured validity, completeness, or thoroughness that is applied to a process. [1]

Risk. Something that might occur, and if it did occur would cause an impact on the project. Occasionally, risks can be opportunities that, when identified and pursued, will bring value to the project. [1]

S

Schedule. An outline of all critical steps in the project to be tracked, along with their time estimate, in the order they will be done. [1]

Schedule Performance Index (SPI). In EVM, an index that can reflect the progress of baselined work within the schedule. [5]

Schedule Variance (SV). The value of the work expected to have been performed during the time period compared to the amount of work that has actually been accomplished. [4]

Scope. A statement of the complete work to be done on a project that is written in plain language and used to align the stakeholders' understanding. [1]

Society. A subset of humanity as it is geographically, culturally, and governmentally categorized. [8]

Specifications. A detailed set of technical descriptions. [1]

Specification Tree. A visualization of the

specifications at each level and for each function in a system. [2]

Spiral. A project management process that incorporates cyclical development with well-defined criteria within multiple phases. [2]

Stakeholders. Customers, project team, organizational leadership, external organizations, and other persons with a vested interest in a project. [1]

Standard Reporting of Schedule/Duration. Tracking of comparisons between planned labor, materials, and services and the actual labor, materials, and services that have been planned during established intervals. [4]

Subjective. Based on opinion, judgments, personal interpretations. [5]

Success. For a project, the subjective assessment by the project team and the stakeholders that the needs of the stakeholders were met. [5]

Synthesis. The process of bringing together architectures and systems so that they behave as an overarching system, or system-of-systems. [1]

System. The collection of interrelated elements, viewed as a whole. [1]

System Definition Review (SDR). A review confirming that the stakeholder-focused view of the system is understood by the project team. [5]

System Integration Review (SIR). A gate review to assess the readiness or completeness of the integration activities. [3]

System Requirements Review (SRR). A review that confirms the solution-focused project scope descriptions against the configuration-controlled system definition. [2, 3]

Systems Analysis. An approach that reviews a problem in logical steps and thoroughly describes the system explicitly. [1]

Systems Engineering. A discipline that provides the structured processes necessary to manage projects that require performance from many disciplines in order to achieve their objectives. [1]

Systems Engineering Management Plan (SEMP). A standard systems engineering document that describes the processes that will be used to manage the technical scope of a project. [2]

Systems Engineering Master Schedule (SEMS). A technical schedule associated with the systems engineering management plan that describes the technical milestones, such as reviews, tests, and other verification events. [2]

Systems Thinking. Methodology that attempts to understand the independent structures associated with the thinking and how they are interrelated. [1]

T

Tailoring. A standard process within both the project management and systems engineering disciplines that provides flexibility in applying rigor to the processes. [1]

Tailoring Down. The act of starting from the full implementation of all standard processes and reducing the level of effort as appropriate. [6]

Tailoring Tool. A model that facilitates tailoring decision making and communication. [6]

Tailoring Up. The act of starting from no process and applying additional processes, as appropriate. [6]

Technical Debt. Used to describe technical work that is undone when the project enters the next phase of its life cycle. [2]

Technical Performance Measures (TPM). Identifies the most vital technical goals needed to be met in order for the project to be successful. [2]

Test. An assessment used to measure something. [1]

Test and Evaluation Master Plan (TEMP). A document that describes the planned test and evaluation activities, and evaluation criteria. [2]

Technical Readiness Levels (TRL). A sequential process used to evaluate technological readiness to move into the next phase of the life cycle. [5]

Technical Risk. The inability to meet technical specifications.

Time Bounding. Establishing time boundaries so that measurements can be taken against a defined time period. [4]

To-Complete Performance Index (TCPI). In earned value management (EVM), a calculation of the actual cost expected to finish the work. [5]

Trade. The exchange of goods and services, typically a commercial transaction that benefits both parties. [8]

Trade Study. A technical risk reduction strategy that provides visibility into options, allowing a solution to be chosen between two or more potentials. [2]

Traditional Method. A method of project management known for its linearity and most-work to least-work anticipated work flow. [2]

Transportation. Refers to the systems that are in place to enable the transfer of persons or objects from one location to another. [8]

Trend Analysis. The review of historical data to forecast future performance. [5]

U

Ubiquitous Technology. Pervasive and omnipresent technology, often embedded into other materials, surfaces, and objects. [8]

Unified Approach. A method to take parts and make them perform as a whole. [2]

Use Case. A text-based story that describes the operating environment that the stakeholders envision in a working system. [2]

V

Validation. The confirmation from the key stakeholders that the system outcomes meet their needs as defined and controlled through change management. [1]

Variance Analysis. A technique for assessing the difference between the baseline and performance. [5]

Verification. The process which confirms that every system element is tested and meets the necessary specifications. [1]

W

Waterfall Method/Approach. An approach to a project that uses a series of sequential phases, aligned in a downward-flowing noniterative representation, like a waterfall. [2]

Work Breakdown Structure (WBS). A logical, all-inclusive representation of all of the elements within a project. [2]

Workforce Management Plan. A document that describes all unique human resource processes that will be used, and identifies all unique skills that will be needed on a project. [4]

Work Package. Represents the smallest unit of work that will be tracked and usually has a product as a deliverable. [2]

Bibliography

Adcock, R., Hutchison, N., and Nielsen, C. (2016). "Defining an Architecture for the Systems Engineering Body of Knowledge." Annual IEEE Systems Converence (SysCon), 2016.

Agile Alliance (2017). *Agile Practice Guide*, PMI Global Standard. Newtown, PA: Project Management Institute.

Alva-GreenCoaching.com. (n.d.) "Pathway to Profit Through Productivity." Accessed January 12, 2018, from www.alva-greencoaching.com

Alva-Green Coaching. (n.d.). "About Alva-Green." Accessed July 9, 2017, from http://www.alva-greencoaching.com/About.htm

Alva-Green Coaching. (n.d.). Accessed January 1, 2018, at https://www.facebook.com/PatAlvaGreen/.

Alva-Green, P. (n.d.) "Patricia Alva-Green." Accessed January 12, 2018, from https://www.linkedin.com/in/patricia-alva-green-437b471

APM. (n.d.) Association for Project Management. Accessed January 8, 2018, from https://www.apm.org.uk/about-us

APM. (n.d.) "Schedule Management." Accessed January 12, 2018, from https://www.apm.org.uk/body-of-knowledge/delivery/schedule-management

ARM. (2017). "Uniting to Deliver Technology for the Global Goals." 2030 Vision Global Goals Technology Forum, SustainAbility, September. Accessed December 21, 2017, from https://www.unglobalcompact.org/library/5491

Beck, S. (2014, December 19). "Snow Art." TEDxKlagenfurt. Accessed January 12, 2018, from https://youtu.be/CfPPZS4IvWs

Beck, S. (2014, October 1). *Simon Beck: Snow Art*. S-Editions.

Beck, S. (n.d.). "Facebook Page: Simon Beck's Snow Art." Accessed January 12, 2018, from https://www.facebook.com/search/top/?q=simon%20beck's%20snow%20art

Beck, S. "Simon Beck: Snow Art Gallery." Accessed January 12, 2018, from http://snowart.gallery/see.php

Bellos, A. (2014, November 6). "Simon Beck's Astonishing Landscape and Snow Art Illustrates the Cold Beauty of Mathematics—In Pictures." *The Guardian*. Accessed January 12, 2018, from https://www.theguardian.com/science/alexs-adventures-in-numberland/gallery/2014/nov/06/simon-becks-snow-art-landscapes-mathematical-designs-drawings-alps

Benjamin, E. (ed.) (2017, July 12). "This Artist Walks 20 Miles to Create Geometric Patterns in the Snow." Culture Trip. Accessed January 12, 2018, from https://theculturetrip.com/europe/articles/simon-beck-artist-snow-art-alps/

Big Think Editors (n.d.) "What Will Life Be Like in 2050?" Big Think. Accessed December 31, 2017, from http://bigthink.com/the-voice-of-big-think/what-will-life-be-like-in-2050

BKCASE Editorial Board. (2017). *The Guide to the Systems Engineering Body of Knowledge (SEBoK)*, v. 1.9. R.D. Adcock (EIC). Hoboken, NJ: The Trustees of the Stevens Institute of Technology. Accessed 1/6/2018. www.sebokwiki.org. BKCASE is managed and maintained by the Stevens Institute of Technology Systems Engineering Research Center, the International Council on Systems Engineering, and the Institute of Electrical and Electronics Engineers Computer Society.

Blanchard, B. S., and Blyler, J. E. (2016). *System Engineering Management*, 5th Edition. Hoboken, NJ: John Wiley & Sons, Inc.

Boehlje, M.J. (2006). "Economics of Animal Agriculture Production, Processing and Marketing." *Choices,* 3rd Q, *21*(3). Accessed January 1, 2018, from http://www.choicesmagazine.org/2006-3/animal/2006-3-08.htm

Bricher, J., and Keener, L. (2007). "Innovations in Technology: Promising Food Safety Technologies." *FoodSafety Magazine,* April/May. Accessed January 1, 2018, from https://www.foodsafetymagazine.com/magazine-archive1/aprilmay-2007/innovations-in-technology-promising-food-safety-technologies/

Brooks, F. P., Jr., and Frederick, P. (2010). *The Design of Design: Essays from a Computer Scientist.* 1st Edition. Indianapolis, IN: Addison-Wesley Professional.

Cann, O. (2017, June 26). "These are the Top 10 Emerging Technologies of 2017." World Economic Forum, Geneva, Accessed December 31, 2017, from https://www.weforum.org/agenda/2017/06/these-are-the-top-10-emerging-technologies-of-2017/

Carolina Soap Market. "Carolina Soap Market." Accessed January 12, 2018, from https://www.carolinasoapmarket.com

Centre for Public Impact (2017, August 4). "The FiReControl Project in the UK", Case Study. Accessed December 31, 2017, from https://www.centreforpublicimpact.org/case-study/firecontrol-project-uk/

Chester, R. (2017). "The Future of Food Safety: The Revolution Is on Our Doorsteps." *New Food,* 05(05). Accessed January 1, 2018, from https://www.newfoodmagazine.com/article/41390/future-food-safety-revolution-doorsteps/

Courbe, J. "Financial Services Technology 2020 and Beyond: Embracing Disruption." pwc|global. Accessed December 31, 2017, from://www.pwc.com/gx/en/industries/financial-services/publications/financial-services-technology-2020-and-beyond-embracing-disruption.html

Crawley, E., Cameron, B., and Selva, D. (2015). *System Architecture: Strategy and Product Development for Complex Systems.* London, UK: Pearson Publishing Company.

Dauphin, C. (2014, May 29). "615 Spotlight: Jill & Julia Talk 'Wildfire' Single, Vegas Life." *Billboard.* Accessed January 12, 2018, from http://www.billboard.com/articles/columns/the-615/6106201/jill-julia-wildfire-single-las-vegas

Defense Acquisition University. (2001). *Systems Engineering Fundamentals.* Fort Belvoir, VA: Defense Acquisition University Press.

Defense Systems Management College. (1997). *Earned Value Management Textbook.* Fort Belvoir, VA: Defense Systems Management College.

Deming, W. E. (1986). *Out of the Crisis.* Cambridge, MA: MIT Press.

Desjardins, J. (2017, December 6). "Next Generation Food Systems." *Visual Capitalist.* Accessed December 31, 2017, from http://www.visualcapitalist.com/future-of-food/

de Weck, O. L., Roos, D., and Magee, C. L. (2011). *Engineering Systems: Meeting Human Needs in a Complex Technological World.* Cambridge, MA: The MIT Press.

Eisenträger, J. (n.d.). Item Interchangeability Rules—Rev. C-2013-08-26. Accessed July 2, 2017, from http://www.joergei.de/cm/interchangeability-guide_en.pdf

EM-DAT: The International Disaster Database. (n.d.) Accessed December 31, 2017, from http://www.emdat.be

FIRSTINSPIRES. (n.d.). "FIRST INSPIRES Organization." Accessed July 4, 2017, from https://www.firstinspires.org

FIRSTINSPIRES. (n.d.). "FIRST INSPIRES Vision and Mission." Accessed July 4, 2017, from https://www.firstinspires.org/about/vision-and-mission

FIRSTRoboticsCompetition. (n.d.). "FIRST STRONGHOLD Game Reveal." Accessed July 4, 2017, from https://youtu.be/VqOKzoHJDjA

Fisher, W. (2017, June 6). "The Future of Food Traceability." *FoodSafety Magazine.* Accessed January 1, 2018, from https://www.foodsafetymagazine.com/enewsletter/the-future-of-food-traceability/

Fleming, Q. W., and Koppelman, J. M. (1996). "Forecasting the Final Cost and Schedule Results." *PM Network®* (v. 10) 1:13–18. Newtown Square, PA: Project Management Institute.

Fordham, D. (2000). Greenland 1999. *Alpine Journal, 105,* 230–234. Accessed March 2, 2017, from https://www.alpinejournal.org.uk/Contents/Contents_2000_files/AJ%20 2000%20230-234%20Greenland.pdf

Fractal Foundation. (n.d.). "Fractals are SMART: Science, Math and Art!" Accessed January 12, 2018, from http://fractalfoundation.org/resources/what-are-fractals/

Harwell, R. (ed.). (1997, August). "Systems Engineering: A Way of Thinking, A Way of Doing Business, Enabling Organized Transition from Need to Product." Brochure prepared as a joint project of the American Institute of Aeronautics and Astronautics (AIAA) Systems Engineering Technical Committee and the International Council on Systems Engineering (INCOSE) Systems Engineering Management Methods Working Group.

Hicks, O. (n.d.). "The Greenland to Scotland Challenge." Accessed January 12, 2018, from http://www.ollyhicks.com/greenland-to-scotland-challenge/

Higginbotham, S. (2017, February 17). "The Future Is Now: Welcome to My (Smart) House." *Fortune.* Accessed December 31, 2017, from http://fortune.com/2017/02/17/smart-home-tech-internet-of-things-connected-home/

Hindy, S. (2015). "The Craft Beer Revolution: How a Band of Microbrewers Is Transforming the World's Favorite Drink." *The Economist.* Accessed January 12, 2018, from https:// www.economist.com/news/business-books-quarterly/21600664-master-microbrewer-analyses-revolution-hops-and-dreams

IAPM. (n.d.) International Association of Project Managers. Accessed January 12, 2018, from https://www.iapm.net/en/start/

IAPPM. (n.d.) International Association of Project and Program Management. Accessed January 8, 2018, from https://www/iapm.net/en/start

Immink, R. (2017). *The Future: Slow Down or Go Faster.* Bookbuzz and Strategy Crowd, Business Book Series, No. 01. Ireland: Oak Tree Press.

INCOSE. (n.d.) "INCOSE Principles." Accessed April 9, 2017, from http://www.incose.org/about/principles

INCOSE. (2015). *Project Manager's Guide to Systems Engineering Measurement for Project Success: A Basic Introduction to Systems Engineering Measures for Use by Project Managers*, v 1.0, INCOSE-TP-2015-001-01

INCOSE. (2015). *Systems Engineering Handbook: A Guide for System Life Cycle Processes and Activities*, 4th Edition. D. Walden, G. Roedler, K. Forsberg, R. Hamelin, and T. Shortell (eds.). Hoboken, NJ: John Wiley & Sons, Inc.

International Council on Systems Engineering (INCOSE). (2014). "A World in Motion: Systems Engineering Vision 2025." International Council on Systems Engineering. Accessed December 17, 2017, from https://www.incose.org/docs/default-source/aboutse/se-vision-2025.pdf

International Organization for Standardization (ISO). (2013). *ISO Standards.* Geneva, Switzerland: International Organization for Standardization. Accessed December 5, 2013, from http://www.iso.org/iso/home/standards.htm

International System of Units. (n.d.). Accessed December 29, 2017, from https://www.britannica.com/science/International-System-of-Units

IPMA. International Project Management Association. Accessed January 8, 2018, from http://www.ipma.world

ISO. (2008). ISO 9000:2005. *Quality Management Systems—Fundamentals and Vocabulary.* Geneva, Switzerland: ISO.

ISO/IEC/IEEE. (2015). *Systems and Software Engineering—System Life Cycle Processes.* Geneva, Switzerland: International Organisation for Standardisation / International Electrotechnical Commissions / Institute of Electrical and Electronics Engineers. ISO/IEC/IEEE 15288:2015.

Jill and Julia. (n.d.). "Jill and Julia." Accessed January 12, 2018, from www.jillandjulia.com.

Jill and Julia. (n.d.). Accessed January 12, 2018, from https://www.linkedin.com/in/jill-and-julia-6b45ba111/?trk=public-profile-join-page

Kaku, M. (2012). *Physics of the Future.* New York, NY: First Anchor Books Edition, Doubleday Imprint.

Kamenetz, A. (2017). Why One Educator Says It's Time to Rethink Higher Eduction. *NPREd: Higher Ed.* 09(13). Accessed December 31, 2017, from https://www.npr.org/sections/ed/2017/09/13/547037211/a-future-forward-look-at-higher-ed

Kokatat. (n.d.) "Wake of the Finnmen." Accessed January 1, 2018, from https://kokatat.com/expeditions/wake-of-the-finnmen

Kossiakoff, A., Sweet, W. N., Seymour, S. J., and Biemer, S. M. (2011). *Systems Engineering: Principles and Practice,* 2nd Edition. Hoboken, NJ: John Wiley & Sons, Inc.

Lamon Records, LRC©. (n.d.) "Sister Duo Jill and Julia Bring Unique Country Sound to Lamon Records." Accessed January 12, 2018, from https://lamonrecords.com/sister-duo-jill-and-julia-bring-unique-country-sound-to-lamon-records

Lano, R. J. (1977). The N^2 Chart. *TRW document #TRW-SS-77-04.* Redondo Beach, CA: TRW, Systems Engineering and Integration Division.

Las Vegas Weekly Staff. (2015, January 22). "The Next Wave of Vegas Music: 10 Acts to Hear This Year." *Las Vegas Weekly.* Accessed January 12, 2018, from https://lasvegasweekly.com/ae/music/2015/jan/22/next-wave-vegas-music-10-acts-local-band-hear-year/?framing=home-news

Lipke, W. (2003). "Schedule Is Different." *The Measurable News.* College of Performance Management® of the Project Management Institute.

Liu, D. (2016). *Systems Engineering: Design Principles and Models.* Boca Raton, FL: CRC Press/Taylor & Francis Group.

Manyika, J. (2017, May). "Technology, Jobs, and the Future of Work." *Executive Briefing,* McKinsey Global Institute. Accessed December 31, 2017, from https://www.mckinsey.com/global-themes/employment-and-growth/technology-jobs-and-the-future-of-work.

Maslow, A. H. (1943). "A Theory of Human Motivation." *Psychological Review, 50*(4): 370–396. Accessed January 12, 2018. doi:10.1037/h0054346–via psychclassics.yorku.ca

McCarthy, E. (2014, November 11). "11 Questions for Snow Artist Simon Beck." *Mental Floss.* Accessed January 12, 2018, from http://mentalfloss.com/article/59958/11-questions-snow-artist-simon-beck

McChrystal, S., Collins, T., Silverman, D., and Fussell, C. (2015). *Team of Teams: New Rules of Engagement for a Complex World.* New York, NY: Portfolio, Penguin Random House.

McKinsey Global Institiute (2017). "Jobs Lost, Jobs Gained: Workforce Transitions in a Time of Automation." McKinsey & Company, December.

NASA. (2007). *NASA Systems Engineering Handbook.* Washington, DC: NASA. Accessed April 24, 2013, from http://www.acq. osd.mil/se/docs/NASA-SP-2007-6105-Rev-1-Final-31Dec2007.pdf

PMI. (2017). *A Guide to the Project Management Body of Knowledge®* (*PMBOK® Guide*), 6th Edition. Newtown Square, PA: Project Management Institute (PMI).

PMI. Project Management Institute. Accessed January 8, 2018, from https://www.pmi.org

Rad, R. F., and Levin, G. (2006). *Metrics for Project Management: Formalized Approaches, Management Concepts.* Vienna, VA: Management Concepts, Inc.

Rainie, L., and Anderson, J. (2017, May 3). "The Future of Jobs and Jobs Training." Pew Research Center: Internet & Technology. Accessed December 31, 2017, from http://www.pewinternet.org/2017/05/03/the-future-of-jobs-and-jobs-training/

Rebentisch, E. (ed.). (2017). *Integrating Program Management and Systems Engineering: Methods, Tools, and Organizational Systems for Improving Performance.* Hoboken, NJ: John Wiley & Sons, Inc.

Rebentisch, E. S., Townsend, S., and Conforto, E. C. (2015, June). "Collaboration Across Linked Disciplines: Skills and Roles for Integrating Systems Engineering and Program Management." Presented at American Society for Engineering Education Annual Conference, Seattle, Washington. Accessed June 30, 2017, from www.asee.org/public/conferences/56/papers/12512/view

Redbull TV. (n.d.) "Voyage of the Finnmen." Accessed January 1, 2018, from https://www.redbull.tv/video/AP-1NBYKYF9W1W11/voyage-of-the-finnmen

ROBOCUBS. (n.d.) "Robocubs." Accessed July 4, 2017, from http://robocubs.com/

Rothman, J. (2016). *Manage Your Project Portfolio: Increase Your Capacity and Finish More Projects.* Raleigh, NC: Pragmatic Bookshelf.

Rothman, J., (2016). *Agile and Lean Program Management: Scaling Collaboration Across the Organization.* Victoria, British Columbia: Practical Ink.

Sillitto, H. (2014). *Architecting Systems—Concepts, Principles and Practice.* London, UK: College Publications.

SEBoK contributors. "Guide to the Systems Engineering Body of Knowledge (SEBoK)." *SEBoK.* http://sebokwiki.org/w/index.php?title=Guide_to_the_Systems_Engineering_Body_of_Knowledge_(SEBoK)&oldid=53122 (accessed March 17, 2018).

Smith, A. (n.d.) "Scot of the Arctic; Sue Conquers the North Pole." *Daily Record.* Accessed January 29, 2014.

Stockdale, S. (n.d.) "What Really Counts." Accessed January 8, 2018, from https://www.sue-stockdale.com/what-really-counts-leader-striving-achieve-tough-goals

"Surprise! Cool Sue Bumps into Fellow Explorer in Arctic Wastes." (1998, May 12). *Oxford Mail.*

Systems Architecture Guild. (n.d.) "FIRST Robotics Strategy Introduction." Accessed July 4, 2017, from https://www.youtube.com/watch?v=NTouZLC8tCk&list=PLDz4YEQgpXfyqUQ90ACBbFJAhi2Ow1hYU

Systems Architecture Guild. "Show Me the Wow." Accessed July 4, 2017, from www.show-methewow.com

Team 1701. "FIRST Robotics Team 1701." Accessed July 4, 2017, from http://robocubs.com/2016/03/

The Scotsman. "No Business Like Snow Business for Motivation." Accessed January 8, 2018, from https://www.scotsman.com/lifestyle/no-business-like-snow-business-for-motivation-1-938670

Walden, D. D., Roedler, G. J., Forsberg, K. J., Hamelin, R. D., and Shortell, T. M. (eds.). (2015). *Systems Engineering Handbook: A Guide for System Life Cycle Processes and Activities,* 4th Edition. INCOSE-TP-2003-002-04. Hoboken, NJ: John Wiley & Sons, Inc.

Wallace, J. (Rev.). (1688). *A Description of the Isles of Orkney,* written by the Rev. James Wallace, A.M., Minister of Kirkwall, about the year 1688.

Wasson, C. S. (2016). *System Engineering: Analysis, Design, and Development,* 2nd Edition. Hoboken, NJ: John Wiley & Sons, Inc.

Wingate, L. M. (2014). *Project Management for Research and Development: Guiding Innovation for Positive R&D Outcome.* Boca Raton, FL: CRC Press/Taylor & Francis Group.

Wingate, L. M. (2016, October 27). *Project Management Tailoring Methods.* Technical Project Management Series, www.itmpi.org.

Index

A

active management, 32, 66, 87, 93, 135, 156–158, 171, 186, 192, 225
Agile, 37, 181, 217, 241, 269
allocated budget, 91, 115
audit, 41, 74, 83, 115, 143, 146, 157–158, 162–163, 175, 178, 185, 192, 195, 216, 224–226, 266

B

baseline, 34–37, 40–42, 46, 49–53, 60, 66–71, 75–78, 82–84, 89–96, 102–107, 111–112, 117, 132, 135–143, 146–155, 160–163, 171–174, 181–182, 188, 191, 209, 216–218, 223–226, 240
baseline execution index, 151
basic research, 148
blocker, 93, 144, 150, 159, 224
bottom-up engineering, 192
boundary, 33, 39, 45, 70, 77–79, 82, 95–96, 108–112, 128–129, 133, 140–144, 169, 180, 205, 215–216, 220, 232, 252–254, 260, 267
budget, 2, 29, 32, 35–40, 45–47, 50–53, 70, 73–75, 81–84, 89–91, 95, 102–103, 111, 114–120, 132–133, 137–138, 141–144, 150–164, 172, 178–186, 189–191, 194, 205, 211, 215–220, 223–224, 257

C

change control, 48, 52–53, 63, 66, 69, 75, 84, 87–89, 93–95, 103, 108, 122, 132, 137, 147, 156–158, 161–163, 172–173, 186, 191–194, 218, 223–226, 240
change management, 10, 51, 60, 84, 93–95, 122, 143, 162–163, 188, 223, 246
charter, 14, 31–33, 36, 39, 42–43, 55, 66, 76–77, 93, 96, 106–108, 112, 126, 132–133, 138, 147, 167, 172, 184–185, 188, 196, 216, 229
collaboration, 7, 15, 31–33, 38–39, 42, 49–52, 58–59, 64, 69–70, 82–85, 101–103, 112, 119, 132, 142–144, 147, 172–173, 184, 189, 193–194, 214–215, 218, 224–227, 234–235, 240, 249, 260, 263
complementarity, 27, 61, 270
complex, 3–17, 22–24, 27–33, 37–39, 54, 61–65, 70, 75, 81–84, 89, 92–96, 99–110, 114, 119–123, 131–132, 135–136, 140–146, 149–150, 155–156, 164, 170–177, 183–192, 195, 205–219, 223–227, 232–233, 237–248, 252–253, 256, 261, 264–265, 268–271
compliance, 6, 13, 22, 46, 56, 74, 79, 86, 147, 157–158, 188, 195, 210, 246
complicated, 3–4, 10, 33, 64, 82–83, 98, 123, 147, 202, 212, 239, 244–245, 252, 260
composite risk index, 87
configuration, 6, 10, 31, 39–41, 47–60, 73–77, 80–81, 86, 91–96, 102–103, 112–113, 123, 142–143, 160–164, 172–174, 182, 186, 191–192, 201, 212–213, 216, 221–226, 236–240

conservation, 249–250
contingency, 11, 29, 32, 35–36, 47, 50,
    83–84, 87–91, 114, 117–119, 143, 150–
    152, 156, 163, 186, 189, 194, 217–226
cost, 2–5, 9–17, 28–38, 47–54, 57, 60, 69,
    73–77, 80–95, 98, 103–107, 111–119,
    122, 128, 131–132, 138, 141–146, 149–
    158, 161, 164, 169–180, 185, 188–194,
    200–203, 207–227, 230–235, 238–240,
    244–246, 258–266
critical path, 29, 35, 83–84, 114–115, 150–151,
    185
cross-disciplinary, 6, 142, 154, 225

D
design synthesis, 188, 198, 230, 234

E
early warning, 139, 143–147, 171, 223, 238,
    241, 249
ecology, 249–250
economic development, 248
economic growth, 136, 248
emergence, 8, 11, 228, 245
energy, 99, 129, 167, 180, 249–251, 255–257,
    260
engagement, 16, 31–33, 41–42, 63–69, 77,
    90, 106–108, 167–168, 183, 187–188,
    192, 196, 214, 229, 237, 246, 255,
    258–259, 264–266
environment, 5, 9, 12–17, 20–21, 39, 42,
    48–49, 57, 65, 71, 78, 90, 100–101, 110,
    122–125, 130, 133, 137, 140, 149, 156,
    160, 167, 176, 179, 197–198, 202–204,
    207–212, 218–219, 222, 227–229,
    235–241, 244, 247–271

F
factor, 9, 14–15, 19, 29, 38, 45, 49–51, 69,
    74, 80–82, 87, 97, 110, 117, 126, 141,
    145, 149–155, 169, 175–179, 183–184,
    204–208, 211, 217, 224, 237, 250–254,
    257–259, 262
fit, 7–8, 14–15, 19–20, 29, 38, 43, 46,
    50–54, 58–63, 70–71, 75, 80–81, 87,

    90, 96–98, 102, 112, 120–127, 133, 138,
    141, 144, 150, 155–160, 164, 174, 177,
    182, 203–204, 208, 212–214, 220–221,
    227–228, 233–239, 245, 250–256,
    260–262, 265, 270
flexible method, 37, 51, 115, 181–182, 191, 212,
    217
float, 29, 35, 83, 114, 150, 153, 185, 197,
    200–203, 224
force multiplier, 198
form, 1–23, 28–96, 100–103, 106–132,
    135–164, 167, 171–202, 205–238,
    241–245, 248–249, 252–254, 257–271
foundation, 4, 55–56, 64, 104–106, 135,
    139–145, 171, 176, 186, 224, 238
function, 4–10, 15–16, 22, 29, 39–45,
    48–50, 53–55, 59, 68–69, 78–81,
    91–92, 96–97, 102–103, 112–115,
    120–125, 132–133, 147, 156, 160–162,
    172–173, 189, 196, 209–210, 220–221,
    239–240, 245, 264
functional architecture, 4, 22, 45, 78–81,
    102–103, 112–113, 132, 160, 172–173,
    196, 220–221, 240

G
governance, 7–8, 21, 31, 37, 46, 50, 53,
    58–60, 63, 68, 75–77, 81, 90–95,
    100–103, 130–133, 140–142, 149, 162,
    170–174, 191–192, 202, 209, 218, 223,
    236, 253

H
holistic, 2–4, 8, 11, 23, 110, 213, 244–245,
    261, 270
hypothesis, 37, 196

I
indicator, 52, 90, 146–147, 152–153, 215,
    223, 226, 236, 247–248, 261
information review, 237
innovation, 24, 40, 54, 61, 117, 137, 148, 174,
    254
interdisciplinary, 4, 9, 53, 60, 82, 95, 114,
    140–141, 163, 209, 250

interface, 6–8, 12–13, 18, 23, 33, 36–40, 43–45, 55–57, 60, 73–74, 80–81, 92, 95, 98, 106, 112–113, 123–127, 131–133, 143, 149, 154, 160, 172, 181, 188–189, 198, 220–221, 225, 245, 254–258, 262–268
interface charter, 39, 112, 188
interface management, 18, 57, 80–81, 106, 112, 131, 188, 221, 254–255, 258
invention, 2–3, 244–246

K
key, 1–4, 13, 20–23, 28–29, 32, 39, 42, 45–57, 60, 64–69, 73–77, 80, 89–95, 98, 101, 106–107, 111–115, 119, 122–123, 127, 131, 135–139, 147–150, 153–162, 168–171, 176–178, 181–184, 187–190, 196–198, 202–209, 214–217, 220–226, 229, 236–238, 244, 248, 255–257, 270

L
leadership, 7, 12–13, 31–33, 53–54, 87, 121, 141, 210
life cycle, 4–12, 22–24, 28–38, 41–43, 47–54, 60–68, 73–95, 106–108, 111–115, 119–125, 135–138, 144–150, 153–162, 170–171, 175–193, 198, 204–213, 216–228, 235–241, 245, 251–253, 258–263, 267–271
life cycle phase, 10, 30–33, 38, 42, 47, 51, 54, 60, 65, 68, 73, 78, 81–82, 93, 113, 149–150, 154, 160, 175–183, 191, 204–205, 208–212, 216–218, 224–225, 228, 237–240, 251–253, 258–262, 267–271

M
management review, 48, 158, 223, 226
matrix management, 120
measurement, 8, 20, 31, 39–41, 48, 52–53, 58, 62, 90–92, 98, 102, 123, 135–139, 143–157, 162–164, 170–178, 186, 190–191, 203, 207–208, 216, 221–228, 236–241, 247–248, 251–255, 259–260
measures, 13, 18, 39–48, 52–53, 62–65, 68–69, 75, 79–81, 84, 89–92, 102, 111, 122–123, 138, 145–147, 151, 154–157, 160–164, 169, 174, 188–190, 203, 215, 221–223, 226, 232, 241, 252
metrics, 31, 52, 90, 102, 144–145, 148, 151, 154, 164, 174, 221–228, 241, 248, 252, 268
mobility, 248–253
modularity, 12, 58

N
N^2 chart/N^2 diagram, 113, 133

O
objective, 1, 13, 22, 37, 50, 54, 58, 64, 70, 73, 85, 89, 106, 110, 120–122, 131, 136, 143–150, 157–160, 170, 204, 207, 215, 218–224, 243
organization culture, 210, 239
organization structure, 7, 22, 70, 83, 139–140, 171, 174, 208–211, 238–239
outcome, 1–2, 5, 8, 11–13, 17, 21–24, 27–33, 36–43, 50–54, 59–68, 74, 77, 81–82, 85–87, 90–93, 99–102, 105–108, 114, 117, 121, 125–127, 130–131, 135–144, 147–149, 152–153, 156–164, 169–171, 174–187, 193–195, 203–216, 222–223, 226–227, 236–238, 241–245, 248, 251, 254, 264–271

P
performance specifications, 12, 58, 78–79, 102, 112, 132, 138, 157, 160–162, 172, 221
physical architecture, 4, 10, 22, 29, 39, 45–47, 78–81, 102–103, 112–113, 132–133, 160, 172–173, 220–221, 240, 245
power, 4, 37, 45, 80, 130, 167, 170, 188, 248–251, 257, 261–262, 265, 268–270
process, 1–16, 19–24, 27–43, 46–54, 59–97, 101–128, 131–143, 146–164, 168–195, 200–201, 204–271
project, 1–16, 19–24, 27–55, 59–127, 130–164, 167–198, 202–229, 237–271

project management, 4–6, 24, 27–32,
    36–43, 47–54, 60–66, 69–71, 76–77,
    85–89, 93, 101, 105–108, 111, 114–117,
    120–122, 131, 135–141, 146–153,
    158, 164, 167, 171, 174–195, 203–217,
    222–223, 226–227, 237–243, 246–247,
    270–271
prototype, 2–3, 56, 59, 81, 89, 95, 121–122,
    132–133, 136, 149, 156, 161–163, 172,
    190, 198
qualitative, 36, 75, 80, 87, 123, 144–147,
    157, 221–223, 227
quality, 5–6, 10, 16, 20–22, 28–32, 36–43,
    48–53, 58–64, 68–69, 73–77, 90–93,
    97–105, 111, 123, 126–133, 143–149,
    152, 157–164, 169–174, 186, 190, 195,
    198, 202, 210, 215, 218, 221–224,
    228–237, 240, 245–246, 250–271
quantitative, 5, 11, 36, 39, 64, 68, 75, 80,
    87, 91, 102, 123, 139, 144–147, 157,
    221–223
rebaseline, 153
requirement, 3–5, 9–13, 16–21, 28–60,
    66–81, 84–86, 89–93, 97–103, 106,
    112–114, 120–127, 132–133, 138–139,
    144–146, 152–155, 159–161, 164,
    168–172, 175–198, 201–207, 210–222,
    225–234, 237–241, 245–250, 253,
    256–257, 262–269
requirements traceability matrix, 33, 123,
    154
rigor, 3–4, 8–13, 22–23, 29, 32, 36–37,
    49, 54, 59–60, 95, 107, 114–115, 121,
    143, 148, 152, 163, 175–180, 183–187,
    190–194, 201, 205, 210–214, 227–228,
    237–238, 245, 251–253, 262
risk, 5–23, 29–53, 57–60, 63–66, 69–70,
    73–77, 81–93, 96, 99–106, 111–115,
    119–123, 127–135, 138–144, 147–164,
    169–190, 193–194, 197–218, 222–230,
    234–240, 245–255, 259–270

S

schedule, 4–5, 9–13, 17–18, 28–43, 46–54,

    60, 63–64, 69–77, 80–98, 101–107,
    111–122, 126–128, 131–133, 137–138,
    141–146, 149–164, 169, 172–194, 199,
    203–225, 234, 237, 240, 246, 250, 257
scope, 4–5, 10–13, 16, 28–47, 50–55, 60,
    63–65, 68–70, 73–97, 101–112, 115–
    117, 120–122, 127, 131–132, 137–138,
    141–144, 147–150, 153–164, 168–173,
    177–191, 194–196, 203–226, 230, 237,
    240, 246, 251–254, 268
society, 6–7, 157, 247–265, 269–270
specification, 3, 9, 12, 33, 36, 39–40, 43–49,
    56–58, 74–75, 78–80, 86, 90–93,
    102, 105, 112–113, 123–125, 132–133,
    136–138, 143, 153–164, 172, 178–180,
    188–191, 197–198, 203, 213, 217–223,
    226, 238, 241
specification tree, 43–45, 79, 160, 219–220
spiral, 37, 181–182
stakeholder, 1–13, 16, 21–23, 28–55, 60,
    63–81, 87–97, 101–114, 117, 122–126,
    131–139, 142, 146–150, 153, 156–164,
    167–196, 203–207, 212–232, 237–241,
    245–246, 250–251, 255, 258–259,
    264–268
Standard Reporting of Schedule, 114
subjective, 129, 136–138, 146–147, 171, 223
success, 1–8, 12–16, 20–22, 27–28, 37, 40,
    43–45, 48–63, 66–70, 76–77, 80–89,
    93, 105–106, 115, 120–123, 126–130,
    135–141, 147–149, 153–154, 160–163,
    167–171, 174–184, 187, 191, 194, 198–
    199, 203–209, 212–216, 222, 225–227,
    234–238, 241–245, 250, 262, 268–271
synthesis, 17, 181, 188, 198, 207, 230, 234
system, 1–13, 16–17, 20–24, 27–69, 73–97,
    101–114, 117–125, 128, 131–164,
    170–183, 186–195, 198, 201–228, 231,
    237–272
systems analysis, 3, 244
systems engineering, 1–13, 16, 20–24,
    27–31, 34–56, 59–69, 73–74, 77–86,
    89–90, 93, 101, 104–108, 111, 121–123,
    131–135, 138–142, 146, 150, 158–161,

164, 170–171, 174–182, 186–194, 204–
214, 219, 223, 227, 237–238, 241–272
systems thinking, 2, 8, 11

T

tailoring, 8, 22, 32, 37–38, 92–93, 105–107,
150, 175–195, 204–207, 224, 227–228,
237–241, 250–251, 265
tailoring down, 175, 180, 183–190, 194, 205,
227, 238
tailoring tool, 206
tailoring up, 180, 184–185, 189–188, 194, 205,
227
technical debt, 47–48, 81–82, 92, 144, 148,
154, 159–161
technical risk, 39–40, 47, 50–52, 64, 75–76,
81, 87–89, 102, 105–106, 121–122, 131–
133, 156, 159, 190, 222–226, 235–237,
245, 253–255, 266
test, 2–7, 10, 17–22, 37–43, 46–53, 56–58,
73, 76, 81–83, 89–90, 99, 102, 105–106,
111, 122–125, 130–133, 143, 146–149,
156–164, 169, 172, 179–182, 189–192,
198–204, 211–213, 217, 221–223, 230–
231, 234–238, 245, 252–255, 258–268
time bound, 29, 110, 114, 128, 151–153
trade, 10, 39, 43–47, 51–52, 75, 80–83,
89, 121–122, 128–129, 132–133, 156,
160–161, 172, 182, 190, 221–222, 225,
248–250, 255
trade study, 122, 161

traditional method, 36–37, 151–153, 181
transportation, 203, 248–253, 256
trend analysis, 146

U

use case, 38, 42, 68, 75, 78, 108–110, 113,
131–133, 137, 160, 172, 216–218

V

validation, 10, 20, 33, 39–42, 46–50, 53,
58, 65, 68–69, 75–76, 79, 83, 90, 96,
102–103, 113, 123–125, 130–132,
138–139, 143–144, 157, 160, 169, 172,
179, 182, 190–191, 198, 202, 211–213,
219–222, 226, 234–238, 245, 253–255,
258–259, 262–266
variance analysis, 146, 152
verification, 3, 9–10, 20, 36, 39–43, 46–50,
53, 57–58, 76, 79–85, 90, 96, 102, 113,
123–125, 130–133, 139, 143, 146–148,
154–160, 169, 172, 179, 182, 189–192,
198, 202–204, 211–213, 219–222,
234–238, 241, 245, 253–255, 258–268

W

waterfall method, 36, 51, 212
workforce management plan, 121, 132–133,
172
work package, 34, 51, 77–78, 81–86, 115,
119, 145, 151–155, 159–161, 164,
216–219, 222–224